AF539395

Principles and Techniques of Biochemical Techniques

Principles and Techniques of Biochemical Techniques

Balaji Yadav Maddina

RANDOM PUBLICATIONS
NEW DELHI (INDIA)

Principles and Techniques of Biochemical Techniques

ISBN 978-93-5111-798-8

Published in 2016 in India by

RANDOM PUBLICATIONS

4376-A/4B, Gali Murari Lal, Ansari Road
New Delhi-110 002
Phone : +9111-43580356, 011-23289044, 011-43142548
e-mail: sales@randompublications.com,
info@randompublications.com, randomexports@gmail.com

Reprinted 2022

Type Setting by : Friends Media, Delhi-110089
Digitally Printed at: Replika Press Pvt. Ltd.

Preface

Biochemical analysis techniques refer to a set of methods, assays, and procedures that enable scientists to analyze the substances found in living organisms and the chemical reactions underlying life processes. The most sophisticated of these techniques are reserved for specialty research and diagnostic laboratories, although simplified sets of these techniques are used in such common events as testing for illegal drug abuse in competitive athletic events and monitoring of blood sugar by diabetic patients. To perform a comprehensive biochemical analysis of a biomolecule in a biological process or system, the biochemist typically needs to design a strategy to detect that biomolecule, isolate it in pure form from among thousands of molecules that can be found in an extracts from a biological sample, characterize it, and analyze its function.

The biochemical test that characterizes a molecule, whether quantitative or semi-quantitative, is important to determine the presence and quantity of a biomolecule at each step of the study. Detection assays may range from the simple type of assays provided by spectrophotometric measurements and gel staining to determine the concentration and purity of proteins and nucleic acids, to long and tedious bioassays that may take days to perform.

– Author

Contents

Preface .... *v-vi*

1. Biochemical and Molecular Methods **1**

Molecular Plant-microbe Interactions 1
Molecular Components of Cells 21
Procaryotic Cell Architecture 24
Cell Membrane 30
Endoplasmic Reticulum 41
Penetration of Ions through Cell Membranes 43
Structure of DNA and RNA 56
Unraveling DNA 60
Enzyme Structure of DNA 61
Synthesis of DNA 63
RNA 67

2. Biochemical and Immunocytological Characterizations of Plant Physiology **91**

Relation between Physiology and Physical Sciences 92
Relation between Plant Physiology and Agricultural Sciences 93
Plant Physiology as a Science 94
Possessions of Solutions 96
Role of Interfacial Phenomena in Plant Physiology 106
Understanding the Colloidal Systems 113
Plant Physiology and Sols and Gels 116

3. Tracer Techniques in Plant Pathology **133**

Objectives of Plant Pathology 133

4. Investigation of Antioxidant Enzymes and Biochemical Techniques in Seed Production **180**

Seed 180
Seed Structure 180
Production 191

Factors Affecting Seed Production 204
Dormancy 206

5. **Protein Purification 210**

Western Blot 210
Medical Diagnostic Applications 218
Chromatography 218
ELISA 229

6. **Structural Determination 237**

X-ray Crystallography 237
Nuclear Magnetic Resonance 265
Electron Microscope 281

Bibliography **283**

Index **285**

1

Biochemical and Molecular Methods

MOLECULAR PLANT-MICROBE INTERACTIONS

PHYTOCHROME GENE

The structures and functions of the phytochrome apoprotein genes (the PHY genes), their diversity across the plant kingdom, and their evolution are central concerns in the study of red-light sensing in plants. We summarize here recent advances in two areas relating to these topics: (1) the characteristics of the PHY gene family in Arabidopsis thaliana, the higher plant species for which the most extensive information on these genes is available, and (2) the similarity relationships, phylogeny, and evolutionary implications of PHY gene sequences and partial sequences which have been described from various plants.

Together, these two areas of study, one directed at understanding in detail the phytochromes present in a single species and the other directed at a much broader understanding of PHY gene relatedness and distribution, are producing an increasingly clear picture of the diversity and evolution of plant red-light photoreceptors. Moreover, they suggest that the complexity of the phytochrome family has increased as land plants have evolved novel morphologies.

The Arabidopsis Phytochrome Gene Family

Arabidopsis PHY gene structures

Sharrock & Quail (1989) and Clack, Mathews & Sharrock (1994) described five phytochrome coding regions derived from Arabidopsis cDNA sequences and designated them PHYA, B, C, D and E. Comparison of the deduced protein sequences of the five phytochromes indicates that four divergent types are encoded, the phyA, phyB/D, phyC and phyE types, while the phyB and phyD proteins are clearly a more closely related subgroup. The phylogenetic analysis of PHY sequences described later in this review addresses more extensively the sequence relationships and possible evolutionary origins of these various receptor types and generally supports the contention that Arabidopsis is representative of other flowering plants in terms of its PHY gene content. The possibility that additional PHY genes, beyond PHYA-E, are present in the Arabidopsis genome cannot be discounted, but low-stringency hybridization and degenerate primer PCR analyses have not uncovered any evidence for this. The five PHY genes have been mapped to four of the five Arabidopsis chromosomes, with the PHYD and PHYE genes both located on chromosome 4 but 5-10 cM apart. Figure shows the structures of the five PHY genes. Intron number and location are highly conserved, with two notable exceptions. All of the genes contain two short introns at homologous positions which divide the genes into coding exons I, II and III in the figure. Four of the five contain a third intron, producing coding exon IV, but this intron is missing in PHYC.

Multiple transcription start sites for the PHYA gene have been mapped and the 5' untranslated region has been shown to contain a large intron. For the other

PHY transcripts, lower limits to the 5' ends are known from cDNA and 5' RACE sequences and, as yet, no evidence for multiple start sites or for a 5' untranslated-region intron in any of the other genes has been reported. Primer extension and S1 nuclease protection analysis of the PHYD and PHYE 5' ends indicate that the transcription start sites for these genes are likely to be close to the 5' RACE products already described. Hence, though in the absence of definitive mapping of the 5' ends of PHYB-E it is not possible to draw final conclusions, it appears that each of the three most divergent phytochrome gene lineages - PHYA, PHYB/D/E and PHYC - is distinguished by the presence or absence of specific introns. Other molecular characteristics shared by the Arabidopsis PHY genes and shown in Fig. include the presence of multiple short upstream open reading frames (URFs) in the 5' untranslated regions and the use of multiple poly(A) addition sites in at least PHYA and PHYB.

Arabidopsis PHY gene Expression and Function

The five Arabidopsis PHY genes are expressed at both the mRNA and protein levels, yielding products of the sizes predicted from gene and cDNA sequences. These gene products are present throughout most stages of plant development and in most plant organs, indicating that, at least to a first approximation, the red-light photoreceptor types overlap extensively in terms of their locations within the plant and constitute a very generally distributed antenna for light. Translational fusions of 2-2·5 kb 5' upstream regions of the PHYA, B, D and E genes to the GUS coding sequence have been introduced into Arabidopsis. The patterns of GUS fluorometric activity and histochemical staining in these transgenic plants confirm that, while there are some distinct differences in the developmental and tissue-specific controls on the activities of these four promoter regions, they are active in fairly general and highly overlapping patterns.

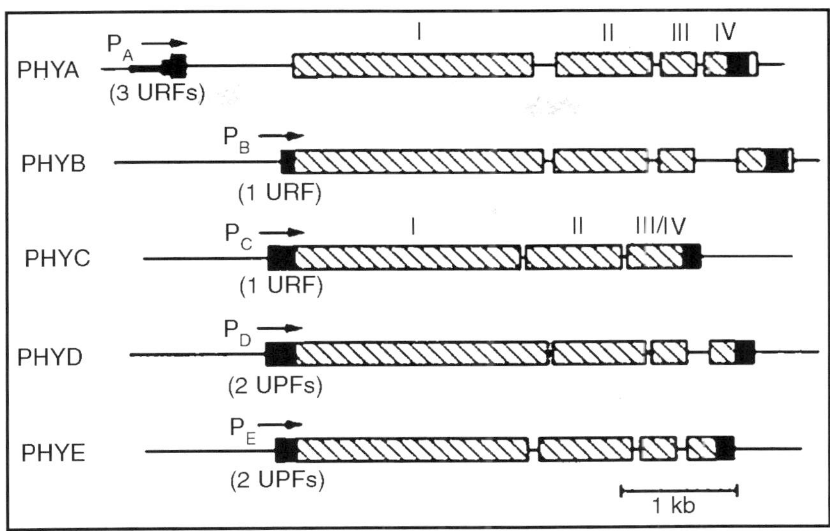

Fig. Structures of the five Arabidopsis PHY genes.

Genomic DNA sequences of PHYA, PHYB, PHYC, PHYD and PHYE and the corresponding cDNA sequences were used to deduce gene structures. Untranslated regions at the 5' and 3' ends of the transcripts are shown in black or white and coding regions are stippled. The number of upstream open reading frames (URFs) present in the 5' untranslated region of each gene is noted and occurrence of multiple poly(A) addition sites is indicated by a white box in the 3' untranslated region. The individual functions of three of the Arabidopsis PHY genes have been identified by isolation of null mutations in these genes. Mutants lacking phyA exhibit loss of far-red light (FR) high irradiance control of hypocotyls elongation, cotyledon expansion and seed germination while mutants lacking phyB show alteration in shade avoidance responses including the effects of red light (R) and of the R:FR ratio on hypocotyl elongation, flowering time and leaf morphology.Aukerman et al. (1997) describe a null mutation in the PHYD gene found as a naturally occurring allele in the Wassilewskija (Ws) ecotype of Arabidopsis.

Loss of phyD causes alteration of many of the same shade avoidance responses which are affected in the phyB mutant, but comparison of the two null mutants shows that phyB plays a much more prominent role than phyD. Hence, diversification of the PHY gene family, at least in the case of the dicot plant Arabidopsis, has allowed the evolution of distinct photosensory roles for the photoreceptor subfamilies, with the most divergent genes, exemplified by PHYA and PHYB, having highly divergent functions and the most closely related genes, PHYB and PHYD, having overlapping or even somewhat redundant roles. Mutants lacking phyC or phyE have not been described and their functions therefore remain to be determined.

Diversification of the Phytochrome Gene Family During the Evolution of Plants

The Origin of Phytochrome

The most ancestral phytochrome that has been fully characterized is that of the green alga Mesotaenium

Evidence of homologues in prokaryotes is limited, though similarity among the C-termini of phytochromes and the histidine kinase domains of bacterial two-component response regulators has been noted. The strongest support for common ancestry comes from characterization of a phytochrome-like putative photoreceptor from the cyanobacterium Fremyella and from sequence analysis of the genome of Synechocystis strain PCC 6803.

The sequence from Fremyella, RcaE, comprises a C-terminus with motifs characteristic of histidine kinase domains and an N-terminus that is weakly similar to the chromophore-binding domain of phytochrome. Synechocystis sequence 1001165 is more similar to phytochrome than is RcaE but, while striking, the degree to which the likeness reflects homology remains unclear because the pattern of similarity among phytochromes and sequences from Synechocystis suggests that the chromophore domain and the C-termini of phytochromes have different evolutionary histories.

For example, an additional sequence from Synechocystis has a putative chromophore-binding domain that is more similar to phytochromes than is RcaE, but the C-terminus is similar to a different class of response regulators, exemplified by PleD from Caulobacter. In addition, the direct repeat in the hinge region of phytochrome may be related to yet other bacterial molecules. Thus, further data are needed to confirm the relationship of phytochromes with specific prokaryotic molecules.

PHY Gene Diversity in Land Plants

During land plant evolution, phytochromes have diversified from a single progenitor into multiple related lineages or subfamilies. Here, we evaluate and review their sequence relationships and evolution by phylogenetically analysing all available full-length phytochromes and two partial sequences using neighbour-joining and parsimony algorithms in PAUP.

Additional PHY gene sequence fragments were not included because, for the most part, they result in phylogenies that are not robust to changes in rooting or method of analysis (e.g. provide little further insight. Results shown in Figure are from analysis of homologous amino acids rather than nucleotides because the divergence (nucleotide substitutions per site) between many sequence pairs is > 1.

Trees were rooted by designating a phytochrome sequence from an ancestral taxon, the green alga Mesotaenium, as the outgroup, and support for phylogenetic groups was evaluated by bootstrap resampling. We addressed the following questions: (1) what is the mode of phytochrome evolution in land plants, and (2) what are the relationships of phytochromes from sporogenous plants and gymnosperms to phytochrome subfamilies in flowering plants.

In sporogenous plants, which diversified before the origin of seeds plants, evidence of multiple divergent phytochromes in any single taxon is limited. Ceratodon has two expressed genes related to phytochrome, designated PhyCer and CpPHY2. Their N-termini are homologous with other phytochromes and share 89% of their amino acids; however, the C-terminus of PhyCer is unrelated to CpPHY2 and to all other phytochromes.

Additional evidence of diversity in a single taxon comprises reports of multiple phytochromes in Psilotum, and Anemia. However, this diversity may result from taxon-specific diversification; in Psilotum, it may reflect a highly duplicated genome. Phylogenetic analyses could suggest the presence of multiple phytochrome lineages in sporogenous plants if branch order of phytochromes clearly conflicted with organismal phylogeny.

For example, the placement of Adiantum phytochrome in trees that include phytochromes from chlorophytes and Psilotum is unexpected. However, this placement may result from attraction of long branches to one another because, in trees from which the outgroup Mesotaenium phytochrome is excluded, the branch order of phytochromes is both well supported and consistent with organismal phylogeny.

Thus, phylogenetic results do not strongly suggest that sporogenous plants have, or had, more than one phytochrome subfamily. Moreover, they suggest no well-supported relationship of phytochromes from sporogenous plants with individual seed plant PHY subfamilies.

Open circles represent gene duplications. Distances in nucleotide substitutions per site are indicated by branch lengths relative to the scale of 0·100 substitutions per site shown. Values from 100 bootstrap replicates are given above branches.

The presence of multiple phytochrome genes in individual species of seed plants is well established, and phylogenetic analyses indicate that, following the major duplication events, these genes have been evolving independently. As shown in Fig., the earliest gene duplication gave rise to the lineage comprising homologues of Arabidopsis PHYA and PHYC on the one hand and the lineage comprising homologues of Arabidopsis PHYE and PHYB/D on the other.

Subsequent duplications resulted in the divergence of PHYA from PHYC and of PHYE from PHYB/D. Divergence of PHYB from PHYD in Arabidopsis followed a recent duplication. This evolutionary pattern is consistent with the functional divergence among phytochromes in flowering plants reviewed in the section 'Arabidopsis PHY gene expression and function' above.

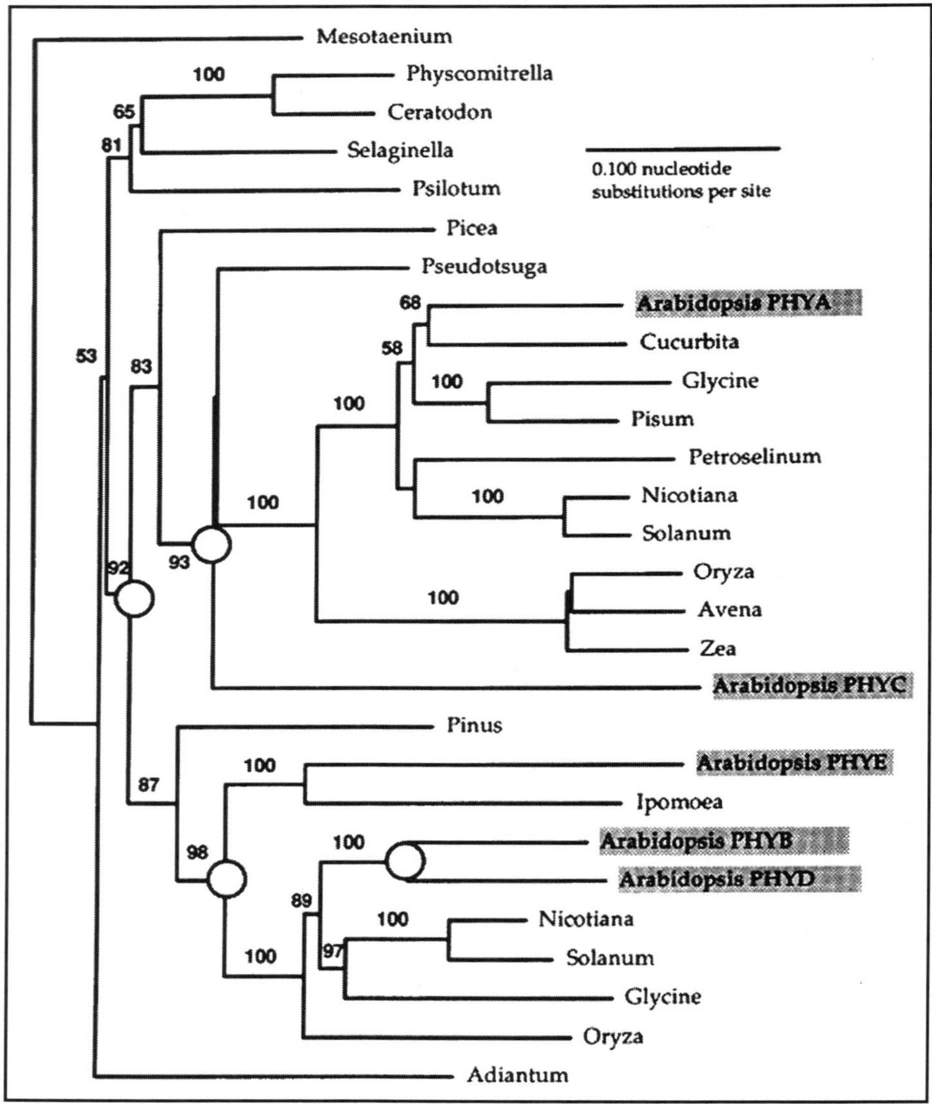

Fig. Neighbour-joining Tree of Phytochrome Amino Acid Sequences.

Furthermore, it is notable that the earliest three diversification events in the phytochrome gene family may have occurred at or near major morphological transitions that mark the evolution of plants: the PHYA/C-PHYB/D/E duplication near the origin of seed plants and the PHYA-PHYC and PHYB/D-PHYE duplications near the origin of flowering plants.

At least two of the phytochrome lineages known from flowering plants appear to have been established before the divergence of conifers from other gymnosperms. For example, the sequence from Pinus is most closely related to the PHYB/D/E subfamily, while the sequence from Picea is most closely related to the PHYA/C subfamily. Furthermore, Pseudotsuga phytochrome does not cluster with Picea phytochrome, as we would expect of homologous sequences from closely related taxa, so it is possible that the phytochrome gene family in conifers comprises three divergent lineages.

This interpretation is consistent with the observation that PHYA, PHYB/D/E and PHYC are well separated from one another in the earliest flowering plant, with the position of the phytochrome fragment from the gymnosperm Ephedra as sister to PHYAs, and with the suggestion that two unpublished sequences from pine are PHYA- and PHYC-like. Together, these observations suggest that three of the four major flowering plant phytochrome lineages may be homologous with phytochromes found in gymnosperms.

Phytochrome Lineages in Flowering Plants

Phytochrome genes have been detected in all major subclasses of flowering plants. Figure summarizes results of phylogenetic analyses of 172 full or partial sequences in GenBank. Data were either partitioned according to sequence length or combined in a matrix in which absent nucleotides were coded as missing data. Each of the sequences belongs unequivocally (91% or higher bootstrap support) to one of the four phytochrome lineages identified in analysis of full-length sequences: PHYA, PHYB/D, PHYC or PHYE. Three of these four gene lineages occur widely in flowering plants but, to date, homologues of PHYE have not been detected in monocots.

This distribution could result from the divergence of PHYE from PHYB/D having occurred after the divergence of monocots from dicots or from PHYE being lost from monocots. There is currently no DNA sequence data suggesting the presence of additional, widely distributed phytochrome lineages in flowering plants. In our analysis, tomato PHYF is a member of the well-supported (91% bootstrap value) cluster including eight PHYC homologues from dicots.

The interpretation of PHYF as a novel PHY subfamily was based on the degree of divergence between Arabidopsis PHYC and tomato PHYF and assumes that rates of nucleotide substitution among phytochrome subfamilies are equal. However, evidence is presented below that the PHYC subfamily is in fact diverging in sequence faster than the PHYA and PHYB subfamilies. Reports of additional genes in tomato (total of 9-12) inferred from Southern analyses have not been substantiated with sequence data. The possibility remains that such very high diversity may be the result of recent duplications within the PHY gene family in Solanum, or in some larger taxonomic group to which tomato belongs.

Numbers in parentheses represent the number of sequences sampled from GenBank; tomato PHYF occurs in the PHYC clade.

Recent diversification in the gene family is evident within dicot flowering plants, exemplified by gene duplications in the PHYA and PHYB/D subfamilies.

Multiple homologues of PHYB/D are found in Arabidopsis, carrot and tomato, while multiple homologues of PHYA are found in carnation, legumes and Ceratophyllum, an aquatic angiosperm. In legumes, the duplicate PHYAs form a discrete lineage that is derived from within their PHYA subfamily.

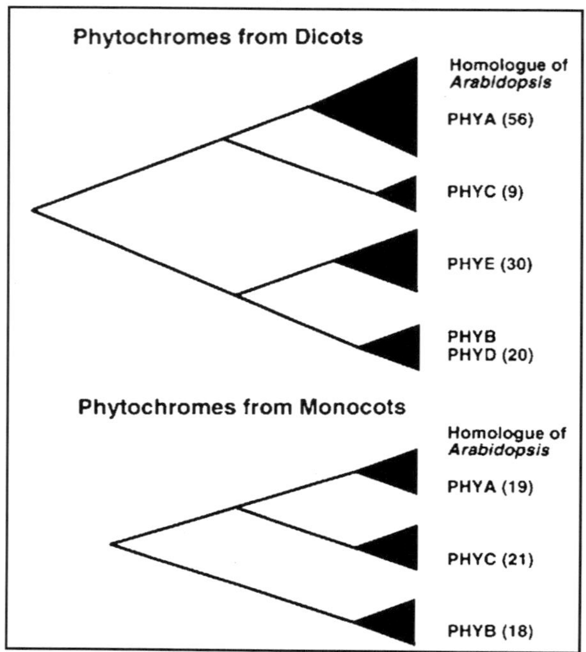

Fig. Distribution and Relationships of 172 Complete or Partial Phytochrome Nucleotide Sequences from Flowering Plants.

In all cases, multiple sequences from a taxon are more divergent than one would expect if they represented merely allelic diversity. Conversely, evidence for recent diversification of PHY genes in monocots is limited. Hershey et al. suggested that four genes encode phytochrome in oat, and reported sequences of two PHYAs and a PHYA fragment that are nearly identical.

This may not be strong evidence of diversification within a gene subfamily because Avena is a polyploid. Similarly, Christensen & Quail estimated the size of the gene family in maize to be from two to four members. They isolated clones representing three loci; one is a homologue of PHYA, another was interpreted as a pseudogene, and the sequence of the third was not reported.

Subsequently, partial sequences homologous with PHYB/D and PHYC have been isolated from maize, possibly accounting for the additional loci. Results from Southern blot analyses of genomic DNA of rice indicate that the three observed hybridization bands correspond to PHYA-, PHYB/D- and PHYC-like sequences. Finally, no evidence of multiple homologues was detected in any of the monocots sampled for phytochrome diversity, except from the grass Panicum which has two homologues of PHYA, one of which is very divergent from other PHYAs from grasses. At the nucleotide level, diversification in the PHY gene family in flowering plants is characterized by unequal rates of divergence. Relative rates of non-synonymous nucleotide substitution between all currently available full length PHY coding sequences and the reference PHY sequence from Selaginella were determined. Table shows that, in all possible pairwise comparisons of PHYA and PHYB sequences from single taxa, PHYAs are significantly more divergent from the common ancestor than are PHYBs.

Furthermore, in Arabidopsis where full-length sequences are available, PHYC is significantly more divergent compared to PHYA, PHYB and PHYD, while PHYE is evolving rapidly relative to PHYB and PHYD. Thus, overall, PHYB/D appears to be the most evolutionarily constrained PHY sequence, followed by PHYA, then PHYC and PHYE. Differences in the rates of divergence for two full-length phytochrome coding sequences, for instance PHYA and PHYC, are largely due to differences in the C-terminal half of the molecule. When the data are partitioned by exon, nucleotide substitution is more highly constrained in exons I, III and IV and is least constrained in exon II (data not shown). It has been proposed from mutational studies that sequences coded in exon II are critical for phytochrome's regulatory activity so it is possible that rapid evolution of PHYC and PHYE in this region reflects functional divergence.

Table. Results from pairwise comparisons of ralative rates of nucleotide substitutions (Wu & Li 1986) among full-lengh PHY sequences. The value d13 - d_{23} is the difference in non- synonymous nucleotide substitution between sequence 1 and sequence 2 ralative error: Single asterisks indicate differences significant at the 0.5 level and double asterisks differences at the 0.01 level. At = Arabidopsis, Nt = Nicotiana, St = Solanum,Os = Oryza, Gm = Glycine

Seq 1	Seq 2	d_{13}-d_{23}	SE	Direction
At A	At B	0·0454*	±0·0183	(A>B)
At A	At C	–0·0429*	±0·0202	(C>A)
At A	At E	–0·0242	±0·0198	
At B	At C	–0·0883**	±0·0254	(C>B)
At B	At E	–0·0696**	±0·0189	(E>B)
At C	At D	0·0770**	±0·0194	(C>D)
At C	At E	0·0187	±0·0207	
At D	At E	–0·0583**	±0·0191	(E>D)
Os A	Os B	0·0732**	±0·0188	(A>B)
Nt A	Nt B	0·0962**	±0·0176	(A>B)
St A	St B	0·0786**	±0·0181	(A>B)
Gm A	Gm B	0·0403*	±0·0192	(A>B)

MOLECULAR MAP OF THE PHYTOCHROMES

Spectral Activity

Chromophore Lyase

The seminal demonstration by Lagarias and coworkers that the PHYA apoprotein can autocatalytically attach the chromophore in vitro as well as in living yeast has been followed by similar demonstrations in vitro for phyB and phyC. Analysis of deletion derivatives of phyA and phyB expressed in yeast or transgenic plants has mapped the chromophore lyase activity sufficient for attachment towithin the NH2-terminal segment between positions 115 and 450. In a more detailed site-directed mutational analysis of five conserved residues (positions 360, 369, 372, 375 and 377) surrounding the chromophore attachment residue (the cysteine at position 374), Song and colleagues have obtained evidence that none is essential for chromophore lyase activity, suggesting a structural rather than a catalytic role.

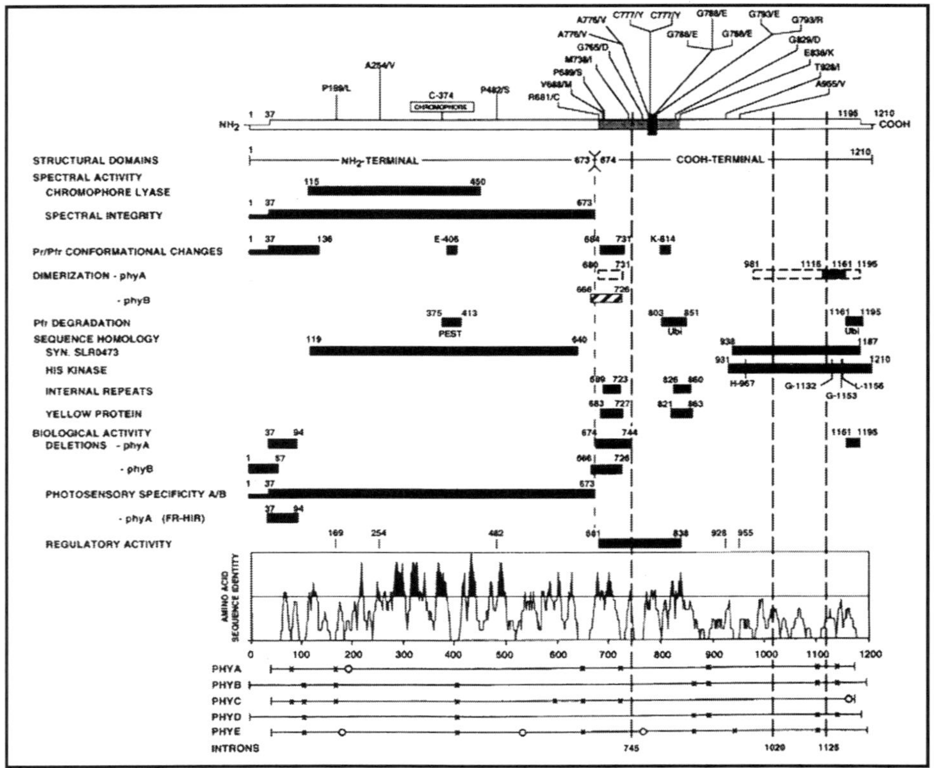

Fig. Molecular map of Phytochrome.

A schematic diagram of a 'consensus' phytochrome molecule derived from the alignment of multiple full-length phytochrome polypeptide sequences by Mathews et al. is shown at the top. The coordinates for this figure and throughout the paper are those for the consensus line of this alignment and are referred to as 'positions' within the consensus sequence. The extremities of the schematic (positions 1 and 1210) correspond to the NH2- and COOH-termini, respectively, of Arabidopsis phyB, the longest sequence in the alignment.

The indented positions at each end (positions 37 and 1195) correspond to the NH2- and COOH-termini, respectively, of Arabidopsis phyA. The location of the tetrapyrrole chromophore covalently linked to a cysteine is indicated at position 374. The residue substitutions indicated at various positions along the consensus molecule are a composite of phyA and phyB missense mutations causing loss of regulatory activity. The stippled area demarcates a region containing a high density of such mutations (the 'core' region) and the small black box, a short segment containing multiple substitutions of individual residues. The horizontal bars in the body of the figure show the locations within the polypeptide of the activity, property, or feature of the molecule shown to the left. At the bottom is a plot of the distribution of amino acid sequence identity across the aligned polypeptide sequences of the five Arabidopsis phytochromes PHYA, B, C, D and E, shown below. The light vertical dashed line at position 673/674 represents the junction of the two major structural domains. The heavy vertical dashed lines represent the positions of the introns within the structural PHY genes of higher plants (except PHYC which lacks the intron at position 1020).

Spectral Integrity

Sequences necessary to sustain complete spectral integrity equivalent to that of the full-length molecule extend in both directions beyond that for chromophore lyase activity to encompass most, if not all, of the NH2-terminal domain (positions 1-673). Spectral integrity here refers to maintenance of photoreversibility and absorption spectra in both Pr and Pfr forms unaltered from that obtained with the native photoreceptor. Analysis of deletion derivatives of phyA in transgenic seedlings has shown that the full NH2-terminal structural domain (positions 1-673) is sufficient for spectral integrity.

A series of deletions covering a variety of segments within the first 115 positions at the NH2-terminus (residues 1-69 of the phyA sequence) has revealed subdomains necessary for stability and spectral integrity of the Pfr form alone (positions 51-94) or of both the Pr and Pfr forms (positions 51-115). These data are consistent with earlier proteolytic studies on native phyA suggesting interaction of this NH2-terminal segment with the chromophore, especially in the Pfr form. Similarly, determinants between positions 524 and 673 at the COOH-terminal end of the NH2-terminal structural domain have been shown to be necessary for spectral normalcy. Preliminary spectral analysis of some comparable deletion derivatives of phyB suggests a similar pattern to that of phyA.

Pr/Pfr Conformational Changes

Conformational changes in the protein moeity induced by photoconversion between the Pr and Pfr forms have been mapped in phyA to the NH2-terminal segment (within positions 37-136), to a segment around position 406 near the chromophore attachment site, to a segment in or near the linker region (positions 684-731), and to a region around position 814. Most of these data have come from older biochemical, immunochemical or physico chemical studies on extracted or purified, non-recombinant phytochrome.

Dimerization

Proteolytic fragmentation coupled with the use of monoclonal antibodies against mapped epitopes provided evidence some years ago that the COOH-terminal structural domain of phyA carries the determinants for dimerization, whereas the NH2-terminal domain is released as a monomer upon cleavage, indicating no stable intra- or intersubunit interactions. Transgene-encoded NH2-terminal domains of both phyA and phyB behave as globular monomers, confirming this latter conclusion, and the transgene-encoded COOH-terminal domain of phyB has been shown to associate into homodimers in Arabidopsis.

Using a conceptually elegant prokaryotic assay system, Edgerton & Jones presented data interpreted to indicate that two regions of the phyA COOH-terminal domain (positions 680-731 and 1116-1195, respectively) are capable of mediating dimerization. However, subsequent analysis of a COOH-terminal deletion series by Cherry et al. (1993) showed that simply removal of residues COOH-terminal of position 981 was sufficient to cause loss of dimerization capacity.

This result indicates that neither the initial (positions 680-731) nor revised (positions 652-741) central segments are sufficient to mediate dimerization in the context of the native molecule. Moreover, since internal deletion of a similar region of phyB (positions 666-726) has been shown not to affect dimerization when expressed in transgenic Arabidopsis, it would appear that this region is neither necessary nor sufficient for dimerization in the context of the native molecule. By contrast, the COOH-terminal deletion analysis by Cherry et al. indicates that residues between positions 981 and 1161 are necessary for dimerization of native phyA in transgenic plants. Combined with the proposal from the prokaryotic assay that positions 1116-1195 are necessary for dimerization, these transgenic plant data might suggest that critical residues for dimerization lie between positions 1116 and 1161.

Pfra Degradation

The rapid intracellular degradation of phyA induced upon its photoconversion to the Pfr form (representing a 100-fold greater turnover rate for PfrA than PrA) is one of the earliest molecular properties of the photoreceptor recorded in the literature. The other phytochromes (phyB, C, D and E) do not display this dramatic isoform specific difference in degradation rate and are therefore considered to be more 'light stable. Recent domain swapping experiments using transgenic Arabidopsis have provided evidence that the determinants specifying recognition of phyA for degradation in the Pfr form reside in the NH2-terminal structural domain.

However, these determinants are alone insufficient for degradation because the NH2-terminal domain alone of phyA is light stable in vivo, requiring a contiguous COOH-terminal domain for degradation to proceed. Indeed, deletion of the extreme COOH-terminus of the phyA polypeptide has been reported to stabilize the molecule as Pfr, localizing one determinant necessary for degradation to a small stretch of amino acids at the extremity of this domain. Because the full COOH-terminal domain from either phyA or phyB supports degradation when fused to the phyA NH2-terminal domain, the necessary determinants are apparently common to both family members.

Proposed mechanisms of selective PfrA degradation have focused on the possible involvement of the ubiquitin- mediated proteolytic pathway and/or the hypothesized PEST-sequence mechanism. There is evidence that phyA is rapidly ubiquinated in vivo upon conversion to Pfr and kinetic data are at least partially consistent with subsequent degradation ensuing via the ubiquitin pathway. The proposed location of the ubiquitin target site (one or more of the three lysine residues between positions 803 and 851) in the COOH-terminal domain, as well as the conservation of these three lysine residues between phyA and phyB sequences would be consistent both with the need for a contiguous COOH-terminal domain for degradation and with the interchangeability of the phyA and phyB COOH-terminal domains in this process.

Unfortunately, site-directed substitution of arginines for the three lysines apparently did not affect the rate of PfrA degradation, indicating that if ubiquitination does occur at these sites it is not necessary for proteolysis. Thus, although still clearly a viable possibility a direct link between ubiquitination of PfrA and degradation remains to be demonstrated. The presence of a PEST sequence and its associated cluster of acidic residues in phyA, but not phyB, C, D or E sequences, as well as its location adjacent to the chromophore (positions 375-413) in the amino terminal domain, and the photoconversion-induced exposure of this region in the Pfr form, are all consistent both with the selective degradation of phyA following Pfr formation and the location of the determinants for this selectivity in the NH2-terminal domain.

One possibility would be that the PEST sequence provides specific recognition of PfrA by the degradative machinery which then initiates degradation via the ubiquitin pathway involving determinants in the COOH-terminal domain.Unfortunately, no direct experimental test of the relevance of the PEST sequence to PfrA degradation has been reported to date and the importance of this sequence to the degradation of several other PEST-containing proteins has been questioned.

Sequence Homology

Schneider-Poetsch and colleagues provided the first provocative evidence from computerized database searches that the distal half of the COOH-terminal domain of the phytochromes (positions 931-1210) exhibits a degree of sequence similarity to the sensor histidine kinase module of bacterial two-component signalling systems, thereby raising the possibility that the photoreceptors function as a family of photoregulated histidine kinases. Subsequently, three considerations tended to detract from the likelihood that this simple extrapolation from the prokaryotic systems is accurate.

First, the histidine residue that is autophosphorylated in the majority of the bacterial kinases is conserved at that site (position 967) only in the monocot phyAs and Arabidopsis phyC, and not any of the other reported phy sequences. Secondly, attempts to detect autophosphorylation of histidine residues in preparations of purified phyA were unsuccessful. Thirdly, site-directed mutagenesis of four single residues [positions 967 (histidine), 1132 (glycine), 1153 (glycine), 1156 (leucine)] in oat phyA (the latter three residues conserved among all phytochromes and the majority of histidine kinases) failed to abrogate the activity of the molecule when tested in transgenic Arabidopsis.

However, the recent sequencing of the entire genome of the cyanobacterium Synechocystis sp. strain PCC6803 has strongly revitalized this proposal. One of the open reading frames (ORFs) in the Synechocystis genome (Cyanobase database), designated ORF SLR0473, has regions of striking sequence similarity both to the phytochromes and to the sensor histidine kinases.

ORF SLR0473, encoding 748 amino acids, consists of an NH2-terminal domain with 36% identity (60% similarity) to the NH2-terminal domain of phyE between positions 119 and 640, and an immediately adjacent COOH-terminal domain with 20% identity (52% similarity) to the COOH terminal domain of phyE between positions 938 and 1187. This latter region of phyE is within the proposed histidine kinase-like domain of the phytochromes.

In turn, the COOH-terminal domain of ORF SLR0473 shows convincing sequence similarity to the histidine kinase module of established bacterial two-component systems (e.g. 31% identity and 55% similarity to B. subtilis KinA), including the presence of the conserved signature motifs, designated H, N, G1, F and G2. These data suggest that the Synechocystis SLR0473 protein may be a photoregulated histidine kinase and that the plant phytochromes may be evolutionary remnants of these prokaryotic molecules.

Strong support for at least part of this notion has been provided recently by Lagarias and coworkers who have shown that the SLR0473 protein can attach tetrapyrrole chromophores, including phytochromobilin, with the formation of a chromoprotein that is fully photoreversible by red and far red light analogous to plant phytochromes. In addition, a genetically defined locus for chromatic adaptation in the cyanobacterium Fremyella has also recently been shown to encode a protein with some sequence similarity to the phytochromes in its NH2-terminal domain and to the histidine kinases in its COOH-terminal domain.

Two other related observations arising from sequence comparisons are that the COOH-terminal domain of the phytochromes contains two repeats of a sequence (positions 689-723 and 826-860, respectively), and that these repeats have weak sequence similarity to a segment of photoactive yellow protein from the purple bacterium Ectothiorhodospira. The significance of these findings is yet to be determined.

Biological Activity

Deletion Analysis

A number of investigations have used overexpression of various deletion derivatives of phyA or phyB in transgenic plants to identify regions of the photoreceptors necessary for biological activity. Although subtleties exist between the effects of some individual constructs and plant systems, the basic conclusion from these combined studies is that deletion of relatively short segments at either terminus of the polypeptide either eliminates or modifies detectable biological activity. The results indicate that determinants within the first 60 residues or less at the NH2-terminus of phyA (positions 37-94) or phyB (positions 1-57), and within as few as 35 residues at the COOH-terminus of phyA (positions 1161-1195) are necessary for normal biological activity.

In addition, internal deletions near the linker region at the proximal end of the COOH-terminal domain of phyA (positions 674-744) or phyB (positions 666-726) also eliminate normal activity, indicating the necessity of this region. The limitation of such data is that is unknown whether these regions are directly involved in photoreceptor activity or the deletions indirectly perturb the function of another critical part of the molecule. One interesting 'gain-of-function' effect is that either deletion of the serine-rich region between residues 6 and 12 of phyA (positions 42-51) or site directed substitution of alanines for all the serines in this region result in enhanced biological activity of the transgene-encoded phyA. This result suggests that the serines in this region may mediate attenuation of phyA activity.

Photosensory Specificity

Like other receptors, the phytochromes can be considered to exhibit dual molecular functions: a sensory function involving perception and interpretation of the incoming light signals, and a regulatory function involving biochemical transfer of the perceived information to downstream transduction chain components. The phytochrome system is well known to monitor multiple parameters of the light environment (wavelength, fluence rate, etc.), to do so in multiple sensory 'modes' (LFR, VLFR, HIR, etc.), and to regulate multiple facets of growth and development throughout the plant life cycle via modulated gene expression.

Studies with photoreceptor mutants of Arabidopsis have shown that some of this sensory complexity, at least in early seedling development, is attributable to the fact that phyA is predominantly, if not exclusively, responsible for FRc perception in the FR-HIR, whereas phyB is predominantly responsible for Rc perception in the R-HIR, and that the two phytochromes transduce mutually antagonistic signals in response to Rc or FRc enrichment.

To begin to map the determinants involved in this photosensory specificity, reciprocal domain-swap experiments have been performed in which the NH2-terminal domain of phyA fused to the COOH terminal domain of phyB at position 673 (phyA/B) and the converse fusion protein (phyB/A) have been overexpressed in transgenic Arabidopsis. The data show that the wavelength-dependent seedling responses to Rc or FRc mediated by the phyA/B fusion resemble those of phyA, whereas the responses mediated by the phyB/A fusion mimic those of phyB. The evidence therefore indicates that the determinants for the photosensory specificity of phyA and phyB to FRc and Rc, respectively, reside in the NH2-terminal domains. Because deletion of the NH2-terminal 52 residues of phyA (positions 37-94) produces a trans gene-encoded molecule defective in FRc perception, but retaining Rc responsiveness, it is possible that one or more of these determinants of phyA photosensory specificity reside in this short-terminal segment.

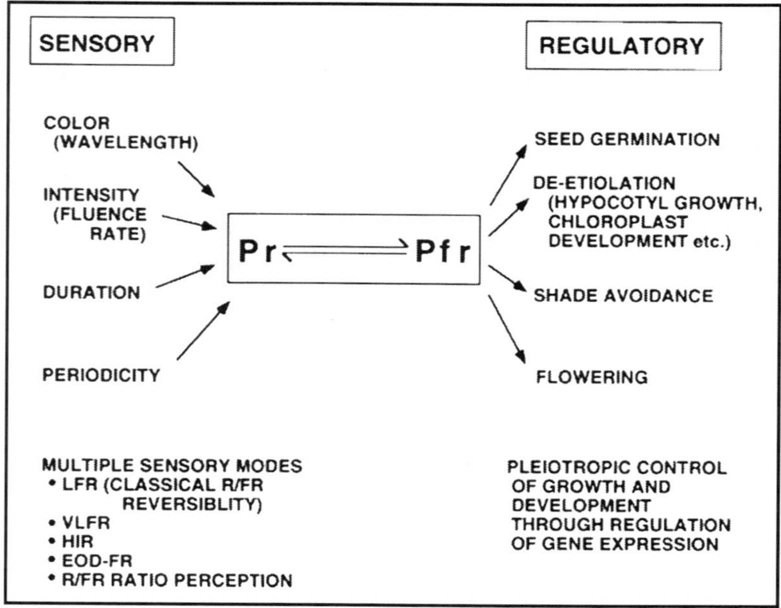

Fig. Dual Molecular Functions of the Phytochromos.

On the other hand, the NH2-terminal domain of neither phyA nor phyB when expressed alone exhibits normal biological activity despite being fully photoactive and having apparently normal spectral properties.

This result indicates that a contiguous COOH-terminal domain provides determinants necessary for transmission of perceived photosignals. Because the COOH-terminal domains of phyA and phyB appear to be fully interchangeable in this function, the critical COOH-terminal determinants appear to be common to both phytochromes. From these data alone, it cannot be determined whether this domain plays a purely structural role or is directly involved in signal transfer. However, at least for phyB, the COOH-terminal domain is not capable of autonomously performing phyB function. This result indicates that the phytochromes appear to be unlike some biological molecules, such as receptor kinases, in which removal of the ligand-binding (receptor) domain results in constitutive activation of the signalling (catalytic) domain.

The photoreceptor molecular can be considered to have two sequential functions: sensory, defined as perception and interpretation of signals from the light environment, and regulatory, defined molecular as biochemical transfer of the perceived signals to transduction pathway components that ultimately regulate morphogenesis through altered gene expression. The principal parameters in the light environment knows to be monitored by the phytochrome system are shown to the left, as are the multiple sensory 'modes' displayed by the photoreceptor: LFR = low fluence response: VLFR = very low fluence response; HIR = high irradiance response; EOD-FR end-of-day far red; R/FR ratio = the ratio of the fluence rates of red to far red light (the parameter used for vegetative shade or proximity detection). Major facets of plant growth and development controlled by the phytochromes are indicated to the light.

Regulatory Activity

One strategy for identifying determinants in the phytochrome protein potentially involved in downstream signal transfer, as distinct from signal perception, is to screen for missense mutations within the polypeptide that do not affect expression level, spectral properties, or gross structural properties (dimer formation, apparent molecular cross-section), but do disrupt biological activity. Such mutant molecules can be considered to be competent in photoperception but defective in regulatory activity.

The mutated resides detected in a series of such screens involving phyA and phyB are compiled at the top of Fig.. A majority of these mutations (76%) cluster at the proximal end of the COOH-terminal domain (positions 681-838), designated the 'core'region, with multiple substitutions occurring at four positions within an 18 residue sub-segment (positions 776-793).

The data indicate therefore that this restricted 'core' region of the polypeptide contains determinants necessary for effective communication of perceived light signals to the cellular transduction circuitry. Because the COOH-terminal domains of phyA and phyB are functionally interchangeable and because the regulatory missense mutations cluster in the same region for both phyA and phyB, it would appear that the samefunction is disrupted in both molecules by these mutations. If the region between positions 681 and 838 is directly involved in the transfer of signalling information from photoreceptor to recipient transduction component, the biochemical mechanism of this transfer would appear to be the same for phyA and phyB.

How could the specificity of signal perception determined by the NH2-terminal domains of phyA and phyB be reconciled with a putative common biochemical mechanism of downstream signal transfer specified by the 'core' region? One possible formal model is depicted in Figure. In this model, the NH2-terminal domains of phyA and phyB carry different determinants that allow each to recognize (bind to) its own cognate reaction partner(s) (target selection) upon appropriate signal perception. The biochemical modification of the bound reaction partner involving the 'core' COOH-terminal region would then be identical in all cases (e.g. phosphorylation of a residue in the reaction partner).

Subsequent steps in the signal cascade could then either converge immediately or remain independent to the point where the common cellular functions controlling the growth and developmental responses to phyA and phyB are affected. This model is analogous to that of many other families of receptor molecules, especially the receptor kinases.

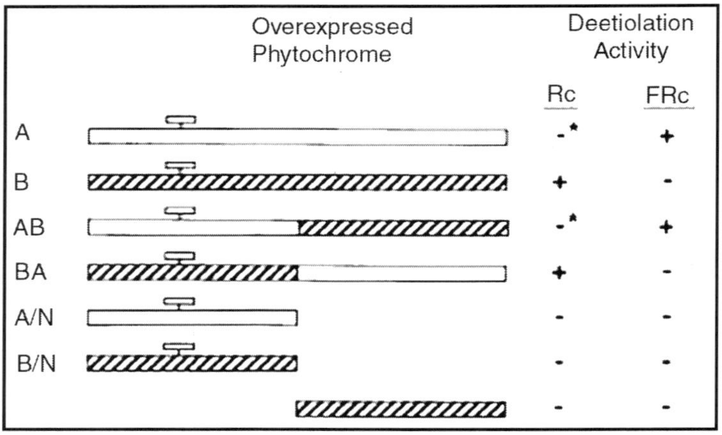

Overexpressed Phytochrome	Deetiolation Activity Rc	Deetiolation Activity FRc
A	-*	+
B	+	-
AB	-*	+
BA	+	-
A/N	-	-
B/N	-	-
	-	-

Fig. Activity of Transgene-encoded Phytochrome Derivatives in Enhancing the Deetiolation process when Overexpressed in Arabidopsis.

Constructs: A, oat phyA; B, rice phyB; AB, chimeric protein with NH2-terminal domain of oat phyA (positions 37-673) fused to COOH-terminal domain of rice phyB (positions 673-1210); BA, chimeric protein with NH2-terminal domain of rice phyB (positions 1-673) fused to COOH-terminal domain of oat phyA (positions 673-1195); A/N, NH2-terminal domain of oat phyA only (positions 37-673); B/N, NH2-terminal domain of rice phyB only (positions 1-666); B/C, COOH-terminal domain of rice phyB only (positions 647-1210).

Deetiolation activity = capacity of transgene encoded phytochrome to enhance deetiolation (suppress hypocotyl elongation) of transgenic Arabidopsis seedlings in Rc or FRc. + = enhanced deetiolation relative to non-transgenic Arabidopsis; - = no enhancement of deetiolation relative to non-transgenic Arabidopsis; -* = quantitatively less effective than rice phyB in enhancement of deetiolation in Rc on a per mole basis of overexpressed phytochrome.

An Integrated Picture

Several interesting insights emerge from considering the various features or activities that have been defined for a given region of the phytochrome molecule.

- The locations of the introns in the higher plant PHY genes do not appear to correlate in any obvious way with the structural domains or various functional activities described. In addition, the algal phytochromes have larger numbers of introns than their higher plant counterparts, and Arabidopsis PHYC lacks the intron at position 1020. The functional significance, if any, of these observations is unknown.
- The region of the higher plant phytochromes that has the highest sequence similarity to the Synechocystis SLR0473 ORF (positions 119-640) corresponds extremely well to the region most highly conserved among the higher plant phytochromes themselves. Indeed 67% of the invariant residues in this region of the plant phytochromes are also invariant in the Synechocystis sequence (data not shown). This observation is consistent with the demonstrated functional homology between the cyanobacterial and plant sequences in chromophore attachment and photoreversibility.
- Each of the regions that undergo photoinduced conformational changes overlaps with other activities. The

NH2-terminus (positions 1-136) and the region surrounding position 406 are each within the larger domain defined as determining the differences in the photosensory specificity and Pfr-specific degradation between phyA and phyB.

As each of these functions requires phototransformation, each of these conformational changes is potentially involved in either or both functions. Similarly, the regions from positions 684-731 and surrounding position 814 both lie within the relatively conserved 'core' region, defined functionally as being necessary for downstream signal transmission. This correlation would be consistent with the notion that photoconversion induced conformational changes expose surfaces in this region necessary for biochemical transfer of perceived informational signals to phytochrome reaction partners.

NH2-terminal domains of phyA and phyB are postulated to contain distinct determinants which recognize specific cognate determinants (protruding triangle or diamond) on separate reaction partners (X1 and X2, respectively) in response to photosignal perception (induced target selection).

Biochemical transfer of the perceived signal to the reaction partner is postulated to be functionally identical for the two photoreceptors and to involve interaction of the 'core' region (stippled area) of the COOH-terminal domain of each with a common determinant (indented triangle) present on all reaction partners (X1, X2, etc.).

Thus, the specificity in signal perception exhibited by phyA and phyB involving the capacity to discriminate between Rc and FRc signals is postulated to be transduced at the first step via selection of different cognate molecular targets that undergo the same local biochemical modification in the process of signal transfer from the photoreceptor.

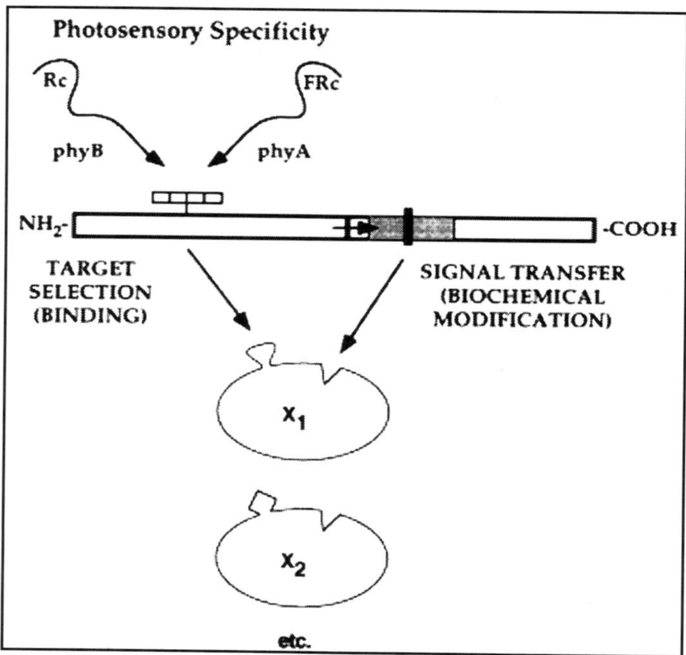

Fig. Two-point Contact model of Phytochrome Action..

It should be noted that, despite the attractiveness of the possibility that the phytochromes may function as photoregulated histidine kinases, none of the mutations affecting regulatory activity thus far reported falls within the postulated kinase domain, and site-directed mutations targeted at conserved residues within this domain failed to disrupt phytochrome activity in vivo. Although there are many possible explanations for these observations, caution is still needed in attempting to transpose the bacterial two-component model directly to the phytochrome system.

Conclusions

Our contemporary view of the phytochrome molecule integrates evidence from earlier biochemical, immunochemical, and spectroscopic studies with that from more recent molecular and genetic studies. Many of the recent advances have come from the capacity afforded by molecular and genetic approaches to perform functional assays on mutagenized molecules in living plants, and to search for similar sequences or structural motifs that have identified functional activities in the rapidly expanding sequence databases. Regions of the molecule potentially involved in reaction partner recognition and in signal transfer have been identified by functional assay, and a domain with tantalizing sequence similarity to the prokaryotic sensor histidine kinases remains the best clue currently available to a potential biochemical mechanism of signal transfer. Despite these advances, however, the picture we have of the photoreceptor molecule remains relatively crude, and the molecular mechanism of phytochrome action remains to be identified. The continued use of the combined power of quantitative photobiological, molecular-genetic and biochemical approaches holds the promise of significant progress on these problems in the near future.

MOLECULAR COMPONENTS OF CELLS

PERMEABILITY BARRIER

The cell membrane is the most dynamic structure in the cell. Its main function is as a permeability barrier that regulates the passage of substances into and out of the cell. The plasma membrane is the definitive structure of a cell since it sequesters the molecules of life in the cytoplasm, separating it from the outside environment. The bacterial membrane freely allows passage of water and a few small uncharged molecules (less than molecular weight of 100 daltons), but it does not allow passage of larger molecules or any charged substances except when monitored by proteins in the membrane called transport systems.

TRANSPORT OF SOLUTES

The presence of transport systems in the membranes allows the bacteria to accumulate solutes and chemical precursors of cell material inside their cytoplasm at concentrations which greatly exceed the concentrations in the environment. Remember, most bacteria live in relatively dilute environments (*e.g.* a lake or stream) where the concentration of the business molecules of life is greater inside of the cell than in the environment. Hence, the bacterial cells must transport their nutrients from the environment and maintain a higher concentration of solutes inside the cell than outside the cell. This comes at a price. To concentrate a substance against the environmental gradient using a membrane transport system always costs energy in one form or another. Bacteria have a variety of types of transport systems which can be used alternatively in various environmental situations. The most important transport systems are called active transport systems since they require energy and concentrate substances inside of the cell. At least 80 per cent of the molecules needed in the cytoplasm are taken up by the process of active transport. Active transport systems are mediated by proteins in the membrane called carrier proteins or "permeases" that are generally quite specific for the substances that they will transport. All active transport systems require energy to operate. Some use chemical energy derived from ATP; others use proton motive force (pmf), which is derived from the establishment of a charge and a pH gradient on opposite sides of the membrane. Besides transport proteins that selectively mediate the passage of substances into and out of the cell, bacterial membranes may also contain sensing proteins that measure concentrations of molecules in the environment or binding proteins that translocate signals from the environment to genetic and metabolic machinery in the cytoplasm.

GENERATION OF ENERGY

Unlike eucaryotes, bacteria don't have intracellular organelles for energy producing processes such as respiration or photosynthesis. Instead, the cytoplasmic membrane carries out these functions. The membrane is the location of electron transport systems (ETS) used to produce energy during photosynthesis and respiration, and it is the location of an enzyme called ATP synthetase (ATPase) which is used to synthesize ATP.

When the electron transport system operates, it establishes a pH gradient across of the membrane due to an accumulation of protons (H^+) outside and hydroxyl ion (OH^-) inside. Thus the outside is acidic and the inside is alkaline. Operation of the ETS also establishes a charge on the membrane called proton motive force (pmf). The outer face of the membrane becomes charged positive while inner face is charged negative, so the membrane has a positive side and a negative side, like a battery. The pmf can be used to do various types of work including the rotation of the flagellum, or active transport as described above. The pmf can also be used to make ATP by the membrane ATPase enzyme which consumes protons when it synthesizes ATP from ADP and phosphate. The connection between electron transport, establishment of pmf, and ATP synthesis during respiration is known as oxidative phosphorylation; during photosynthesis, it is called phosphorylation.

Figure below illustrates the membrane of *E. coli*. The topographical features of the membrane from top to bottom are

1. Lactose transport system;
2. The flagellar motor coupled to the hook and filament;
3. Na^+ transport (export) system;

4. Ca^{++} transport (export) system;
5. Electron transport system;
6. ATPase enzyme;
7. Proline transport system.

The operation ot the electron transport system during respiration produces the H^+ charge on the membrane (pmf). The pmf (H^+) is used by the transport systems to move molecules from one side of the membrane to the other; by the flagellar motor ring to rotate the flagellar filament; and by the ATPase enzyme to synthesize ATP.

THE CYTOPLASMIC MEMBRANE

The cytoplasmic membrane of bacterial cells is a delicate and plastic structure that completely encloses the cell cytoplasm (or protoplasm). The bacterial membrane is composed of 40 per cent phospholipid and 60 per cent protein. The phospholipids are amphoteric molecules, meaning they have a water-soluble hydrophilic region (the glycerol "head") attached to two insoluble hydrophobic fatty acid "tails". In water, such molecules naturally form the molecular bilayer characteristic of membranes. The fatty acid tails from one layer face towards the fatty acid tails of the second layer ("likes dissolve like"), and the glycerol heads naturally turn towards the water (Figure). Dispersed throughout the bilayer are various structural and enzymatic proteins which carry out most membrane functions. This arrangement of proteins and phospholipids forms what is called the fluid mosaic membrane as illustrated in Figure.

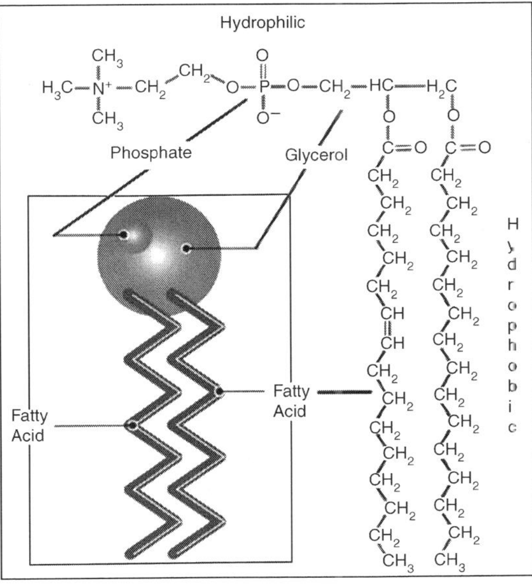

Fig. Molecular Structure of a Phospholipid, the Building block of Membranes. Inset - the usual Depiction of a Membrane Phospholipid Containing a Phosphatidyl-glycerol "head" Attached to two Fatty acid "tails".

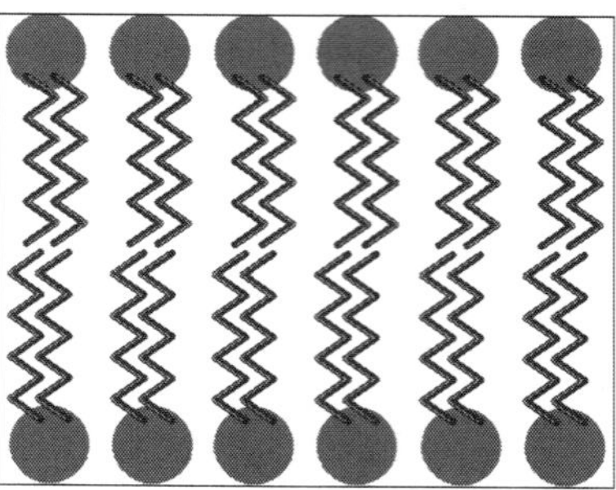

Fig. Organization of Phospholipids in Aqueous Solution to form a Bilayer. The Hydrophilic Phosphatidyl Glycerols form the inner and outer Faces of the Membrane. The Fatty Acids Orient Themselves Towards one Another to form the Hydrophobic Interior of the Membrane.

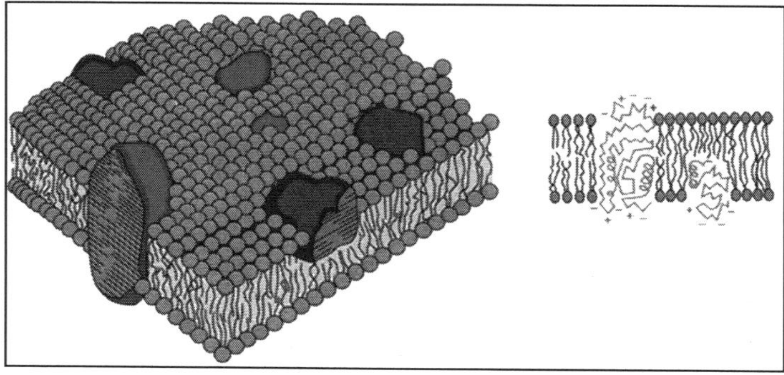

Fig. Fluid Mosaic Model of a Biological Membrane. In Aqueous Environments Membrane Phospholipids Arrange themselves in Such a way that they Spontaneously form a Fluid Bilayer. Membrane Protein may be Either Structural or Functional. Proteins may be Permanently or Transiently Associated with one Side or the other of the Membrane or built into the Bilayer, or they may Span the Bilayer Forming Transport Channels through the Membrane.

PROCARYOTIC CELL ARCHITECTURE

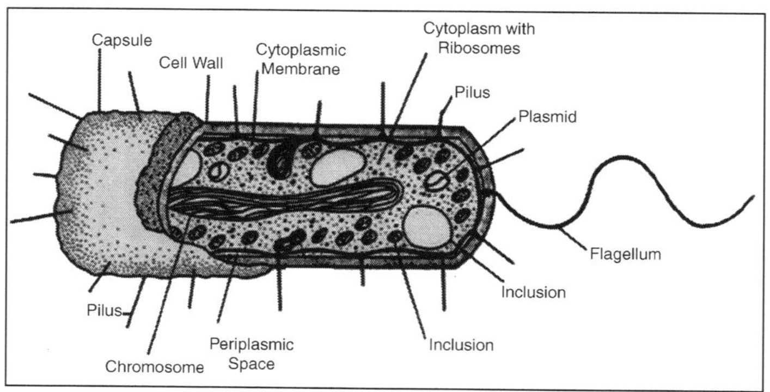

Fig. Schematic Drawing of a Typical Bacterial Cell.

At one time it was thought that bacteria were essentially "bags of enzymes" with no inherent cellular architecture. The development of the electron microscope in the 1950s revealed the distinct anatomical features of bacteria and confirmed the suspicion that they lacked a nuclear membrane. Structurally, a bacterial cell (Figure) has three architectural regions: appendages (attachments to the cell surface) in the form of flagella and pili (or fimbriae); a cell envelope consisting of acapsule, cell wall and plasma membrane; and a cytoplasmic region that contains the cell chromosome (DNA) and ribosomes and various sorts of inclusions. In this lecture, we will discuss the anatomical structures of procaryotic cells in relation to their adaptation, function and behaviour in natural environments.

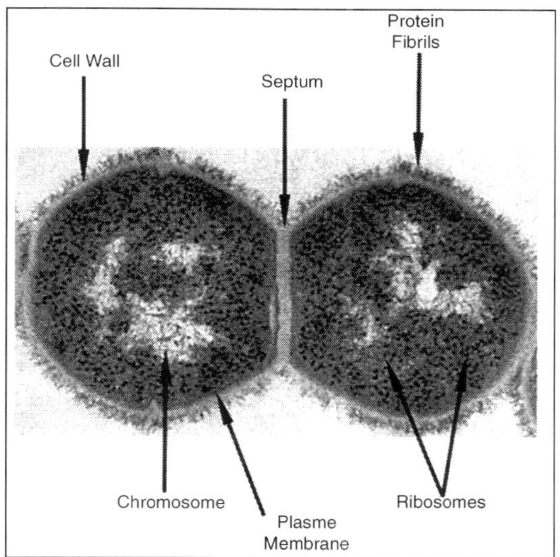

Fig. Electron Micrograph of an Ultra-thin Section of a Dividing Pair of Group A Streptococci (20,000X).

The Cell Surface Fibrils, Consisting Primarily of Protein, are Evident. The Bacterial Cell Wall, to which the Fibrils are Attached, is also Clearly seen as the Light Staining Region between the Fibrils and the Dark Staining cell Interior.

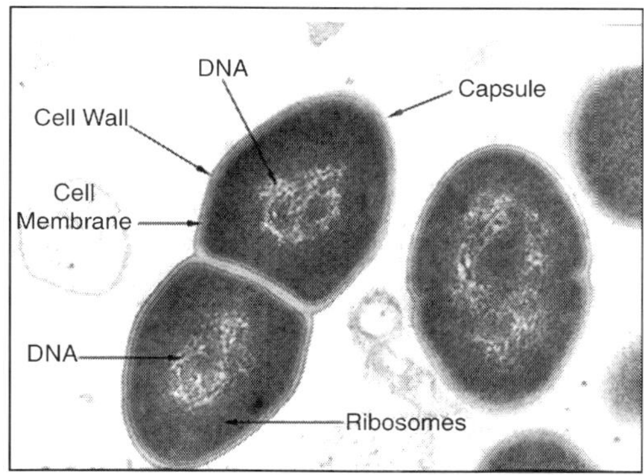

Fig. Electron Micrograph of an Ultra-thin Section of a Dividing Pair of Enterococci

Table. Characteristics of Typical Bacterial Cell Structures.

Structure	Function(s)	Predominant Chemical Composition
Flagella	Swimming movement	Protein
Pili		
Sex pilus	Mediates DNA transfer during conjugation	Protein
Common pili or fimbriae	Attachment to surfaces; protection against phagotrophic engulfment	Protein
Capsules (includes "slime layers" and glycocalyx)	Attachment to surfaces; protection against phagocytic engulfment, occasionally killing or digestion; reserve of nutrients or protection against desiccation	Usually polysaccharide; occasionally polypeptide
Cell wall		
Gram-positive bacteria	Prevents osmotic lysis of cell protoplast and confers rigidity and shape on cells	Peptidoglycan (murein) complexed with teichoic acids
Gram-negative bacteria	Peptidoglycan prevents osmotic lysis and confers rigidity and shape; outer membrane is permeability barrier; associated LPS and proteins have various functions	Peptidoglycan (murein) surrounded by phospholipid protein-lipopolysaccharide "outer membrane"
Plasma membrane	Permeability barrier; transport of solutes; energy generation; location of numerous enzyme systems	Phospholipid and protein
Ribosomes	Sites of translation (protein synthesis)	RNA and protein
Inclusions	Often reserves of nutrients; additional specialized functions	Highly variable; carbohydrate, lipid, protein or inorganic
Chromosome	Genetic material of cell	DNA
Plasmid	Extrachromosomal genetic material	DNA

APPENDAGES

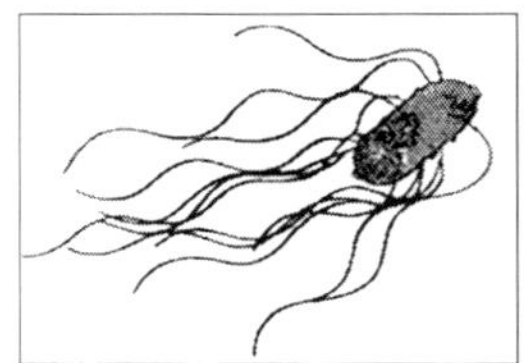

Fig. *Salmonella Enterica. Salmonella* is an Enteric Bacterium Related to *E. coli.* The Enterics are Motile by Means of Peritrichous Flagella.

FLAGELLA

Flagella are filamentous protein structures attached to the cell surface that provide the swimming movement for most motile procaryotes. Procaryotic flagella are much thinner than eucaryotic flagella; the diameter of a procaryotic flagellum is about 20 nanometers, well-below the resolving power of the light microscope. The flagellar filament is rotated by a motor apparatus in the plasma membrane allowing the cell to swim in fluid environments.

The ultrastructure of the flagellum of *E. coli* is illustrated in Figure below. The flagellar apparatus consists of several distinct proteins: a system of rings imbedded in the cell envelope (the basal body), a hook-like structure near the cell surface, and the flagellar filament. The innermost rings, the M and S rings, located in the plasma membrane, comprise the motor apparatus. As the M ring turns, powered by an influx of protons, the rotary motion is transferred to the filament which rotates thereby propelling the bacterium.

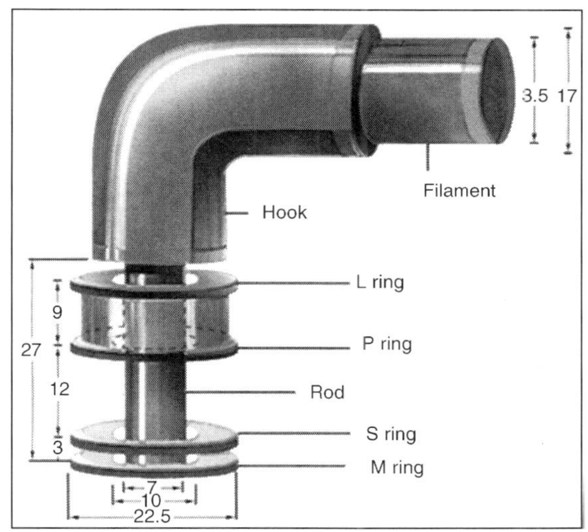

Fig. The Ultrastructure of a Bacterial Flagellum.

Measurements are in Nanometers. The Flagellum of *E. coli* Consists of three Parts, Filament, hook and Basal body, all composed of Different Proteins. The Basal body and hook Anchor the Whip-like Filament to the cell Surface. The Basal body Consists of four ring-shaped Proteins Stacked like Donuts around a Central rod in the cell Envelope.

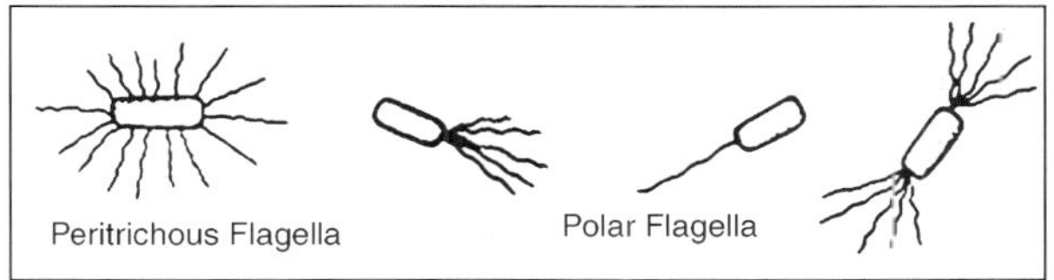

Fig. Different Arrangements of Bacterial Flagella. Swimming Motility, Powered by Flagella, Occurs in Half the Bacilli and Most of the Spirilla.

Flagella may be variously distributed over the surface of bacterial cells in distinguishing patterns, but basically flagella are either polar (one or more flagella arising from one or both poles of the cell) or peritrichous (lateral flagella distributed over the entire cell surface). Flagellar distribution is a genetically-distinct trait that is occasionally used to characterize or distinguish bacteria. For example, among Gram-negative rods, pseudomonads have polar flagella to distinguish them from enteric bacteria, which have peritrichous flagella. Procaryotes are known to exhibit a variety of types of tactic behaviour, *i.e.*, the ability to move (swim) in response to environmental stimuli. For example, duringchemotaxis a bacterium can sense the quality and quantity of certain chemicals in its environment and swim towards them (if they are useful nutrients) or away from them (if they are harmful substances). Other types of tactic response in procaryotes include phototaxis, aerotaxis and magnetotaxis. The occurrence of tactic behaviour provides evidence for the ecological (survival) advantage of flagella in bacteria and other procaryotes.

DETECTING BACTERIAL MOTILITY

Since motility is a primary criterion for the diagnosis and identification of bacteria, several techniques have been developed to demonstrate bacterial motility, directly or indirectly. Flagellar stains outline flagella and show their pattern of distribution. If a bacterium possesses flagella, it is presumed to be motile.

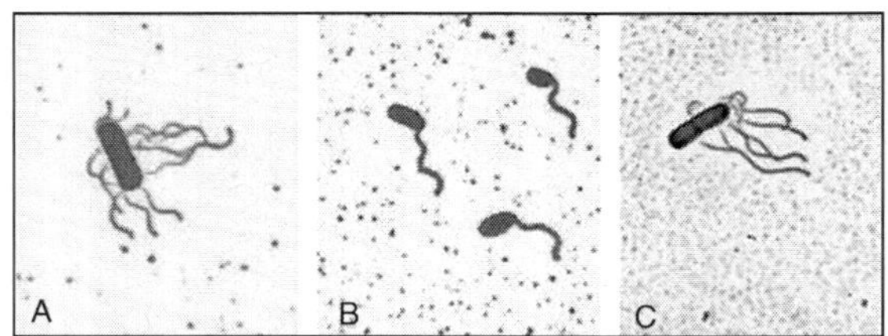

Fig. Flagellar Stains of Three Bacteria a. *Bacillus Cereus* b. *Vibrio Cholerae* c. *Bacillus brevis*. CDC. Since the Bacterial Flagellum is below the Resolving power of the Light Microscope, although Bacteria can be seen Swimming in a Microscope Field, the Organelles of Movement cannot be Detected.

Motility test medium demonstrates if cells can swim in a semisolid medium. A semisolid medium such as 0.75% agar is inoculated with the bacteria in a straight-line stab with a needle. After incubation, if turbidity (cloudiness) due to bacterial growth can be observed away from the line of the stab, it is evidence that the bacteria were able to swim through the medium.

Fig. Bacterial Cultures Grown in Motility test medium. The tube on left is a non Motile Organism; the tube on Right is a Motile Organism. Motility test medium is a semi-soft medium that is Inoculated with a Straight Needle.

Direct microscopic observation of living bacteria in a wet mount. One must look for transient movement of swimming bacteria. Most unicellular bacteria, because of their small size, will shake back and forth in a wet mount observed at 400X or 1000X. This is Brownian movement, due to random collisions between water molecules and bacterial cells. True motility is confirmed by observing a bacterium swim from one side of the microscope field to the other side.

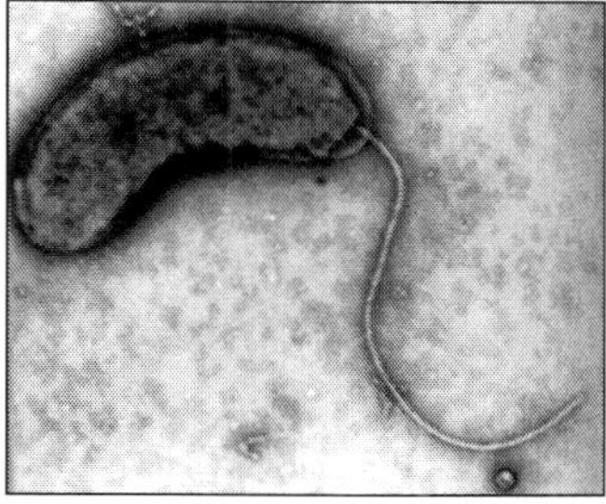

Fig. A *Desulfovibrio* Species. TEM. About 15,000X. The Bacterium is Motile by Means of a Single Polar Flagellum. Of Course, one can Detect Flagella by Means of Electron Microscopy. Perhaps this is an Alternative way to Determine Bacterial Motility, if you happen to have an Electron Microscope.

FIMBRIAE AND PILI

Fimbriae and pili are interchangeable terms used to designate short, hair-like structures on the surfaces of procaryotic cells. Like flagella, they are composed of protein. Fimbriae are shorter and stiffer than flagella, and slightly smaller in diameter. Generally, fimbriae have nothing to do with bacterial movement (there are exceptions, *e.g.* twitching movement on *Pseudomonas*). Fimbriae are very common in Gram-negative bacteria, but occur in some archaea and Gram-positive bacteria as well. Fimbriae are most often involved in adherence of bacteria to surfaces, substrates and other cells or tissues in nature. In *E. coli*, a specialized type of pilus, the F or sex pilus, apparently stabilizes mating bacteria during the process of conjugation, but the function of the smaller, more numerous common pili is quite different

Common pili (often called fimbriae) are usually involved in specific adherence (attachment) of procaryotes to surfaces in nature. In medical situations, they are major determinants of bacterial virulence because they allow pathogens to attach to (colonize) tissues and/or to resist attack by phagocytic white blood cells. For example, pathogenic Neisseria gonorrhoeae adheres specifically to the human cervical or urethral epithelium by means of its fimbriae; enterotoxigenic strains of E. coli adhere to the mucosal epithelium of the intestine by means of specific fimbriae; the M-protein and associated fimbriae of *Streptococcus pyogenes* are involved in adherence and to resistance to engulfment by phagocytes.

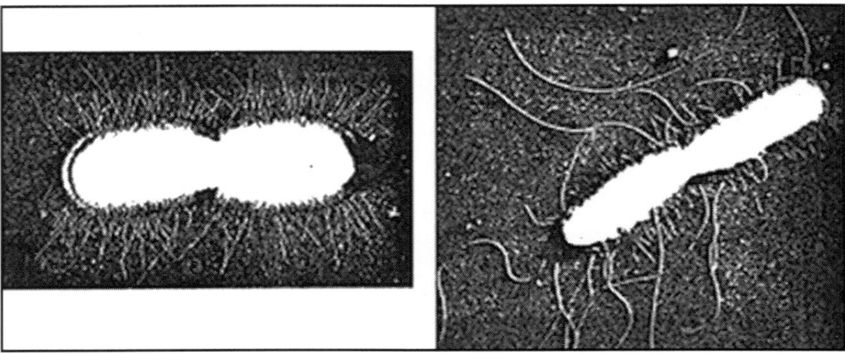

Fig. Fimbriae (Common Pili) and Flagella on the Surface of Bacterial cells. Left: Dividing *Shigella* Enclosed in Fimbriae. The Structures are Probably Involved in the Bacterium's Ability to Adhere to the Intestinal Surface.

Table. Some Properties of Pili and Fimbriae.

Bacterial Species Where Observed	Typical Number on Cell	Distribution on Cell Surface	Function
Escherichia coli (F or sex pilus)	1-4	Uniform	Stabilizes mating during conjugation
Escherichia coli (common pili or Type 1 fimbriae)	100-200	Uniform	Surface adherence to epithelial cells of the GI tract
Neisseria gonorrhoeae	100-200	Uniform	Surface adherence to epithelial cells of the urogenital tract
Streptococcus pyogenes (fimbriae plus the M-protein)	?	Uniform	Adherence, resistance to phagocytosis; antigenic variability
Pseudomonas aeruginosa	10-20	Polar	Surface adherence

CELL MEMBRANE

The membrane that surrounds a cell is made up of proteins and lipids. Depending on the membrane's location and role in the body, lipids can make up anywhere from 20 to 80 per cent of the membrane, with the remainder being proteins.

Cholesterol, which is not found in plant cells, is a type of lipid that helps stiffen the membrane. You may not remember it, but you crossed a membrane to get in here. Every cell is contained within a membrane punctuated with special gates, channels, and pumps. These gadgets let in—or force out—selected molecules. Their purpose is to carefully protect the cell's internal environment, a thick brew (called the cytosol) of salts, nutrients, and proteins that accounts for about 50 per cent of the cell's volume (organelles make up the rest).

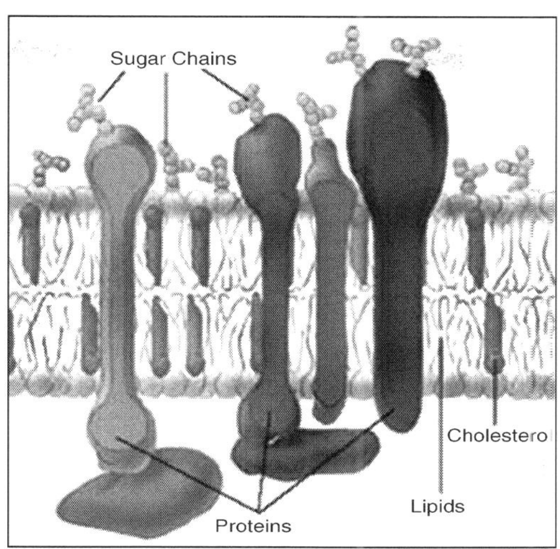

The cell's outer membrane is made up of a mix of proteins and lipids (fats). Lipids give membranes their flexibility. Proteins transmit chemical messages into the cell, and they also monitor and maintain the cell's chemical climate. On the outside of cell membranes, attached to some of the proteins and lipids, are chains of sugar molecules that help each cell type do its job. If you tried to bounce on the cell's outer surface as you did on the nuclear membrane, all these sugar molecules and protruding proteins would make it rather tricky (and sticky).

CHEMICAL COMPOSITION

Regarding chemical composition of the plasma membrane of the human red cell is more than about that of any other cell. Protein represents approximately 52 per cent of its mass, *lipids* 40 per cent, and *carbohydrates* 8 per cent. Oligosaccharides are bound to lipids for example *glycolipids* and, mainly, to proteins for example *glycoproteins).*

There is a wide variation in the lipids-protein ratio between different cell membranes. Myelin is an exception, in the sense that the lipid predominates; in the other cell membranes there is higher protein lipid ratio. It may be observed that in myelin the area occupied by the protein is insufficient to cover that of the lipids, whereas in a red cell ghost the opposite situation is found.

The main lipid components of the plasma membrane are phospholipids, cholesterol, and galactolipids; their proportion varies in different cell membranes. The major proportion of membrane phospholipids is represented by phosphatidyl-choline, phosphatidylethanolamine and sphingomyelin, all of which have no net charge at neutral pH (*i.e., neutral phospholipids)* and tend to pack tightly in the bilayer. This property is also shared by cholesterol.) Five to 20 per cent of the phospholipids are acidic, including: phosphatidylinositol, phosphatidylserine, cardiolipin, phosphatidylglycerol, and sulfolipids. *Acidic phospholipids* are negatively charged and in the membrane are associated principally with proteins by way of lipid-protein interactions. One of the main characteristics of the molecular organization of the plasma membrane is the asymmetry of all of its chemical components.

This refers to their non-uniform distribution between the two surfaces, *i.e.*, the inner or *protoplasmic surface* (P_s), in contact with the ground cytoplasm, and the outer or *external surface* (E_s) in contact with the surrounding fluid medium.

Regarding non-permeant reagents and various *phospholipases (i.e.,* enzymes that hydrolyze different parts of the phospholipid molecule), it has been demonstrated that the distribution of the; *phospholipids is* highly asymmetrical. While the *outer layer* consists mainly of lecithin and sphingomyelin, the *inner layer* is composed mainly of phosphatidylethano-lamine and phosphatidylserine. Furthermore, the *glycolipids* are mainly in the outer half of the bilayer. It is assumed that this asymmetry is rather stable and that there is no exchange of lipids across the bilayer.

Carbohydrates

The distribution of the oligosaccharides is also highly asymmetrical. In both *glycolipids* and *glycoproteins* they are confined exclusively to the external membrane surface. In erythrocyte membranes, hexose, hexosamine, fucose, and sialic acid are bound mainly to proteins. In fact all The proteins present at the outer surface are glycosylated.

Because of the presence of sialic acid residues, as well as carboxyl and phosphate groups, the outer surface of the membrane is negatively charged; consequently, positively charged proteins may be bound by electrostatic interactions to the plasma membrane. Only a small amount of sialic acid exists in the form of *gangliosides* (*i.e.*, glycolipids) in the plasma membrane of liver. However, gangliosides are important constituents of the neuronal surface and are probably involved in ion transfers.

Membrane Proteins

Proteins represent the main component of most biological membranes. They play an important role, not only in the mechanical structure of the membrane, but also as carriers or channels, serving for transport) they may also be involved in regulatory or ligand-recognition properties. In addition, numerous enzymes, antigens, and various kinds of receptor molecules, are present in plasma membranes.

Membrane proteins have been classified as *integral (intrinsic) or peripheral (extrinsic) according* to the degree of their association with the membrane and the methods by which they can be solubilized.

Peripheral proteins are separated by mild treatment, are soluble in aqueous solutions, and are usually free of lipids. Examples include: the above-mentioned *spectrin,* which may be removed from red cell ghosts by chelating agents; cytochrome *c,* found in mitochondria; and acetylcholinesterase, in electroplax membranes, which are easily removed in high salt solutions.

Integral proteins represent more than 70 per cent of the two protein types and require drastic procedures for isolation. Usually they are insoluble in water solutions and need the presence of detergents to be maintained in a non-aggregated form. The study of integral proteins from different membranes has shown that they are rather heterogeneous in relation to molecular weight. These proteins may be attached to oligosaccharides, thus forming glycoproteins.

Polypeptides of the Red Cell Membrane

The usual method used to separate the membrane polypeptides is to dissolve the erythrocyte ghosts in sodium dodecyl sulfate (SDS) (an ionic detergent) and then to use polyacrylamide gel electrophoresis (PAGE). The gel can be stained for proteins (with Coomassie Blue) or for carbohydrates. Densitometric scan of the PAGE which, by the position of the bands, permits estimation of the molecular weight of the polypeptides and, by the surface of the profile, enables determination of its relative mass. For example, *polypeptides 1* and *2* corresponding to *spectrin* (*i.e.*, myosin), represent about 30 per cent of the total protein. *Polypeptide* 3 is the major intrinsic protein and represents about 25 per cent.

Protiens molecular organization is highly asymmetrical, and this can be demonstrated by the use of reagents that are unable to cross the membrane. The reagent is first applied to the *intact erythrocyte* and then to the *white ghost* the difference between the two preparations may give information about the position of a particular protein with respect to the outer or inner surface. Some specific labels can be detected on either side of the membrane by using cytochemical and electron microscopic techniques. For example, using antibodies against spectrin or actin (polypeptide 5) it is possible to show that each is in the cytoplasmic surface of the red cell membrane. On the other hand, acetylcholinesterase can be shown to be in the outer surface by its inactivation with proteolytic enzymes.

It can be said with certainty that *every protein constituent is asymmetrically distributed in the red cell membrane.* All readily soluble polypeptides are localized at the cytoplasmic surface, and those that are at the outer surface are tightly bound to the lipid structure of the membrane.

Major Polypeptides of the Red Cell Membrane

On the cytoplasmic side of the red cell membrane there are *polypeptides 1 and 2* (the myosinlike *spectrin*) and *polypeptide 5 (actin).* These proteins are associated into supramolecular structures forming *microfilaments* which can be identified by electron microscopy, especially by scanning electron microscopy. Under the red cell membrane these structures form a filamentous network that gives stability to the membrane by providing a kind of skeletal support to the fluid lipid bilayer, with its intrinsic proteins. This system of microfilaments may also control the characteristic shape of the erythrocyte. Thus, a possible alteration of the microfilaments has been implicated in a hereditary abnormality of the erythrocytes in which they are spheroidal instead of biconcave discs *(hereditary spherocytosis).*

The molecular weight of the major intrinsic protein is 90,000 daltons, spans the thickness of the membrane and has a small amount of carbohydrate on the pole at the outer surface. This polypeptide is present in the membrane as a dimer held together by S-S bonds. There are between 500,000 and 600,000 such dimers per cell, enough to account for the 8 nm particles observed in freeze-fractured membranes. This dimeric protein appears to be involved in the facilitated *diffusion of anions* (*i.e.*, chloride, bicarbonate) across the membrane.

As per the glycophorin has a molecular weight of 55,000, of which 60 per cent is carbohydrate. Near the COOH end of the molecule there is a region that is very hydrophobic and which interacts with the lipids of the membrane. The COOH end is probably exposed to the interior of the red cell. The NH_2 end is more hydrophilic, is exposed to the external environment, and has the attached oligosaccharides that are at the outer surface of the membrane.

At this surface glycoproteins contain protein-bound antigens of the ABO blood groups and other antigens, such as: the MN groups reacting with rabbit antisera, the influenza virus, phytohemagglutinin, and wheat germ agglutinin. It has been calculated that there are some 700,000 copies of glycophorin per human red cell. This protein accounts for 80 per cent of the carbohydrate and 90 per cent of the negatively charged sialic acid present in the cell surface.

Proteolipids among the most hydrophobic integral proteins of the membrane, are characterized by their strong association with lipids and the fact that they are, soluble in organic solvents. First isolated by Folch and Lees from myelin, proteolipids are found in practically all cell membranes, and in many of them they represent receptor proteins for synaptic transmitters or form channels across those membranes.

Asymmetrical Distribution of Enzymes

More than 30 enzymes have been detected in isolated plasma membranes. Those most constantly found are 5'-nucleotidase, Mg^{2+} ATPase, Na^+—K^+ activated-Mg^{2+} ATPase, alkaline phosphatase adenyl, cyclase, acid phosphomono-esterase, and RNAse. Some enzymes have a preferential localization; for example, alkaline phosphatase and ATPase are more abundant at the bile capillaries, while disaccharidases are present in microvilli of the intestine. A specific localization with a mosaic arrangement has been postulated for some of these enzymes. Disaccharidase forms 5 to 6 nm globular units coating the membrane of the microvilli. The plasma membrane lacks the respiratory chain and glycolytic activity.

Of all the membrane-associated enzymes, Na^+—K^+ activated-Mg^{2+} ATPase is one of the most important because of its role in ion transfer across the plasma membrane. This enzyme is dependent on the presence of lipids and is inactivated when all lipids are extracted. Enzymes also show an *asymmetrical distribution*. Such as in the outer surface of erythrocytes there are acetylcholinesterase, nicotinamide-adenine dinucleotidase, and the ouabain binding site of the Na^+ K^+ ATPase. In the inner surface there is NADH-diaphorase, G3PD, adenylate cyclase, protein kinase, and ATPase.

MOLECULAR MODELS OF THE CELL MEMBRANE

On the molecular structure of the membrane before the isolation of plasma membranes, theories were generally based on indirect information. Since substances soluble in lipid solvents penetrate the plasma membrane easily, Overton postulated in. 1902 that the plasma membrane is composed of a thin layer of lipid. In 1926 Gorter and Grendell found that the lipid content of hemolyzed erythrocytes was sufficient to form a double layer of lipid molecules over the entire cell surface. This theory was also supported by electrical measurements that indicated a high impedance at the plasma membrane. The high impedance is because of the faex fact that it is difficult for ions to penetrate a lipid layer.

Tension at a water-oil interface is about 10 to 15 dynes per centimeter, whereas surface tension of cells is almost nil. It has been postulated that the low tension is due to the presence of protein layers on the lipid components. In fact, when a very small amount of protein is added to a model lipid-water system, the surface tension is lowered comparably.

In 1935 Danielli and Davson explained about all these properties. They proposed that the plasma membrane contained a *lipid bilayer,* with protein adhering to both lipid-aqueous interfaces. Other models containing globular lipid micelles, globular proteins, or a combination of both have been proposed a globular-bilayer transition has also been postulated. These globular models do not account satisfactorily for the high electrical impedance.

On the fine structure of the plasma membrane, electron microscopy has thrown some light and has revealed the numerous structural differentiations that this membrane and the underlying cell cytoplasm have in different cell types. To resolve the structure of the plasma membrane, extremely thin sections (~ 20 nm) must be used, otherwise one would observe different orientations of the plasma membrane with respect to the plane of the section.

The membrane appears most thin when it is exactly perpendicular to the plane of the section. Membranes of *6* to 10 mm have been observed at the surface of all cells. The membranes of two cells that are in close contact appear as dense lines separated by a space of 11 to 15 nm, which is strikingly uniform and contains a material of low electron density. This intercellular component can be considered as a kind of cementing substance.

The Unit Membrane–Re evaluation of the EM Image

There are three layer in cell membrane outer dense layers of 2.0 nm and a middle clear one of about 3.5 nm led Robertson to postulate the so-called *unit membrane* model. The electron microscopic image was interpreted along the lines of the Danielli-Davson model, so that the clear layer corresponded to the hydrocarbon chains of the lipids, and the dense layers, to the proteins. This unit membrane model was also extended to all kinds of intracellular membranes. That this concept is an oversimplication which does not account for the many proteins traversing the membrane can be readily seen by observing some high resolution micrographs in which fine bridges cross the unit membrane.

On the basis of some experiments the unit membrane has been re-evaluated recently methods designed to avoid the removal of proteins from membranes during the preparation for conventional electron microscopy or freeze-fracturing. With both methods a granular appearance of the membrane is obtained, which is more in keeping with the fact that a high proportion of the proteins traverse the membrane. According to these findings the unit membrane image is, to a great extent, artifactual.

The Fluid Mosaic Model

The main source of the knowledge of the molecular organization of biological membrane comes from an integration of the data on chemical analysis and from the application of several biophysical techniques.

The important concepts are:

- That the lipid and integral proteins are disposed in a kind of mosaic arrangement;
- That biological membranes are quasi-fluid structures in which both the lipids and the integral proteins are able to perform translational movements within the overall bilayer. The concept of fluidity implies that the main components of the membrane are held in place only by means of non-covalent interactions.

For better study of the molecular organization of the membrane, it is necessary to remember that, not only the lipids, but also many of the intrinsic proteins and glycoproteins of the membrane, are amphipathic molecules. The term *amphipathy,* coined by Hartley in 1936, refers to the presence, within the same molecule, of hydrophilic and hydrophobic groups these amphipathic molecules constitute liquid crystalline aggregates in which the polar groups are directed towards the water phase and the non-polar groups are situated inside the bilayer.

The integral proteins of the membrane are intercalated to a greater or lesser extent into a rather continuous lipid bilayer. This arrangement is based on the fact that these integral proteins are also amphipathic, with polar regions protruding from the surface and non-polar regions embedded in the hydrophobic interior of the membrane.

It is well recognized that a protein of appropriate size or a cluster of protein subunits may pass across the entire membrane (transmembrane proteins). Such traversing proteins could be in contact with the aqueous solvent on both sides of the membrane. A similar model has been postulated for the cholinergic receptor. This model also reflects the generally accepted view that the major proportion of the phospholipids in the membrane are arranged in a bilayer form.

This mosaic model of the membrane, with intercalating proteins, comes from the use of freeze-fracturing techniques in erythrocytes and other cell membranes. The red cell ghosts show a large number of particles, about 8 nm in diameter, that have been interpreted as representing proteins embedded within the plane of cleavage which passes through the middle of the lipid bilayer. There are between 500,000 and 600,000 such particles in a cell and more are attached to the inner half of the membrane (PF) than to the outer half (EF). We mentioned earlier that these particles represent dimers of the polypeptide found in *band 3,* and that they may also correspond to anionic permeation channels. The degree of dispersion or aggregation of these particles within the membrane appears to be influenced by the *spectrin-actin system.* This suggests that some link between this system and the dimers of band 3 could exist in the living membrane. The mosaic model allows for the characteristic *asymmetry* of the membrane, in which specific components predominate in the outer half or the inner half of the membrane. This implies that there are constraints on the transmembrane rotation of those macromolecules that are well oriented for the carrying of information across the cell membrane. In this model the fluidity of the membrane also implies that both the lipids and the proteins have considerable freedom of movement within the bilayer.

The fluidity of the lipids depends:

- On the degree of saturation of the hydrocarbon chains
- On the ambient temperature. A considerable proportion of the lipids in the membrane are unsaturated, so that the melting point of the bilayer is below body temperature.

Membrane Fluidity

For studying purpose, it can be classified as physical and biological.

The physical techniques are of two main classes:

1. Those that involve a minimal perturbation of the membrane, such as *X-ray diffraction* and *nuclear magnetic resonance (NMR) spectroscopy,*
2. Those that use certain added molecules to monitor specific sites of the membrane. Into this second class fall *fluorescence microscopy,* which uses fluorescent probes and *electron spin resonance (ESR)* spectroscopy.

In practice, these physical techniques can be subdivided on the basis of their ability to examine mainly either the lipid or the protein components, or both. ESR spectroscopy uses paramagnetic probes (*e.g.*, nitroxide-containing amphipathic molecules) which are introduced into the lipid bilayer. This can be done with natural and artificial membranes The data obtained provide information on lipid-protein packing, lateral diffusion of lipids, lipid-protein interactions, the fluidity of the membrane, and the rate of transmembrane rotation (so-called flip-flop) of molecules across the bilayer.

The *biological techniques* involve light and fluorescence microscopy and electron microscopy, including freeze-fracturing and radioisotope labeling methods. One of the simplest methods consists of binding gold or carbon particles to the cell surface and observing under the light microscope the movement of those particles on the surface.

More information has come from studies in which different ligands, such as antibodies and plant lectins, interact with cell surface receptors. If these ligands are labeled with fluorescent dyes their movement can be followed by fluorescence microscopy.

For example, in a classical experiment by Frye and Edidin mouse L cells and human transformed cells which have different surface antigens, were marked with the corresponding antibodies labeled with two distinct fluorescent dyes. These cells were then induced to fuse under the influence of Sendai viruses.

While at the beginning both cell surfaces could be recognized by their differing labels, after 40 minutes considerable intermixing of the antigens had occurred, so that the two labels could no longer be recognized. This intermixing was retarded by temperatures below 20° C.

In surface antigens of cultured muscle cells, a diffusion rate of 1×10^{-9} cm per second was calculated. Also, the so-called clustering or capping effect observed in lymphocytes may result in the displacement of antigenic molecules within the surface membrane. In this case, upon treatment with a labeled antibody, the antigens are first randomly distributed, but after some time they become clustered, forming patches, and agglomerate at one pole of the cell (*i.e.*, capping) where pinocytosis of the antigen-antibody complexes takes place. This process is also inhibited by temperatures that produce solidification of the lipid bilayer.

A number of histochemical techniques using the electron microscope may give information about the fluidity of the membrane. For example, antigens labeled with ferritin or peroxidase can be used. Another extremely valuable tool is freeze-fracturing with which it is possible to demonstrate the movement of the particles in response to different experimental conditions that produce changes in the molecular structure of the membrane.

Membrane Fluidity

The fluid mosaic model has been used to explain the fact that many receptors present in the cell membrane are coupled to enzymes such as adenylate and guanylate cyclase. We can say that they are macromolecules, in most cases of a protein nature, that act at the interface between the cell surface and the environment.

A receptor has the dual function of recognizing a chemical signal (*i.e.*, one of the many regulatory agents, such as peptide hormones, neurotransmitters, prostaglandins, antigens plant lectins, and some bacterial toxins) and of initiating a biological response. The function of recognition is determined by the presence of a site that binds specifically to the ligand (*i.e.*, to the binding portion of a particular regulatory agent). This interaction produces a conformational change that may induce the translocation of ions or the activation of membrane enzymes, such as adenylate cyclase, thus producing cyclic AMP (cAMP).

In a cell there may be several kinds of receptors (at least eight in a fat cell) coupled to the same adenylate cyclase. The ligands do not compete for each specific receptor when binding, and the effect of multiple ligand-receptor bindings on adenylate cyclase is not additive. These findings have been interpreted as indicating that each ligand reacts with a unique receptor and that several receptors may be coupled to a certain number of enzyme molecules in the membrane.

According to the *mobile hypothesis* both the receptors and the enzymes can diffuse independently within the plane of the membrane The greater degree of freedom that this model includes within the lipid realm of the membrane would allow different receptors to couple to a single enzyme.

Fusing cells in which the adenylate cyclase was inactivated but the receptor remained intact, with other cells which contain only the enzyme, resulted in receptor-enzyme coupling. In other words, the receptor of one cell (*e.g.*, the b-adrenergic receptor of an erythrocyte) was coupled to the adenylate cyclase of the other cell (*e.g.*, a tumor cell lacking the receptor). The coupling is manifested by a great increase in cAMP after receptor interaction with the specific ligand (in this case isoproterenol).

The diagram shown below interprets in a highly simplified manner the fusion of two different cells that results in coupling between a receptor from one cell and adenylate cyclase from the other.

The Myelin Sheath and the Photoreceptors

Important information about the molecular structure of the cell membrane is derived from studies of natural multi-layered lipoprotein systems, such as the myelin sheath and the photoreceptors.

The Myelin Sheath: The myelin sheath is lipoprotein membrane that surrounds the axon of the nerve fibre. In peripheral nerves this sheath is formed by the Schwann cells. In central nerves the myelin sheath is produced by the activity of the oligodendroglial cells. It is a very special membrane structure in which the lipids are more abundant than the protein. In central myelin there are two main protein moieties: the proteolipid and the basic protein.

The Myelin Sheath has been known for over a century that the myelin sheath has a strong birefringence, which indicates a high degree of organization at a submicroscopic level. Studies with x-ray diffraction have revealed a spacing of 17 nm in amphibian and 18.0 to 18.5 nm in mammalian peripheral nerves. Within this period, the proportion corresponding to the lipid, protein, and water content has been estimated.

The existence of lecithin-cholesterol and Sphingomyelin-cholesterol complexes is postulated. The first type can be accommodated within the thickness of the lipid layer (L), but in the second, the longer sphingomyelin molecules must interdigitate in order to fill the same space. This model also accounts for the localization of protein (HP) and water (HL) and is in accord with the view that each x-ray diffraction period of 18 nm corresponds to two unit membranes, *i.e.*, to two lipid bilayers separated by a watery space (H), that originally corresponds to extracellular space.

It has confirmed by electron microscopic studies that myelin has a multi-layered membranous structure. In most cases, however, the x-ray diffraction period is reduced to 10 to 12 nm, thereby introducing a great many artifacts. Recently, with a highly polar embedding medium based on polymerized glutaraldehyde-urea, the periodicity obtained was of the same order as by x-ray diffraction. With this technique the lipids are not extracted; this explains the lacks of collapse of the structure.

In nerve conduction to function as an insulator, preventing the dissipation of energy into the surrounding medium. It might act not only as a dielectrics (insulating) material but also as a kind of resonant conductor in which the energy waves resonating in the lipid layers between the protein membranes could pass with maximum speed and minimum loss of energy.

Photoreceptors

The retinal rods and cones are highly differentiated cells that have at their outermost segment a lipoprotein structure that is specialized for photo-reception. Studies with the polarization microscope suggest a submicroscopic organization consisting of transversely oriented protein layers alternating with lipid molecules arranged longitudinally along the axis of the photoreceptor. This type of layered organization has been demonstrated by electron microscopy in fragmented rod outer segments and in thin sections of the retina. These observations indicate that the rod consists of a stack of superimposed disks (several hundred) along the axis. These disks are really flattened sacs made of two membranes, which surround a thin space of 3 nm and become continuous at the edges. The space between the rod sacs is 5 to 12 nm. The cone outer segments, with minor differences, have a similar structure.

To osmotic change rod sacs are highly sensitive. In hypotonic solutions they swell considerably, and the inner space between the membranes becomes very large. The rod outer segments can easily be isolated in the frog and be submitted to chemical analysis. The protein composition is simpler than in other membranes.

The photopigment protein, *rhodopsin,* represents 40 per cent of the total mass of the rod sacs; another 40 per cent is lipid. Rhodopsin is an integral type of protein requiring detergents for extraction. In the pline of the membrane the rhodopsin molecules have a liquid-like distribution. X-ray diffraction studies have led to the interesting conclusion that this molecule may change its position within the lipid layer.

In the dark (non-stimulated condition) the molecule is about one third embedded in the lipid and two thirds exposed to the water surface. Upon activation by light (bleached rhodopsin) the molecule becomes more deeply embedded within the lipid bilayer. This change is presumably caused by changes in surface charges of the molecule.

The process of photoreceptors transform light energy into another type of energy that can be conducted as nerve impulse. It is based on a cycle of chemical reactions which involve the visual pigments present in the protein membranes of the rod and cone sacs. This multilayered structure is a very effective system that facilitates the maximum absorption and utilization of light by the chromophoric groups present in the visual pigments (retinenes). The acute sensitivity of the photoreceptors, which can react to a single photon, can be explained by the fact that the possibility of striking a sensitive molecule is increased by a factor of hundreds or thousands by the molecular organization of the photoreceptor.

It is known that the interaction with photons of light produces liberation of Ca^{2+} ions which, in turn, inhibit the sodium current that is characteristic of the photoreceptor in the dark condition. In this way photoreceptors act as energy transducers, transforming light energy into electrical signals.

ENDOPLASMIC RETICULUM

If you peer over the side of the nucleus, you'll notice groups of enormous, interconnected sacs snuggling close by. Each sac is only a few inches across but can extend to lengths of 100 feet or more. This network of sacs, the endoplasmic reticulum (ER), often makes up more than 10 per cent of a cell's total volume.

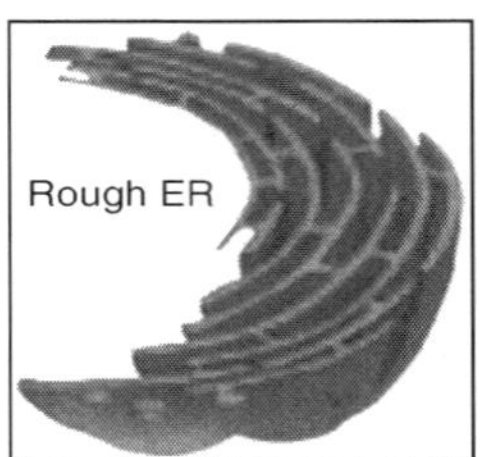

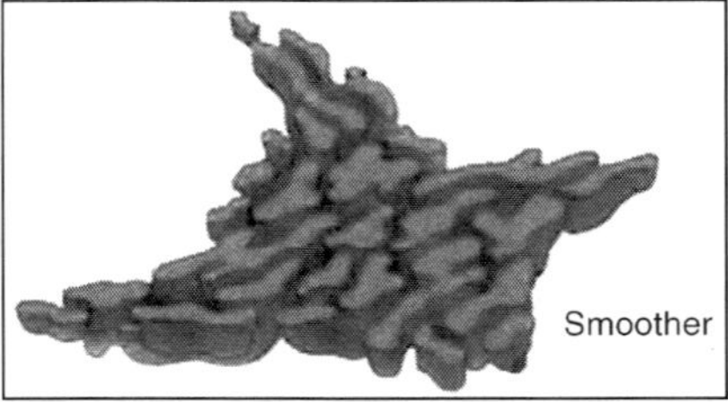

The endoplasmic reticulum comes in two types: Rough ER is covered with ribosomes and prepares newly made proteins; smooth ER specializes in making lipids and breaking down toxic molecules.

Take a closer look, and you'll see that the sacs are covered with bumps about 2 inches wide. Those bumps, called ribosome's, are sophisticated molecular machines made up of more than 70 proteins and 4 strands of RNA, a chemical relative of DNA. Ribosomes have a critical job: assembling all the cell's proteins. Without ribosomes, life as we know it would cease to exist. To make a protein, ribosomes weld together chemical building blocks one by one. As naked, infant protein chains begin to curl out of ribosomes, they thread directly into the ER. There, hard–working enzymes clothe them with specialized strands of sugars.

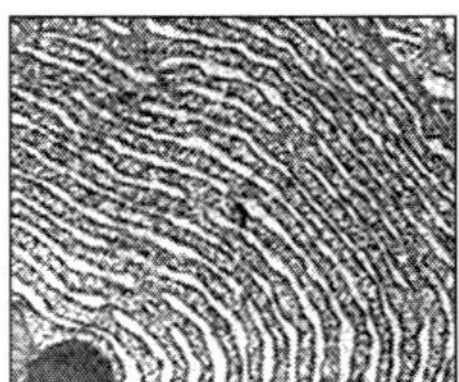

Fig. Rough ER

Now, climb off the nucleus and out onto the ER. As you venture farther from the nucleus, you'll notice the ribosomes start to thin out. Be careful! Those ribosomes serve as nice hand—and footholds now. But as they become scarce or disappear, you could slide into the smooth ER, unable to climb out.

In addition to having few or no ribosomes, the smooth ER has a different shape and function than the ribosome–studded rough ER. A labyrinth of branched tubules, the smooth ER specializes in synthesizing lipids and also contains enzymes that break down harmful substances. Most cell types have very little smooth ER, but some cells—like those in the liver, which are responsible for neutralizing toxins—contain lots of it.

Next, look out into the cytosol. Do you see some free–floating ribosomes? The proteins made on those ribosomes stay in the cytosol. In contrast, proteins made on the rough ER's ribosomes end up in other organelles or are sent out of the cell to function elsewhere in the body. A few examples of proteins that leave the cell (called secreted proteins) are antibodies, insulin, digestive enzymes, and many hormones.

PENETRATION OF IONS THROUGH CELL MEMBRANES

Penetration of ions through cell membranes is of special interest because for instance, Na^+ ions are more concentrated outside the cell (in the body fluids) and K^+ ions are more concentrated inside the cell, yet we know that K ions will pass into the cell and Na ions will pass out. Both these passages thus take place against a concentration gradient and this can only occur by the application of energy. It is claimed that oxidative enzymes and phosphatases play a part in the movement of ions in and out of membranes; a process which is known as "active transport."

Substances which interfere with the activity of these enzymes have been shown in many cases to interfere with the movement in or out of the cell of both sodium and potassium. There are certain specialized cell membranes, *e.g.*, the cell membrane of the intestinal cells, in which the distal edge of the cell is folded to produce a larger number of extremely small fingerlike processes called microvilli; these microvilli are covered with a typical cell membrane. There are a large number of dephosphorylating enzymes localized around these microvilli. Their precise function in this position is not known, but it is almost certain, however, that they play a part in active transport by utilizing the energy derived from the hydrolysis of high-energy phosphates.

It is obvious from what we have said, therefore, that the cell membrane, in general, is a very selective structure anti exercises a good deal of control over the cell by deciding What goes in or out and by enabling the cell to be isolated from its environment. In this way it permits labour to go on in the cell which may be quite different from that going on in a neighbouring cell, or in the body fluids surrounding the cell. Here then is the first division of labour in the cell to be described by us. However the story is not yet ended. In 1952-54, Danielli proposed the structure of the cell membrane which explained the way in which large protein molecules in the cell membrane could permit the passage of ions through the membrane.

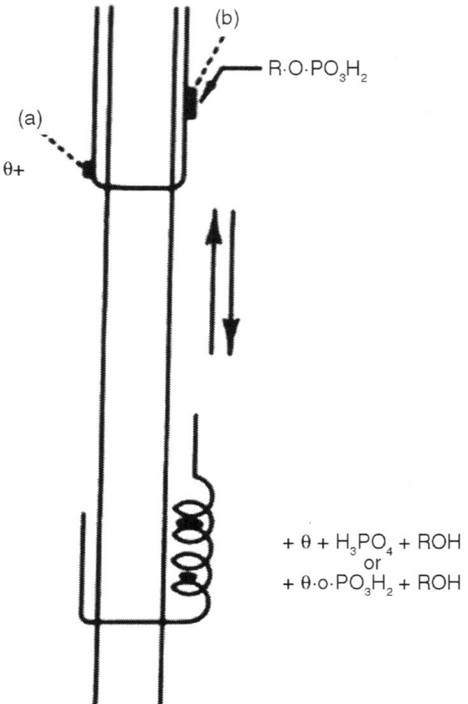

Fig. Diagram of Danielli's Theory of Secretion by a Contractile Protein

Danielli conceived of a long protein molecule extending right through the membrane and passing outside it. This end became attached to an ion and then by contraction pulled the latter through into the interior of the membrane. According to Lundgard and Hodgkin, a cytochrome oxidase energy-producing system is involved in the penetration of ions. They claim that their ion moving system requires adenosine triphosphatase (ATPase), creatine phosphatase and cytochrome oxidase.

Conway has produced a redoxpump hypothesis that attempts to explain the penetration of ions through cell membranes by a mechanism which involves the successive oxidation and reduction of a compound in the membranes; possibly a phospholipid provides the energy for the movement of ions in this way. Phosphatidic acid is now thought to be concerned with ion movements through membranes.

The possible relationship of phosphatases to cell membrane permeability has also been mentioned in connection with the microvilli (brush borders) of intestinal epithelium. The passage of glucose across the mebranes of gut cells could be explained by a mechanism involving these enzymes. It is of interest that in yeast cells too there is some evidence that phosphatases are localized on the membrane and that these enzymes are also involved in the permeability not only to glucose but also to phosphates. Phosphatases are also present on the brush borders of the kidney tubule cells where a good deal of absorption takes place.

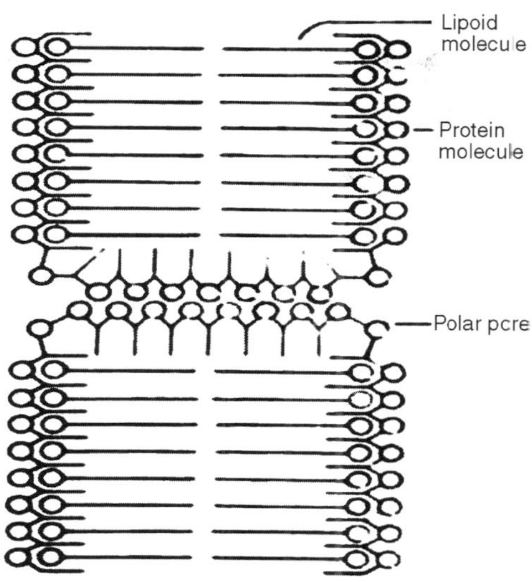

Fig. Diagram of Cell Membrane as Suggested

More recent work has indicated that ribonuclease may play a part in the penetration of cell membranes by ions. Lansing and Rosenthal and Tenardo, for instance, showed that when ribonuclease was added from the outside to certain cells, it modified their permeability to ions; Brachet and Leduc found exactly the same thing for amphibian eggs. According to Brachet "amphibian eggs swell very much when they are immersed in a ribonuclease solution and since cell membranes often give strong cytochemical tests for ribonucleic acid it might well be that the integrity of this nucleic acid is of importance for normal permeability." Danielli made a suggestion that the cell membrane consisted, as suggested before, of bimolecular leaflets of lipoid with protein molecules stretched both inside and outside the membrane, but that it also contains a series of pores which he calls polar pores. In other words, groups of protein molecules are oriented radially with their polar groups directed toward the interior of the pore and such molecules are thus able to control the penetration of the compounds or substances through this pore according to their polar group affinities.

Danielli discusses the association with all membranes of a group of enzymes, which are called "permeases." These compounds he says "permit either facilitated diffusion or active transfer of 'signal' molecules across the membranes." It is possible that insulin may function as a permease for cells of certain organs. To quote Danielli again "Once permease becomes incorporated in the plasma membrane the situation is transformed: specific substrates can penetrate, more permease will become available by induced synthesis, induced enzyme will appear and a whole range of further induced enzymes may appear in response to the action of the induced enzymes of the inducing substrate.

Thus a transient infection with permease may potentially change drastically the state of differentiation of a cell—possibly even result in carcinogenesis." It is very likely that there will be considerable further developments in this subject of permeases and their relation to cell membrane permeability and these will be awaited with interest.

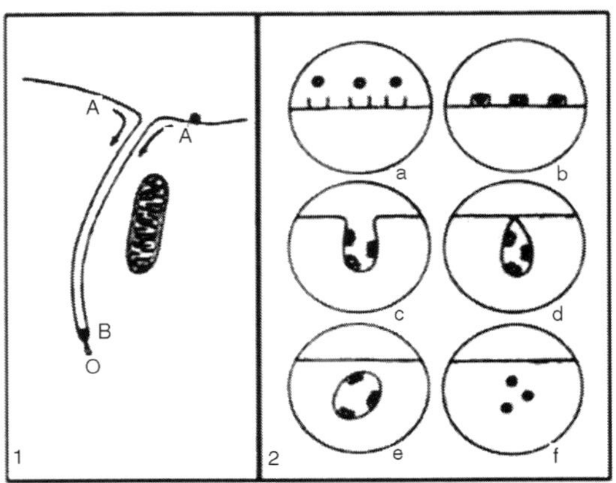

Fig. Diagrammatic Representation of Pinocytosis

A further development which should be noted when considering problems of penetration of compounds into cells is that of pinocytosis. This literally means "drinking by cells" and was first described by Warren Lewis many years ago in tissue culture cells. In this process, these cells were seen by a pseudopodial-like movement to engulf a droplet of culture fluid in which they were lying. The process was equivalent and similar to the phagocytosis of solid food by the *Amoeba*, only in the case of the tissue culture cell it is a droplet of fluid that is engulfed.

Now it seems that the cells of multicellular animals even in vivo can do the same sort of thing though there may be specialized parts of the cell membrane which carry out the process. A case in point is the kidney tubule cell. From the basal part of the cell, long internal extensions of the cell membrane pass a considerable distance up into the interior of the cytoplasm. They are present in some other cells too and are called "caveolae intracellulares."

The cell appears to be able to imbibe droplets of fluid by an engulfing movement of the membranes of these caveolae. It is possible too that the cell membranes of a variety of cells may be able to carry out pinocytosis even without the presence of these intracellular membranous extensions. Thus we see that, if a cell really "wants" to take in a few macromolecules that pass its membranes with difficulty, it can always take them by "the scruff of the neck" and pull them pinocytotically into its interior.

PERMEABILITY OF CELL MEMBRANES

Membrane permeability is a quality of a cell's plasma membrane that allows substances to pass in and out of the cell, so that the cell can expel waste products and ship out the chemicals it assembles for the body. At the same time, the nutrients that the cell needs can pass through the membrane to the inside. Cell membranes have selective permeability, meaning that the membrane will allow certain substances to pass while forming a barrier against others. Cells are like microscopic factories: they design, produce, and package the substances the body needs to survive each day. Just like a factory, a cell needs a way to bring the raw materials for its products — such as nutrients from food — inside its workshop. Once the cell has assembled and packaged a substance, the cell needs a way to ship the finished product out into the bloodstream so that the body can make use of it. The cell membrane is a flexible plasma that envelopes the exterior boundaries of the cell. It separates the intracellular fluid — the fluid within the cells — from the extracellular fluid, which is the fluid outside of the cells. The cell membrane is not a passive or insurmountable wall however, as there is a constant and dynamic exchange of substances between the two fluids.

The various problems associated with the permeability of natural membranes and the account which follows is taken largely from his writings.

PENETRATION OF COMPOUNDS

Danielli points out that there are three sites of resistance to free diffusion in cell membranes:

- The membrane/water interface for diffusion into the membrane;
- The membrane/water interface for diffusion from the membrane into the water;
- The interior of the membrane.

Every molecule which requires to pass into the interior of the cell has to get through these three sites of resistance. A molecule which has polar groups (-OH groups), as opposed, for instance, to nonpolar groups such as methylene, forms at least one hydrogen bond with water for each polar group; all these hydrogen bonds must be fractured simultaneously if the molecule is to penetrate even into the lipoid membrane.

Glycerol, having three OH groups, must acquire sufficient kinetic energy to break three hydrogen bonds simultaneously before it can penetrate into the membrane. This involves a large amount of energy, consequently resistance 1 (diffusion from water to membrane) is so high for glycerol that resistances 2 and 3 are reduced to insignificance.

On the other hand, a molecule such as methyl alcohol which has only one OH group and one CH_3 group penetrates easily into the membrane and can also pass easily out of the lipoid layer into water and presumably ethyl alcohol (C_2H_5OH) can function similarly. Thus we find that for such molecules the rate of diffusion across the interfaces is very large compared with rate of diffusion across the membrane.

The interior of the membrane in the case of these compounds is the most important factor controlling penetration. Differing from these examples are molecules which are predominantly hydrocarbon in nature such as carotene ($C_{40}H_{56}$). With a molecule such as this, resistance I is insignificant; the molecule has no polar groups and thus even resistance 3 is not enormous; however, resistance 2 is very large indeed, because the hydrocarbon groups are hydrophobic and a considerable amount of kinetic energy is required to transfer CH_2 groups from lipid into water. When many such groups are present they must all be transferred simultaneously from the lipoid layer into water, since otherwise the molecule remains substantially part of the lipoid layer and cannot diffuse away into the aqueous phase. With these examples in mind, Danielli classified penetrating molecules into four groups.

- Molecules with few polar and few nonpolar groups —for these resistance 3 is most important and they can penetrate comparatively rapidly *e.g.*, oxygen and methyl alcohol.
- Molecules having a predominantly polar character-for these resistance 1 is most important and penetration is slow, included are glycerols, sugar and glycogen.
- Molecules having few polar and many nonpolar groups—resistance 2 is most important, penetration is slow *e.g.*, carotene, vitamin A and fat.
- Molecules having many polar and many nonpolar groups—resistances 1 anti 2 are both important and penetration is slow *e.g.*, polyhydroxylic bile acids, the glucuronide of oestrin and proteins.

Danielli has pointed out that these principles of membrane permeability have a distinct physiological significance. For instance, oxygen is required in large amounts by the cell and carbon dioxide must be disposed of rapidly and both of these substances can penetrate the cell membrane rapidly. On the other hand, the first products of glucose utilization, for example, by muscle cells are glycerol derivatives, these are valuable and the cell membrane does not let them get through too easily and so they escape only slowly. During sustained work lactic acid is formed—this would be toxic if it accumulated. The cell membrane is relatively permeable to lactic acid which thus can escape into the blood and this permits violent exercise to be maintained for a much longer period than would be the case if the cell membrane was impermeable to lactic acid.

Amino acids penetrate all membranes moderately well but protein penetrates badly and, therefore, amino acids are stored as protein. Fatty acids will penetrate moderately well, but neutral fat very poorly, therefore fatty acids are stored as neutral fat. Glucose which can get in and out of cells fairly rapidly is polymerized in liver cells to form glycogen which passes the cell membrane only with great difficulty. Each of these three latter products, to which the membranes are impermeable, is the results of polymerization of simpler compounds.

It is of interest that, when compounds which are diffusing through a cell membrane from outside to inside are being incorporated into some compound within the cell, they appear to penetrate the membrane more easily. A typical example of this can be drawn from amino acids. If they are being actively incorporated into proteins, the rate of diffusion through the membrane will be much more rapid than if they were not being incorporated. Presumably, if they are being built into proteins, they cease to accumulate and so cease to build up a concentration gradient against the passage of further amino acids.

Detoxification mechanisms also take advantage of the principles of the cell membrane permeability. For instance, toxic substances which might penetrate the cell membrane become conjugated with amino acids, sulfuric acid, or glucuronic acid. Thus toxic cell-penetrating substances such as bromobenzene or menthol are converted into new molecules which penetrate the cells with great difficulty and once in the bloodstream tend to be filtered off by the glomeruli of the kidneys and cannot be reabsorbed from the urine and returned to the bloodstream by the kidney tubules.

CELL PERMEABILITY

The permitting or activating of the passage of substances into, out of, or through cells, or from one cell to another. These materials traverse either the cell surface that demarcates the living cytoplasm from the extracellular space or the boundaries between adjacent cells. In many cases the materials also traverse the cell wall.

For functioning of the living cell and maintaining of satisfactory intercular physiologic conditions permeability is fundamental This function determines which substances can enter the cell; many of which may be necessary to maintain its vital processes and the synthesis of living substances. It also regulates the outflow of excretory material and water from the cell.

The presence of a membrane establishes a net difference between the *intracellular* fluid and the *extracellular* fluid in which the cell is bathed This may be fresh or salt water in unicellular organisms grown in ponds or the sea, but in multicellular organisms the internal fluid, *i.e.*, the blood, the lymph, and especially the *interstitial* fluid, is in contact with the outer surface of the cell membrane one of the functions of the cell membrane is to maintain a balance between the osmotic pressure of the intracellular fluid and that of the interstitial fluid. In higher organisms, the osmotic pressure of the body as a, whole is regulated principally by the kidneys, and the osmotic pressure of the interstitial fluid is about the same as that of the intracellular fluid. In plants, the intracellular fluid has a higher osmotic pressure than the extracellular fluid. The cell is protected from bursting by a rigid cellulose wall. Animal cells generally lack the turgidity that characterizes plant cells. The unfertilized eggs of some marine animals, such as the sea urchin, behave like genuine osmometers. Since they are spheroid, one can, by measuring the diameter, determine the volume and the changes that the egg undergoes with changes in the osmotic pressure of the medium.

The osmotic equilibrium is maintained by means of a contractile vacuole, in many unicellular This "organelle" extracts water from the protoplasm, releasing its contents into the external medium.

Electrical Potentials acroes

In all cells there is a difference in ionic concentration with the extracellular medium and an electrical potential across the membrane. These two properties are intimately related, since the electrical potential depends on an unequal distribution of the ions on both sides of the membrane. The interstitial fluid has a high concentration of Na^+ and Cl^-; and the intracellular fluid, a high concentration of K^+ and of larger organic anions (A^-). Using fine microelectrodes with a tip of 1 mm or less, investigators are able to penetrate through the membrane into a cell and also into the cell nucleus and to detect an *electrical potential* (also called the *resting,* or *steady,* potential), which is always negative inside. The values of the membrane potential vary in different tissues between -20 and -100 millivolts or mv.

Passive Permeability

The passage of molecules and ions across membranes may be the result of two principal mechanisms. Permeability may be *passive,* if it occurs only because of physical laws such as *diffusion* or *active* if it requires energy for transport. There are several kinds of permeability which will be considered in the following sections. Without an intervening membrane, when two solutions of different concentration are mixed a process of intermixing called *diffusion* occurs. For example, if a concentrated solution of sugar is placed in contact with water, there will be a net movement (*i.e.*, flux) of the solute from the region of higher concentration to that of a lower concentration. In this case the higher the difference in concentration between the two solutions (*i.e.*, the *concentration gradient),* the more rapid the rate of diffusion.

The presence of a lipoprotein membrane, such is the plasma membrane, greatly modifies this diffusion or passive permeability. Overton demonstrated that substances that dissolve in lipids pass more easily into the cell, and Collander and Barlund, in their classic experiments with the cells of the plant *Chara,* demonstrated that the rate at which substances penetrate depends on their solubility in lipids and the size of the molecule. The more soluble they are, the more rapidly they penetrate, and with equal solubility in lipids the smaller molecules penetrate at a faster rate.

The permeability (P) of molecules across the membrane is:

$$P = \frac{KD}{t}$$

with K, the partition coefficient; D, the diffusion coefficient, which depends on the molecular weight; and t, the thickness of the membrane The partition coefficient in most cell membranes is similar to that of olive oil and water.

The *partition coefficient* can be measured by mixing the solute with an oil-water mixture and then waiting until the phases are separated. The coefficient is the concentration of the solute in the oil, divided by the concentration in the aqueous phase. Today, the *diffusion coefficient* can be easily determined by using radioactive solutes and measuring their rate of entry into the cytoplasm at various external concentrations. From the experiments. It is evident that many molecules may enter the cell by simple diffusion if the external concentration is higher than the internal. On the other hand, waste products accumulated in the cytoplasm may be eliminated by diffusion.

DEPENDENT ON THE CONCENTRATION AND ELECTRICAL GRADIENTS

The diffusion of ions across membranes is even more difficult because the same is depends not only on the concentration gradient but also on the electrical gradient present in the system. Within the cell there is a large concentration of non-diffusible anions (A^-) and that an electrical gradient across the membrane is established.

Donnan predicted in 1911 that if a theoretical cell, having a non-diffusible negative charge inside, is put in a solution of KC1, K^+ will be driven into the cell by both the concentration and the electrical gradients; Cl^-, on the other hand, will be driven inside by the concentration gradient, but will be repelled by the electrical gradient. As shown by Donnan, the equilibrium concentrations will be exactly reciprocal:

$$\frac{[K^+_{in}]}{[K^+_{out}]} = \frac{[Cl^-_{out}]}{[Cl^-_{in}]}$$

According to Donnan equilibrium involving only physical forces (*i.e.*, without expenditure of energy by the membrane) was confirmed by the demonstration that the membrane potential was negative on the inside and was accompanied by a high K^+ and a low Cl^- concentration. The relationship between the concentration gradient and the resting membrane potential is given by the Nernst equation:

$$E = RT \log \frac{C_1}{C_2}$$

where E is given in millivolts, R is the universal gas constant, and T is the absolute temperature. Donnan equilibrium for KG can now be expressed as follows:

$$E = RT \log \frac{[K^+{}_{in}]}{[K^+{}_{out}]} = RT \log \frac{[Cl^-{}_{out}]}{[Cl^-{}_{in}]}$$

As per (4) any increase in the membrane potential will cause an increase in the ion asymmetry across the membrane, and *vice-versa*. While the first measurements of membrane potentials and ion concentration seemed to confirm this type of *passive* or *diffusion equilibrium,* more precise determinations in different cell types demonstrated that this was not the case.

Active Transport of Ions

Extra than the diffusion movement of neutral molecules and ions across membranes, cell permeability includes a series of mechanisms that require energy. These medianisms are generally described as *active transport.* Adenosine triphosphate (ATP), which is produced mainly by oxidative phosphorylation in mitochondria, is generally used as the source of energy. For this reason, active transport is generally related to, or coupled with cell respiration.

There is explained active transport against a concentration gradient in which water has to be moved upstream (*i.e.*, against gravity). The osmotic work to be done is expressed by the Nernst equation, ft charged molecule crossing through an electrochemical gradient also may imply expenditure of energy. For example, to maintain a low intracellular concentration (Na^+, the cell must extrude sodium against a gradient (*i.e.*, higher Na^+ concentration outside). In addition, it must do this against an electrochemical barrier since the membrane is negative inside and positive outside.

A "Sodium Pump"

During transportation of ion against an electrochemical gradient an extra consumption of oxygen is required. It is calculated that 10 per cent of the resting metabolism of a frog muscle is used for transport of sodium ions. This consumption may increase to 50 per cent in some experimental conditions in which the muscle is stimulated. The resting membrane potential is due to active transport may be demonstrated in plant and animal cells that have been metabolically blocked by anoxia or specific poisons. In this case leakage of K^+ occurs and the potential may decrease to zero. This is clearly observed when anoxia is combined with the poisoning of glycolysis. As suggested by Krogh in 1946. The membrane potential is not really at equilibrium but in a "steady state" involving the constant expenditure of energy.

There is a example of active transport has been provided by experiments with isolated frog skin. The epithelium is specialized to transport Na^+ from the pond water to the interstitial fluids, and by this mechanism the frog can trap this essential ion for use in different tissues. The isolated skin can be kept alive for many hours and used as a wall between two chambers in which the ionic concentration and other factors are changed experimentally.

By means of this preparation, a difference in potential across the skin has been demonstrated: the inside surface is positive with respect to the outside. The sodium ions are transported from the outer toward the inner surface, and the current produced is due to the flux of sodium. At was also observed that the antidiuretic hormone of the neurohypophysis stimulates the transport of sodium and water. Similar findings have been observed in the isolated toad bladder and the kidney tubules of *Necturus*.

Ionic transport is intense in the salivary and sweat-producing glands, and even more so in the glands of the stomach that produce H^+ and Cl^- that must be replaced by the blood. Transport of ions is fundamental to maintenance of the osmotic equilibrium of the cell, the required concentration of anions and cations and the special ions needed for the functioning of the cell. Together with the extrusion of Na^+, which is continuously pumped out by the cell, there is an exit of water molecules.

In this way the cell keeps its osmotic pressure constant. Potassium ions, which are concentrated inside the cell, must pass against a concentration gradient. This can be achieved by a "pumping" mechanism at the expense of energy.

As explained earlier, Na^+ also is transported by an active process, which is sometimes called the "sodium pump." The passive (downhill) fluxes are distinguished from the active (uphill) fluxes. Notice that the active pumping out of Na^+ is the main mechanism for maintaining a negative potential inside the membrane of -50 mv. This diagram demonstrates that the distribution of ions across the membrane depends on the summation of two distinct processes: (1) simple electrochemical diffusion forces which tend to establish a Donnan equilibrium (*i.e.*, passive transport) and (2) energy-dependent ion transport processes (*i.e.*, active transport)

It may be observed that the Na^+— K^+ pump drives ions extruding Na^+ and taking in K^+ (active fluxes). At the same time the passive Na^+ influx depends on a large driving force (D.F.), resulting from the concentration and voltage gradients, and only a slight permeability of the membrane to Na^+ (P_{Na}). Similarly, the net passive K^+ efflux results not from a small driving force, but from a greater permeability to K^+ (P_k).

Ionic Transport through Charge Pores in the Membrane

The molecular machinery involved in ionic transport is located within the cell membrane. This has been demonstrated in two key materials. For example, if red blood cells are hemolyzed so that only the cell membrane remains, they can be filled again with appropriate solutions containing ions and ATP, and Na^+ is transported and K^+ is taken up as in a normal cell. The giant axon of the squid, which has a diameter of about 0.5 mm, can be emptied of the axoplasm and then refilled with solutions of different electrolytes. The transport of ions against a concentration gradient, steady potentials, and even action potentials with the conduction of impulses can be obtained in this preparation in which most of the axoplasm is lacking and the excitable membrane left alone. The utilization of radioscopes demonstrated that ions can enter into the cell rapidly without obvious osmotic effects. It was then suggested that an ionic interchange across the membrane could take place through electrically charged pores. Knowledge about the diameter of the different ions in the hydrated state is particularly pertinent. In this respect it is interesting to remember that the sodium ion, although smaller than K^+ and CI^- in weight, is large in the hydrated condition and enters with more difficulty into the cell.

The pore could be envisioned as the interstice between four adjacent protein subunits which could form a hydrophilic channel across the membrane. The ion channels for Na^+ and K^+ are represented in the polarized or resting condition, or in the state of depolarization that occurs during the action potential. It is postulated that the Na^+ and K^+ channels are closed in the polarized resting membrane.

The rest of the diagram shows the opening of the Na^+ pores at the peak of depolarization of the action potential and the opening of the K^+ pores during repolarization. In red blood cell the total area of the pores has been estimated to be on the order of 0.06 per cent of the surface area. This means that a 0.7 nm pore would be surrounded by a nonporous square 20 ´ 20 nm. These findings indicate that the cell uses only a minute fraction of its surface area for ionic interchange.

Anion Transport

From tissues to the lung the red cell membrane is endowed with an extremely active anion permeation mechanism to transport CO_2 This system can exchange chloride and bicarbonate ions across the membrane. Using several chemical probes that inhibit anion permeability, knowledge has been gained about the molecular localization of this mechanism. Some of the probes can be transported by the anion system and, under certain conditions, can be fixed covalently to it. In this way it is possible to determine not only the number of permeation sites but also the polypeptide in which they are localized. This approach has permitted the identification of the *polypeptide of band 3* with the site involved in the transport. This protein spans the membrane and may exist as a dimer (or tetramer). (We have seen earlier that it also corresponds to the 8 nm particles observed in freezefractured membranes.

The proposed model for anion transport is that of a continuous proteinaceous aqueous channel across the membrane having, near the outer surface, one anion binding site with three positive charges, and a hydrophobic barrier to limit the free diffusion of anions. Passage through the barrier could occur only as a consequence of the binding of the anion. In the above model it is assumed that the segment containing the binding site could exist in two conformations, one facing outward, the other inward. In this way the segment could act as a gate in the proteinaceous channel, swinging between the two positions; this would permit the binding site to interact with anions coming from the outside or from the interior of the cell. In this model it is observed that the anion binding site is located in a 65,000 daltons segment of the protein which remains in the membrane after a 35,000 daltons segment has been separated by a proteolytic enzyme.

The Vectorial Function of N^+ K^+ ATPase

Previously we have studied that the so-called *sodium pump* is a mechanism of active transport by which Na^+ is eliminated from the cell. This mechanism, discovered by Hodkin and Keynes in 1955, was soon associated by Skou with the Na^+ K^+ ATPase. This enzyme is able to couple the hydrolysis of ATP with the removal of Na^+ from the cytoplasm against an unfavourable electrochemical gradient.

It may be observed that the hydrolysis of one ATP provides the energy for the linked transport of 2 K^+ ions toward the inside, and 3 Na^+ ions toward the outside of the cell. This diagram shows the vectorial characteristics of the enzyme which is sensitive to ATP on the inside of the membrane, but not on the outside. This ATPase is stimulated by a mixture of both Na^+ and K^+ and is inhibited specifically by the cardiotonic glycoside *ouabain.* The Na^+ and K^+ sites are independent and are competitively inhibited by K^+ and Na^+, respectively.

The Na^+ K^+ ATPase has been found to be concentrated in membranes of nerves, brain, and kidney and is particularly rich in the electric organ of the eel and the salt gland of certain marine birds (*e.g.*, albatross) in which the transport of ions is extremely active. The vectorial properties of the enzyme can be demonstrated cytochemically; thus, by electron microscopy the liberation of phosphate has been shown to occur exclusively on the inner side of the red cell ghost. On the other hand, ouabain binds exclusively to the outer surface of the membrane. The vectorial function of Na^+K^+ ATPase can be best explained by the so-called "carrier" hypothesis. In its most simple form a carrier consists of. (1) a membrane protein having a specific site to combine with the ligand to be translocated. (2) a mechanism of translocation Which moves the ligand from one side of the membrane to the other. (3) a mechanism for the release of the ligand. Most carrier models involve some kind of conformational change of the protein by which the affinity of the binding site for the ligand changes during the translocation. The case of Na^+ K^+ ATPase it is possible to envision that the enzyme has binding sites for Na^+ and K^+ and a complex carrier mechanism to deliver and release these ions at opposite sides of the membrane. The coupling of the barrier mechanism to the hydrolysis of ATP generates the energy needed to move the ions against the concentration gradients.

There are two steps of transportation. Step. I consists of the formation of a *covalent phosphoenzyme intermediate* (E ~ P) on the inner side of the membrane and in the presence of Na^+. This first reaction is inhibited by Ca^{2+}, but not by ouabain. In Step II the [Na^+ E ~ P] complex is hydrolyzed to form free enzyme and Pi. This reaction requires K^+, added to the outside of the membrane, and is inhibited by ouabain, which competes with K^+. Steps I and II can be represented as given below:

$$\text{Step I } Na^+_{in} + ATP + E \longrightarrow [Na^+ \cdot E \sim P] + ADP$$

$$\text{Step II } [Na^+ \cdot E \sim P] + K^+_{out} \longrightarrow Na^+_{out} + K^+_{in} + Pi + E$$

Na^+ and K^+ are ultimately translocated in the opposite directions from which they were bound. The Na^+ K^+ ATPase is tightly bound to the cell membrane, from which it may be released by the use of detergents. The molecular weight of the enzyme has been estimated to be about 670,000, and it probably contains several polypeptides. Certain lipids are needed for the activity of this enzyme. In a red cell there are about 5000 enzyme molecules, each of which may extrude 20 Na^+ ions per second.

STRUCTURE OF DNA AND RNA

STRUCTURE OF DNA

Most biological experiments are done with samples of living matter- cells, tissues, whole organisms or extracts of these materials. Only a few experiments are done with mathematical or physical models of biological components. Ocassionally, however, a model experiment gives an insight that would be difficult to obtain in any other way. This was true of the discovery of the structure of DNA. A crucial experiment was done by James Watson in 1952 using nothing more complicated than chemical models cut out of cardboard.

INHERITANCE

While the period from the early 1900s to World War II has been considered the "golden age" of genetics, scientists still had not determined that DNA, and not protein, was the hereditary material. However, during this time a great many genetic discoveries were made and the link between genetics and evolution was made. Friedrich Meischer in 1869 isolated DNA from fish sperm and the pus of open wounds. Since it came from nuclei, Meischer named this new chemical, nuclein. Subsequently the name was changed to nucleic acid and lastly to deoxyribonucleic acid (DNA).

Robert Feulgen, in 1914, discovered that fuchsin dye stained DNA. DNA was then found in the nucleus of all eukaryotic cells. During the 1920s, biochemist P.A. Levene analysed the components of the DNA molecule. He found it contained four nitrogenous bases: cytosine, thymine, adenine, and guanine; deoxyribose sugar; and a phosphate group. He concluded that the basic unit (nucleotide) was composed of a base attached to a sugar and that the phosphate also attached to the sugar. He (unfortunately) also erroneously concluded that the proportions of bases were equal and that there was a tetranucleotide that was the repeating structure of the molecule. The nucleotide, however, remains as the fundemantal unit (monomer) of the nucleic acid polymer. There are four nucleotides: those with cytosine (C), those with guanine (G), those with adenine (A), and those with thymine (T). During the early 1900s, the study of genetics began in earnest: the link between Mendel's work and that of cell biologists resulted in the chromosomal theory of inheritance; Garrod proposed the link between genes and "inborn errors of metabolism"; and the question was formed: what is a gene?

The answer came from the study of a deadly infectious disease: *pneumonia*. During the 1920s Frederick Griffith studied the difference between a disease-causing strain of the pneumonia causing bacteria (*Streptococcus peumoniae*) and a strain that did not cause pneumonia. The pneumonia-causing strain (the S strain) was surrounded by a capsule. The other strain (the R strain) did not have a capsule and also did not cause pneumonia. Frederick Griffith was able to induce a nonpathogenic strain of the bacterium *Streptococcus pneumoniae* to become pathogenic. Griffith referred to a transforming factor that caused the non-pathogenic bacteria to become pathogenic.

Griffith injected the different strains of bacteria into mice. The S strain killed the mice; the R strain did not. He further noted that if heat killed S strain was injected into a mouse, it did not cause pneumonia. When he combined heat-killed S with Live R and injected the mixture into a mouse (remember neither alone will kill the mouse) that the mouse developed pneumonia and died. Bacteria recovered from the mouse had a capsule and killed other mice when injected into them!

Hypotheses:

- The dead S strain had been reanimated/resurrected.
- The Live R had been transformed into Live S by some "transforming factor".

Further experiments led Griffith to conclude that number 2 was correct. In 1944, Oswald Avery, Colin MacLeod, and Maclyn McCarty revisited Griffith's experiment and concluded the transforming factor was DNA. Their evidence was strong but not totally conclusive. The then-current favourite for the hereditary material was protein; DNA was not considered by many scientists to be a strong candidate.

The breakthrough in the quest to determine the hereditary material came from the work of Max Delbruck and Salvador Luria in the 1940s. Bacteriophage are a type of virus that attacks bacteria, the viruses that Delbruck and Luria worked with were those attacking *Escherichia coli*, a bacterium found in human intestines. Bacteriophages consist of protein coats covering DNA. Bacteriophages infect a cell by injecting DNA into the host cell. This viral DNA then "disappears" while taking over the bacterial machinery and beginning to make new virus instead of new bacteria. After 25 minutes the host cell bursts, releasing hundreds of new bacteriophage.

Phages have DNA and protein, making them ideal to resolve the nature of the hereditary material. In 1952, Alfred D. Hershey and Martha Chase conducted a series of experiments to determine whether protein or DNA was the hereditary material. By labeling the DNA and protein with different (and mutually exclusive) radioisotopes, they would be able to determine which chemical (DNA or protein) was getting into the bacteria.

Such material must be the hereditary material (Griffith's transforming agent). Since DNA contains Phosphorous (P) but no Sulpher (S), they tagged the DNA with radioactive Phosphorous-32. Conversely, protein lacks P but does have S, thus it could be tagged with radioactive Sulpher-35. Hershey and Chase found that the radioactive S remained outside the cell while the radioactive P was found inside the cell, indicating that DNA was the physical carrier of heredity.

Erwin Chargaff analysed the nitrogenous bases in many different forms of life, concluding that the amount of purines does not always equal the amount of pyrimidines (as proposed by Levene). DNA had been proven as the genetic material by the Hershey-Chase experiments, but how DNA served as genes was not yet certain. DNA must carry information from parent cell to daughter cell. It must contain information for replicating itself. It must be chemically stable, relatively unchanging. However, it must be capable of mutational change. Without mutations there would be no process of evolution.

Many scientists were interested in deciphering the structure of DNA, among them were Francis Crick, James Watson, Rosalind Franklin, and Maurice Wilkens. Watson and Crick gathered all available data in an attempt to develop a model of DNA structure. Franklin took X-ray diffraction photomicrographs of crystalline DNA extract, the key to the puzzle. The data known at the time was that DNA was a long molecule, proteins were helically coiled (as determined by the work of Linus Pauling), Chargaff's base data, and the x-ray diffraction data of Franklin and Wilkens.

DNA is a double helix, with bases to the centre (like rungs on a ladder) and sugar-phosphate units along the sides of the helix (like the sides of a twisted ladder). The strands are complementary (deduced by Watson and Crick from Chargaff's data, A pairs with T and C pairs with G, the pairs held together by hydrogen bonds).

Notice that a double-ringed purine is always bonded to a single ring pyrimidine. Purines are Adenine (A) and Guanine (G). We have encountered Adenosine triphosphate (ATP) before, although in that case the sugar was ribose, whereas in DNA it is deoxyribose. Pyrimidines are Cytosine (C) and Thymine (T). The bases are complementary, with A on one side of the molecule we only get T on the other side, similarly with G and C. If we know the base sequence of one strand we know its complement.

HEREDITARY MATERIAL

DNA was proven as the hereditary material and Watson et al. had deciphered its structure. What remained was to determine how DNA copied its information and how that was expressed in the phenotype. Matthew Meselson and Franklin W. Stahl designed an experiment to determine the method of DNA replication. Three models of replication were considered likely.

Dispersive Replication

Involved the breaking of the parental strands during replication, and somehow, a reassembly of molecules that were a mix of old and new fragments on each strand of DNA. The Meselson-Stahl experiment involved the growth of *E. coli* bacteria on a growth medium containing heavy nitrogen (Nitrogen-15 as opposed to the more common, but lighter molecular weight isotope, Nitrogen-14). The first generation of bacteria was grown on a medium where the sole source of N was Nitrogen-15. The bacteria were then transferred to a medium with light (Nitrogen-14) medium. Watson and Crick had predicted that DNA replication was semi-conservative. If it was, then the DNA produced by bacteria grown on light medium would be intermediate between heavy and light.

It was. DNA replication involves a great many building blocks, enzymes and a great deal of ATP energy (remember that after the S phase of the cell cycle cells have a G phase to regenerate energy for cell division). Only occurring in a cell once per (cell) generation, DNA replication in humans occurs at a rate of 50 nucleotides per second, 500/second in prokaryotes. Nucleotides have to be assembled and available in the nucleus, along with energy to make bonds between nucleotides. DNA polymerases unzip the helix by breaking the H-bonds between bases. Once the polymerases have opened the molecule, an area known as the replication bubble forms (always initiated at a certain set of nucleotides, the origin of replication). New nucleotides are placed in the fork and link to the corresponding parental nucleotide already there (A with T, C with G). Prokaryotes open a single replication bubble, while eukaryotes have multiple bubbles. The entire length of the DNA molecule is replicated as the bubbles meet. Since the DNA strands are antiparallel, and replication proceeds in thje 5' to 3' direction on EACH strand, one strand will form a continuous copy, while the other will form a series of short Okazaki fragments.

UNRAVELING DNA

Encoded into the double-helical strands of DNA, the human genome is the complete set of instructions required to make a human being. While the mapping of the human genome (roughly three billion components) is certainly a Herculean accomplishment, it is only the first step towards realizing the full potential of genomic medicine in the diagnosis, monitoring, and treatment of disease. Beyond knowing what the genetic blueprint says, scientists must understand how that blueprint gets interpreted, or "expressed" as an individual with unique traits. The pol II enzyme is the catalyst for a major step in this process.

As a pol II molecule slides along a DNA molecule, it "unzips" the strands of the DNA double helix, synthesizes a complementary strand of RNA (which will carry the genetic information to where it is needed), and verifies that no mistakes have occurred. This process is regulated by transcription factors—separate molecules that bind to pol II and determine which genes are expressed, at what stage of development, and in which tissue.

Done correctly, this process results in healthy cell growth and differentiation; otherwise, aberrations such as cancer can be the result. Thus, details of the structure of pol II, including information about its binding sites and how they interact with transcription factors, will provide valuable insight into the detailed mechanisms underlying the flow of genetic information from DNA to RNA to protein, which is necessary for life and health.

Without hydrogen bonds there could be no life because they hold the double helix of DNA together, and this they do by charge attractions. A hydrogen bonded to oxygen, or nitrogen, becomes slightly positively charged which enables it to attract a centre of negative charge on another molecule, such as another oxygen or nitrogen atom. The hydrogen bond is then written, *e.g.*, O–H...N, with the dotted line signifying the hydrogen bond. There are also O–H...O, N–H...O and N–H...N bonds, the last being among the weakest.

The secondary effects they have on structures, molecular vibrations, etc., can be used to infer hydrogen bonding, but there is no primary way of observing them because of their inherent weakness. NMR appears to be the least useful technique because neither of the common isotopes, oxygen-16 or nitrogen-14, has a magnetic nucleus. However, nitrogen-15 has a magnetic moment, and by replacing ^{14}N by N, Grzesiek has opened up a new area of investigating these enigmatic bonds. Paper reports for the first time the direct observation by NMR of an N–H...N hydrogen bond between nucleic acids enriched with ^{15}N, by measuring the coupling of the nitrogen atoms. Grzesiek, working with Andrew Dingley of the Heinrich-Heine University in Düsseldorf, has been able to do this and show that the coupling is surprisingly large. Normally atomic nuclei only couple with each other if they are linked by normal chemical bonds, and in theory hydrogen bonds have neither the strength nor stability for this to occur.

The German researchers studied an ^{15}N enriched sample of the T1 domain of the potato spindle tuber viroid and were able to prove that N–H...N hydrogen bonding was present between the base pairs, uridine...adenosine and guanosine...cytidine, with couplings of approximately 7 Hz. How could they be certain that the signals they were observing are due to N–H...N hydrogen bonds?

The answer was to use triple resonance techniques to examine base pairs that hydrogen bond only via O–H...N hydrogen bonds, and show that the signal they had previously observed was absent. Grzesiek's second paper on hydrogen bonds, coauthored by Florence Cordier, Heinrich-Heine University, extends the work in an even more remarkable way by measuring the NMR coupling between nitrogens and carbons in the backbone hydrogen bonds of the human protein ubiquitin.

The carbons are part of a carbonyl (C=O) group, so are one removed from the hydrogen bond, *i.e.*, N–H...O=C. This time they used material enriched with ^{15}N and ^{13}C (normal ^{12}C has no nuclear magnet), and there too was the evidence for these hydrogen bonds, albeit with an interaction an order of magnitude weaker (at –0.25 to –0.9 Hz) than the N–H...N coupling. Nevertheless, the couplings correlate with the strength of the hydrogen bond, being stronger in the stronger bonds. Grzesiek's third paper, was done in conjunction with Dingley and researchers at UCLA. Together they studied not only the hydrogen bonding of Watson-Crick base pairs but also of Hoogsten base pairs within a DNA triplex consisting of one purine and two pyrimidine strands.

Four different base pairs were identified, their various couplings distinguished–including those of the weaker interactions at the "frayed ends" of the DNA chains–and relationships with other hydrogen bonding parameters, such as bond length, were established. In addition they were able to show that density functional computer simulations by computer could reproduce these findings exactly.

ENZYME STRUCTURE OF DNA

Before a cell can begin to divide or differentiate, the genetic information within the cell's DNA must be copied, or "transcribed," onto complementary strands of RNA. RNA polymerase II (pol II) is an enzyme that, by itself, can unwind the DNA double helix, synthesize RNA, and proofread the result. When combined with other molecules that regulate and control the transcription process, pol II is the key to successful interpretion of an organism's genetic code.

However, the size, complexity, scarcity, and fragility of pol II complexes have made analysis of these macromolecules by x-ray crystallography a formidable challenge. A team of structural biologists has met this challenge using data obtained from both the Stanford Synchrotron Radiation Laboratory and the Macromolecular Crystallography Facility at the ALS. The resultant high-resolution model of a 10-subunit pol II complex suggests roles for each of the subunits and will allow researchers to begin unraveling the intricacies of DNA transcription and its role in gene expression.

In this work, the researchers studied the pol II enzyme from the yeast *Saccharomyces cerevisiae*, which is likely to be an excellent model for the human enzyme in light of its highly similar gene sequences.

It is also the best-characterized form of the pol II enzyme, having been the subject of many biochemical and low-resolution structural studies in the past. To obtain a high-resolution structure, the research team drew on its considerable expertise in the preparation of protein crystals: two-dimensional crystals of pol II (minus two small subunits found to impede crystal growth) were used as seeds for growing three-dimensional crystals.

These crystals, when produced in an inert atmosphere to prevent oxidation, enabled the collection of data to 3.5-angstrom resolution. The addition of a final soaking procedure to produce uniform crystals, combined with high-brightness x-ray sources, resulted in a resolution of 3.0 angstroms. The current results bring into focus the somewhat fuzzy features previously observed in or inferred from earlier experiments. More importantly, the structural details suggest possible explanations for some of the unusual characteristics of this enzyme, which include a high processivity (the ability to synthesize very long strands of RNA) and the tendency to work in periodic spurts separated by pauses.

While it is known that additional proteins (transcription factors) play a role in controlling the activity of pol II, scientists have yet to understand how such proteins interact with pol II binding sites to perform their various functions.

The pol II model reported here establishes the positions of the various subunits and provides detailed information about the DNA/RNA binding domains. The data reveal two main subunits separated by a deep cleft where DNA can enter the complex. At the end of the cleft is the active site, where the DNA can be unwound for a short distance (the "transcription bubble") and a DNA/RNA hybrid can be produced. Two prominent grooves lead away from the active site, either of which could accommodate the exiting RNA transcript. An opening below the active site may allow the entry of nucleotides (for manufacturing RNA) and transcription factors (for regulating the process).

The same opening may provide room for the leading end of the RNA strand during "backtracking" maneuvers, which are important for proofreading and for traversing obstacles such as DNA damage.

Other notable features that might help account for the great stability of this transcribing complex include a pair of "jaws" that appear to grip the DNA strands as they enter the complex and, closer to the active site, a clamp on the DNA that could possibly be locked in the closed position by the presence of RNA.

The high-resolution pol II structure reported here is a landmark achievement, pulling together threads from numerous diverse research efforts into a cohesive whole. Further study should yield many new insights into the detailed mechanisms of pol II and its transcription factors. Construction of an atomic model is already well underway.

SYNTHESIS OF DNA

DNA Replication (also known as DNA synthesis) is a process where the double stranded Deoxyribonucleic Acid (DNA) is copied.

Replication is an important first step in cell division, as cells must duplicate their entire genetic constitution before they can divide into two daughter cells. DNA replication is also a critical requirement for DNA repair.

REQUIREMENTS OF DNA REPLICATION

Replication requires three important components in order for this process to work.

Template Strand

The DNA serves as a template to guide the incoming nucleotides. The template strand attaches to the RNA primer strand in order to replicate the strand of DNA. The template just guides the nucleotides to the place that they need to go in.

This is another component that is needed in order for DNA replication to even take place. If the DNA only had the template strand and none of the rest of the things then the strand could not replicate. The template starnd is very thin. Through an enzyme called primase the DNA template strand is synthesized one nucleotide at a time.

DNA Polymerase

There are many DNA polymerases within a cell, but only one of which is used in the replication of DNA. In order for DNA to replicate, it must have a primer, which is a small strand of RNA. The DNA template strand synthesizes one nucleotide at a time by the use of primase. The adding of nucleotides to the 3' end, which is continued until that section of DNA is replicated. The other

DNA polymerases that are within the cell are used in other ways besides replication. DNA polymerase III is one certain polymerase, which aids in replication of what is known as the "lagging strand". DNA ligase is the final enzyme, which combines the lagging strand

Free 3' Hydroxyl

The free 3' hydroxyl is the starter strand called a primer which is required in order for DNA to be replicated. The primer strand is not very big, it is shorter then a single strand of RNA. The primer is the complementary strand to the template strand of DNA. This is another one of the things that are needed in order to make the process of DNA replication even possible. The DNA polymerase is added to the 3' end of the primer, which keeps growing until the section is finally complete. After a while the RNA primer is degraded and disappears.

Types of Replication

During the discovery of DNA replication they also discovered that there are three modes of replication that a DNA strand can take. The three types of DNA replication are semiconservative, conservative, and dispersive. These three types of replication were tested on in order to see the possible patterns that would result in the complementary base pairing.

Semiconservative Replication

Through this type of replication the parent strand serves as a template for the new strand. The two offsprings would have one of the parent strand and one new strand.

Conservative Replication

Through this type of replication the double helix serves as a template. This although does not contribute to the new double helix.

Dispersive Replication

In this type of replication the fragments from the parent DNA molecule. The DNA serves as a template for the assembly of the new molecule. The new double helix contains old and new parts of the DNA strands.

DNA Repairing

After the DNA is replicated, there are three DNA repair mechanisms. They are to be used when they feel necessary. This will lessen the chance of human mutation. The rate of error in which the DNA polymerases makes a mistake is very high. This is why the DNA has mechanisms to cause fewer mutations. The three mechanisms are:

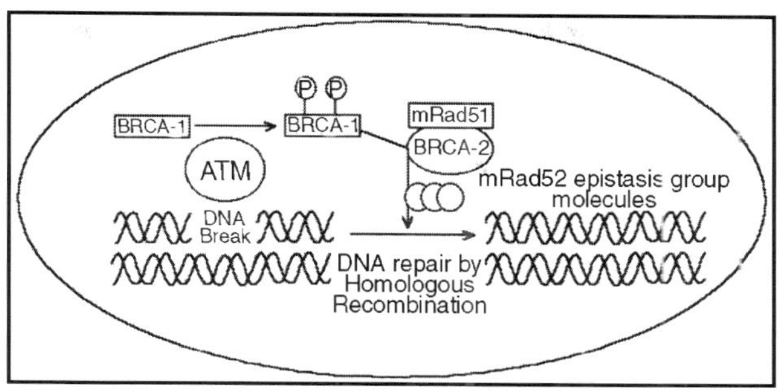

Fig. DNA Repairing

Proofreading

Every time a nucleotide is introduced to an already growing chain, the DNA polymerase checks the connection. If the pair is mismatched, it will be removed and redone. This lowers the overall rate of mutations.

Mismatch Pair

After the replication takes place, another set of proteins check to make sure that there are no more mismatched base pairs and also which strand is the wrong one.

Excision Pair

There can still be damage while the cell is living. Because of this, there are enzymes constantly checking the cell. If they detect a problem, the enzyme cuts the strand and also cuts away the opposite base. DNA polymerase and ligase then fix this base sequence.

LABORATORY TECHNIQUE

Scientists use two techniques in a laboratory that involve DNA replication. They use these techniques to observe genes and genomes. One of the techniques can make multiple copies of DNA from only a short piece. Another technique allows scientists to determine the base sequence of DNA. The two techniques are:

Polymerase Chain Reaction (PCR)

Through this process, a short strand of DNA is copied repetitively. There are three steps that are repeated in the process. The DNA strands are denatured or heated in order to separate the strand. Next, they add a primer, which was artificially made. They also add DNA polymerase and the four deoxyribonucleotide triphosphates. These three are needed in order to replicate DNA. The final step is then the DNA polymerase catalyzes the production of a complementary strand.

DNA Sequencing

This allows scientists to determines the base sequence of a DNA. The process is the DNA is first denatured and a single strand is placed in a test tube. They then add DNA polymerase, primer, the four dNTP's, and small amounts of ddNTP's. The DNA replicates within the tube and after the DNA fragments are denatured.

The fragments then go through electrophoresis, which sorts the lengths of DNA fragments. The fragments then pass through a laser beam which excites the fluorescent tags. This information is fed into a computer at which tells the sequence.

Mendelson-Stahl Experiment

Through this experiment Meselson- Stahl convinced the scientists that the correct model for DNA replication was the semiconservative replication. They used density labeling in order to distinguish the old and new strands of DNA. In their experiment they used "heavy" isotopes of nitrogen. This nitrogen is a nonradioactive isotope that made the molecule more dense.

The chemically identical molecules containing an isotope known as ^{14}N. They needed a way to distinguish the different DNA densities.

Because of this they needed to come up with something to measure the densities of solutions. Meselson, Stahl, and Jerome Vanguard came up with centrifuge, which is a procedure that involves the spinning of solutions at high speed, which causes the particles to separate and form gradient according to the different densities of the solution. The first part of their experiment they grew cultures of Escherichia coli. One of the cultures was grown in ^{15}N, which made the DNA "heavy".

The other culture was placed in a medium using ^{14}N rather than ^{15}N, which made all the DNA "light". In order to see if the use of centrifuge worked they combined the two cultures and when centrifuged it formed two DNA bands. This showed them the different densities with in the two cultures. The next part of their experiment they grew E. Coli in the ^{15}N medium and then transferred the bacteria from that medium to the ^{14}N.

Through this they came up with the conclusion that through E. Coli DNA replicates every 20 minutes. They then took DNA samples from each generation through, which they found that the density gradient was different each generation.

Their observations could only be explained with the semiconservative model of DNA replication. Their results showed that when the DNNA first underwent replication, it was in 15N, which caused the DNA to be heavy. Because in the semiconservative model of DNA replication, the one strand acts as a template for a second strand which the DNA was in a ^{14}N DNA strand and a ^{15}N and they were of intermediate density.

RNA

The high molecular weight nucleic acid, DNA, is found chiefly in the nuclei of complex cells, known as eucaryotic cells, or in the nucleoid regions ofprocaryotic cells, such as bacteria. It is often associated with proteins that help to pack it in a usable fashion.

In contrast, a lower molecular weight, but much more abundant nucleic acid, RNA, is distributed throughout the cell, most commonly in small numerous organelles called ribosomes. Three kinds of RNA are identified, the largest subgroup (85 to 90%) being ribosomal RNA, rRNA, the major component of ribosomes, together with proteins. The size of rRNA molecules varies, but is generally less than a thousandth the size of DNA. The other forms of RNA are messenger RNA, mRNA, and transfer RNA, tRNA. Both have a more transient existence and are smaller than rRNA. All these RNA's have similar constitutions, and differ from DNA in two important respects. As shown in the following diagram, the sugar component of RNA is ribose, and the pyrimidine base uracil replaces the thymine base of DNA. The RNA's play a vital role in the transfer of information (transcription) from the DNA library to the protein factories called ribosomes, and in the interpretation of that information (translation) for the synthesis of specific polypeptides. These functions will be described later.

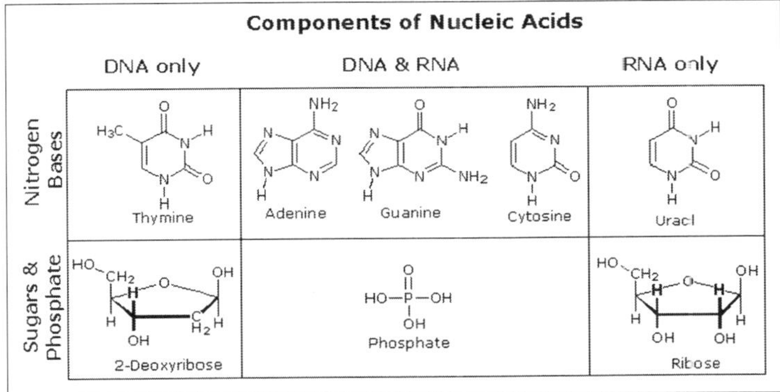

SECONDARY STRUCTURE

In the early 1950's the primary structure of DNA was well established, but a firm understanding of its secondary structure was lacking. Indeed, the situation was similar to that occupied by the proteins a decade earlier, before the alpha helix and pleated sheet structures were proposed by Linus Pauling.

Many researchers grappled with this problem, and it was generally conceded that the molar equivalences of base pairs (A & T and C & G) discovered by Chargaff would be an important factor. Rosalind Franklin, working at King's College, London, obtained X-ray diffraction evidence that suggested a long helical structure of uniform thickness. Francis Crick and James Watson, at Cambridge University, considered hydrogen bonded base pairing interactions, and arrived at a double stranded helical model that satisfied most of the known facts, and has been confirmed by subsequent findings.

Base Pairing

Careful examination of the purine and pyrimidine base components of the nucleotides reveals that three of them could exist as hydroxy pyrimidine or purine tautomers, having an aromatic heterocyclic ring. Despite the added stabilization of an aromatic ring, these compounds prefer to adopt amide-like structures.

These options are shown in the following diagram, with the more stable tautomer drawn in blue.

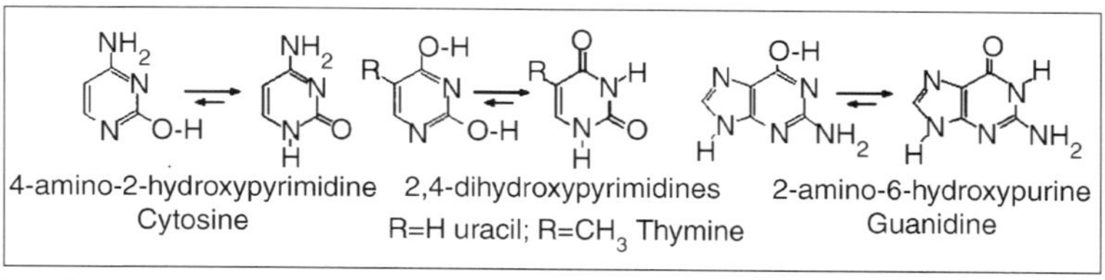

4-amino-2-hydroxypyrimidine Cytosine   2,4-dihydroxypyrimidines R=H uracil; R=CH_3 Thymine   2-amino-6-hydroxypurine Guanidine

A simple model for this tautomerism is provided by 2-hydroxypyridine. As shown on the left below, a compound having this structure might be expected to have phenol-like characteristics, such as an acidic hydroxyl group. However, the boiling point of the actual substance is 100° C greater than phenol and its acidity is 100 times less than expected (pKa = 11.7). These differences agree with the 2-pyridone tautomer, the stable form of the zwitterionic internal salt. Further evidence supporting this assignment will be displayed by clicking on the diagram.

Note that this tautomerism reverses the hydrogen bonding behaviour of the nitrogen and oxygen functions (the N-H group of the pyridone becomes a hydrogen bond donor and the carbonyl oxygen an acceptor).

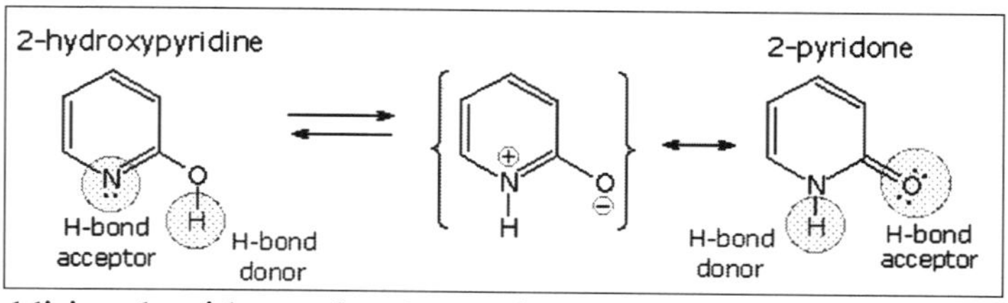

2-hydroxypyridine   2-pyridone

H-bond acceptor   H-bond donor   H-bond donor   H-bond acceptor

The additional evidence for the pyridone tautomer, that appears above by clicking on the diagram, consists of infrared and carbon nmr absorptions associated with and characteristic of the amide group. The data for 2-pyridone is given on the left. Similar data for the N-methyl derivative, which cannot tautomerize to a pyridine derivative, is presented on the right.

Once they had identified the favoured base tautomers in the nucleosides, Watson and Crick were able to propose a complementary pairing, via hydrogen bonding, of guanosine (G) with cytidine (C) and adenosine (A) with thymidine (T). This pairing, which is shown in the following diagram, explained Chargaff's findings beautifully, and led them to suggest a double helix structure for DNA. Before viewing this double helix structure itself, it is instructive to examine the base pairing interactions in greater detail. The G#C association involves three hydrogen bonds (coloured pink), and is therefore stronger than the two-hydrogen bond association of A#T.

These base pairings might appear to be arbitrary, but other possibilities suffer destabilizing steric or electronic interactions. By clicking on the diagram two such alternative couplings will be shown. The C#T pairing on the left suffers from carbonyl dipole repulsion, as well as steric crowding of the oxygens. The G#A pairing on the right is also destabilized by steric crowding (circled hydrogens).

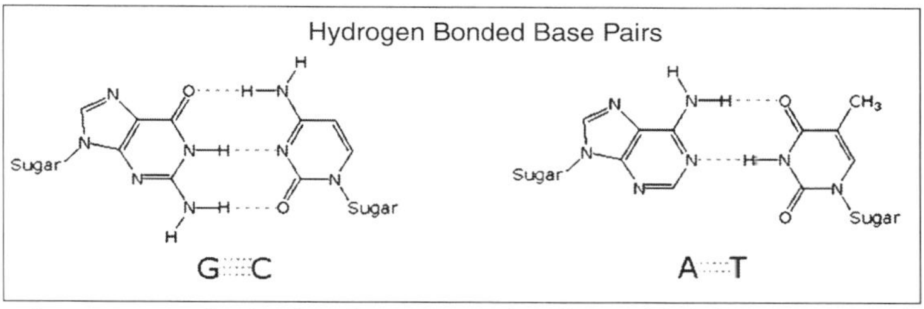

Hydrogen Bonded Base Pairs

G:::C　　A:::T

A simple mnemonic device for remembering which bases are paired comes from the line construction of the capital letters used to identify the bases. A and T are made up of intersecting straight lines. In contrast, C and G are largely composed of curved lines. The RNA base uracil corresponds to thymine, since U follows T in the alphabet.

Double Helix

After many trials and modifications, Watson and Crick conceived an ingenious double helix model for the secondary structure of DNA. Two strands of DNA were aligned anti-parallel to each other, *i.e.* with opposite 3' and 5' ends, as shown in part a of the following diagram. Complementary primary nucleotide structures for each strand allowed intra-strand hydrogen bonding between each pair of bases. These complementary strands are coloured red and green in the diagram.

Coiling these coupled strands then leads to a double helix structure, shown as cross-linked ribbons in part b of the diagram. The double helix is further stabilized by hydrophobic attractions and pi-stacking of the bases. A space-filling molecular model of a short segment is displayed in part c on the right.

The helix shown here has ten base pairs per turn, and rises 3.4 Å in each turn. This right-handed helix is the favoured conformation in aqueous systems, and has been termed the B-helix.

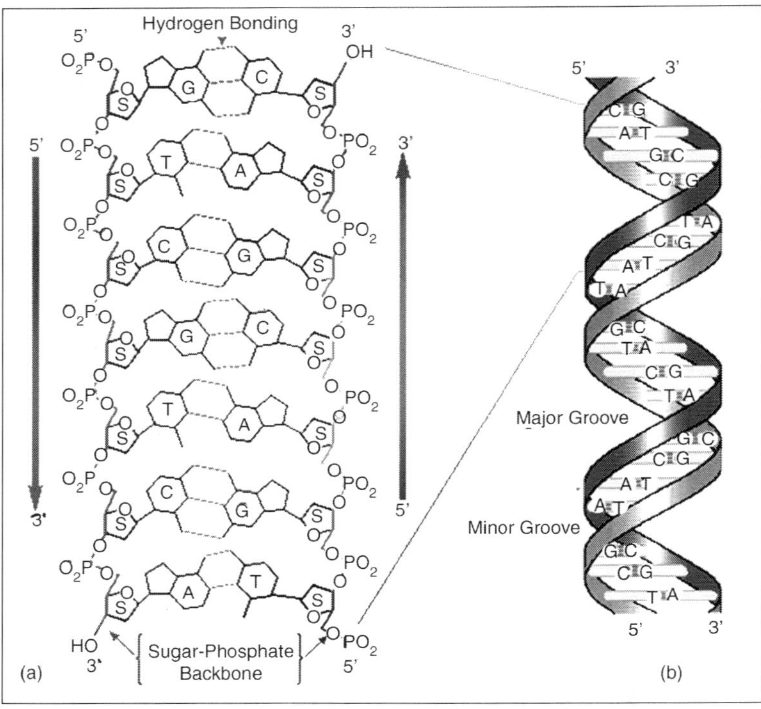

Fig. The Double Helix Structure for DNA

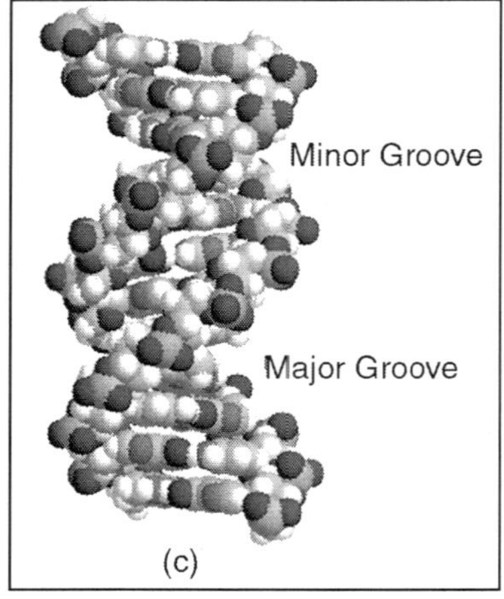

Fig. Space-Filling Molecular Model

As the DNA strands wind around each other, they leave gaps between each set of phosphate backbones. Two alternating grooves result, a wide and deep major groove (ca. 22Å wide), and a shallow and narrow minor groove (ca. 12Å wide). Other molecules, including polypeptides, may insert into these grooves, and in so doing perturb the chemistry of DNA. Other helical structures of DNA have also been observed, and are designated by letters (*e.g.* A and Z).

RNA AND ANTIBODY MOLECULES

CATALYTIC RNA MOLECULES: RIBOZYMES

It was long assumed that all enzymes are proteins. However, in recent years, more and more instances of biological catalysis by RNA molecules have been discovered. These catalytic RNAs, or ribozymes, satisfy several enzymatic criteria: They are substrate–specific, they enhance the reaction rate, and they emerge from the reaction unchanged. For example, RNase P, an enzyme responsible for the formation of mature tRNA molecules from tRNA precursors, requires an RNA component as well as a protein subunit for its activity in the cell.

In vitro, the protein alone is incapable of catalyzing the maturation reaction, but the RNA component by itself can carry out the reaction under appropriate conditions. In another case, in the ciliated protozoan *Tetrahymena*, formation of mature ribosomal RNA from a pre–rRNA precursor involves the removal of an internal RNA segment and the joining of the two ends in a process known as splicing out.

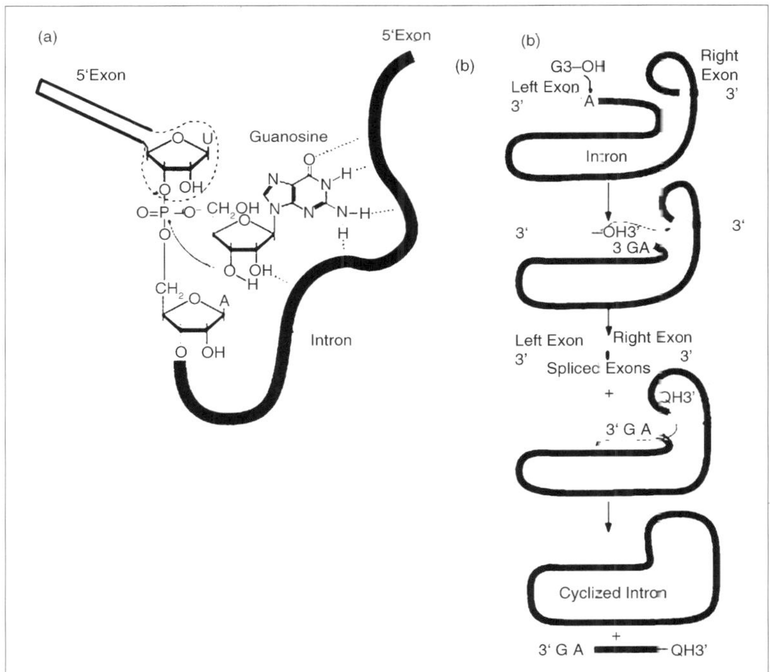

Fig. RNA splicing in *Tetrahymena* rRNA Maturation.
(a) the Guanosine–mediated Reaction Involved in the Autocatalytic Excision of the *Tetrahymena* rRNA Intron, and (b) the Overall Splicing Process

The excision of this intervening internal sequence of RNA and ligation of the ends is, remarkably, catalyzed by the intervening sequence of RNA itself, in the presence of Mg^{2+} and a free molecule of guanosine nucleo–side or nucleotide (Figure). *In vivo*, the intervening sequence RNA probably acts only in splicing itself out; *in vitro*, however, it can act many times, turning over like a true enzyme.

The cyclized intron is formed via nucleophilic attack of the 3'–OH on the phosphodiester bond that is 15 nucleotides from the 5'–GA end of the spliced–out intron. Cyclization frees a linear 15–mer with a 5'–GA end.

PROTEIN–FREE 50S RIBOSOMAL SUBUNITS CATALYZE

Perhaps the most significant case of catalysis by RNA occurs in protein synthesis. Harry F. Noller and his colleagues have found that the peptidyl transferase reaction, which is the reaction of peptide bond formation during protein synthesis (Figure), can be catalyzed by 50S ribosomal subunits from which virtually all of the protein has been removed. These experiments imply that just the 23S rRNA by itself is capable of catalyzing peptide bond formation. Also, the laboratory of Thomas Cech has created a synthetic 196–nucleotide–long ribozyme capable of performing the peptidyl transferase reaction.

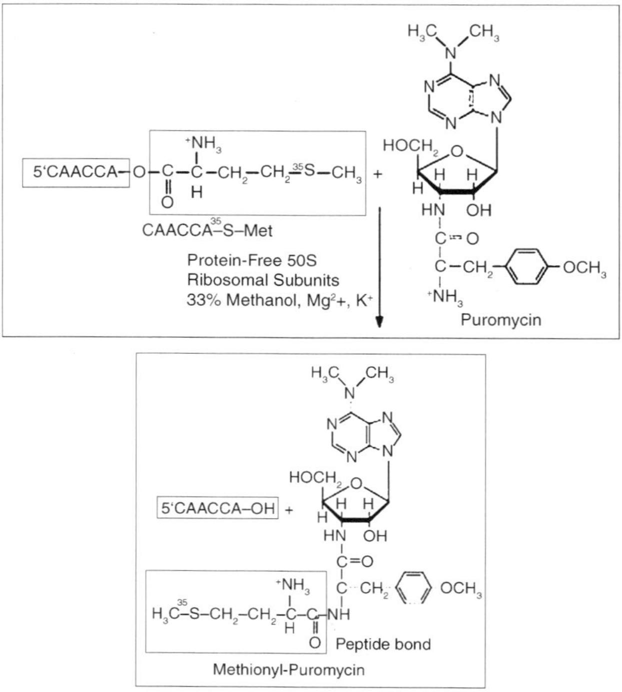

Fig. Protein–free 50S Ribosomal Subunits have Peptidyl Transferase Activity

Peptidyl transferase is the name of the enzymatic function that catalyzes peptide bond formation. The presence of this activity in protein–free 50S ribosomal subunits was demonstrated using a model assay for peptide bond formation in which an aminoacyl–tRNA analog (a short RNA oligonucleotide of sequence CAACCA carrying S–labeled methionine attached at its 3'–OH end) served as the peptidyl donor and puromycin (another amino–acyl–tRNA analog) served as the peptidyl acceptor. Activity was measured by monitoring the formation of S–labeled methioninyl–puromycin.

Several features of these "RNA enzymes," or ribozymes, lead to the realization that their biological efficiency does not challenge that achieved by proteins. First, RNA enzymes often do not fulfill the criterion of catalysis *in vivo* because they act only once in intramolecular events such as self–splicing. Second, the catalytic rates achieved by RNA enzymes *in vivo* and *in vitro* are significantly enhanced by the participation of protein subunits. Nevertheless, the fact that RNA can catalyze certain reactions is experimental support for the idea that a primordial world dominated by RNA molecules existed before the evolution of DNA and proteins.

CATALYTIC ANTIBODIES: ABZYMES

Antibodies are *immunoglobulins*, which, of course, are proteins. Like other antibodies,catalytic antibodies, so–called abzymes, are elicited in an organism in response to immunological challenge by a foreign molecule called an antigen. In this case, however, the antigen is purposefully engineered to be *an analog of the transition–state intermediate in a reaction*. The rationale is that a protein specific for binding the transition–state intermediate of a reaction will promote entry of the normal reactant into the reactive, transition–state conformation. Thus, a catalytic antibody facilitates, or catalyzes, a reaction by forcing the conformation of its substrate in the direction of its transition state.

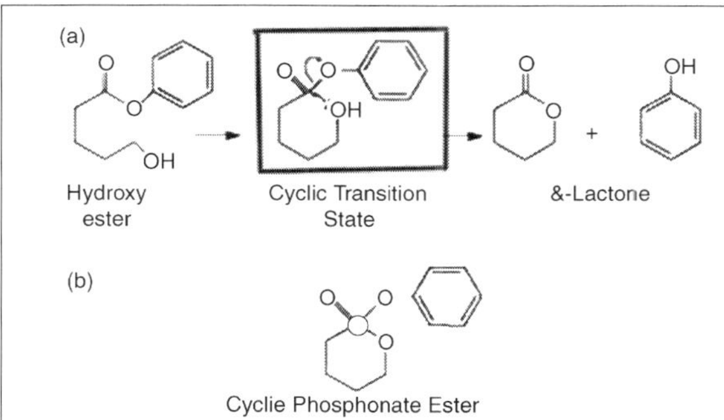

Fig. Catalytic Antibodies are Designed to Specifically Bind the Transition–state Intermediate in a Chemical Reaction

(a) The intramolecular hydrolysis of a hydroxy ester to yield as products a d–lactone and the alcohol phenol.Note the cyclic transition state.

(b) The cyclic phosphonate ester analog of the cyclic transition state. Antibodies raised against this phosphonate ester act as *enzymes*: they are catalysts that markedly accelerate the rate of ester hydrolysis.

RNA AND PROTEIN SYNTHESIS

The genetic information stored in DNA molecules is used as a blueprint for making proteins. Why proteins? Because these macromolecules have diverse primary, secondary and tertiary structures that equip them to carry out the numerous functions necessary to maintain a living organism. As noted in theprotein chapter, these functions include:

- Structural integrity (hair, horn, eye lenses etc.).
- Molecular recognition and signaling (antibodies and hormones).
- Catalysis of reactions (enzymes)..
- Molecular transport (hemoglobin transports oxygen).
- Movement (pumps and motors).

The critical importance of proteins in life processes is demonstrated by numerous genetic diseases, in which small modifications in primary structure produce debilitating and often disastrous consequences. Such genetic diseases include Tay-Sachs, phenylketonuria (PKU), sickel cell anemia, achondroplasia, and Parkinson disease. The unavoidable conclusion is that proteins are of central importance in living cells, and that proteins must therefore be continuously prepared with high structural fidelity by appropriate cellular chemistry.

Early geneticists identified genes as hereditary units that determined the appearance and/ or function of an organism (*i.e.* its phenotype). We now define genes as sequences of DNA that occupy specific locations on a chromosome. The original proposal that each gene controlled the formation of a single enzyme has since been modified as: one gene = one polypeptide.

The intriguing question of how the information encoded in DNA is converted to the actual construction of a specific polypeptide has been the subject of numerous studies, which have created the modern field of Molecular Biology.

DOGMA AND TRANSCRIPTION

Francis Crick proposed that information flows from DNA to RNA in a process called transcription, and is then used to synthesize polypeptides by a process called translation. Transcription takes place in a manner similar to DNA replication. A characteristic sequence of nucleotides marks the beginning of a gene on the DNA strand, and this region binds to a promoter protein that initiates RNA synthesis. The double stranded structure unwinds at the promoter site., and one of the strands serves as a template for RNA formation, as depicted in the following diagram. The RNA molecule thus formed is single stranded, and serves to carry information from DNA to the protein synthesis machinery called ribosomes. These RNA molecules are therefore called messenger-RNA (mRNA).

To summarize: a gene is a stretch of DNA that contains a pattern for the amino acid sequence of a protein. In order to actually make this protein, the relevant DNA segment is first copied into messenger-RNA. The cell then synthesizes the protein, using the mRNA as a template.

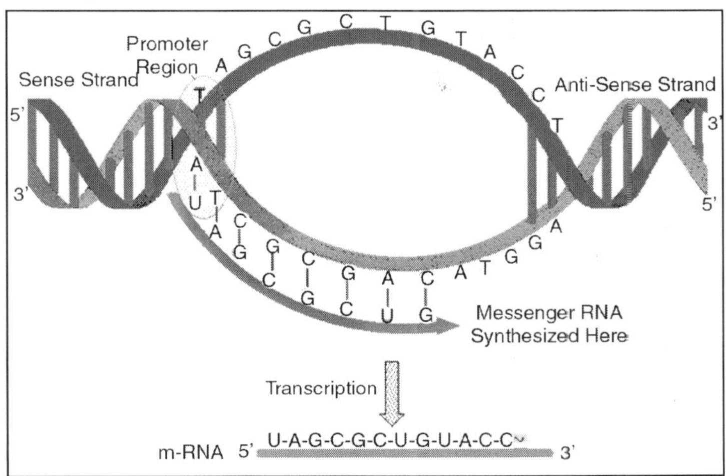

An important distinction must be made here. One of the DNA strands in the double helix holds the genetic information used for protein synthesis. This is called the sense strand, or information strand (coloured red above). The complementary strand that binds to the sense strand is called the anti-sense strand (coloured green), and it serves as a template for generating a mRNA molecule that delivers a copy of the sense strand information to a ribosome. The promoter protein binds to a specific nucleotide sequence that identifies the sense strand, relative to the anti-sense strand. RNA synthesis is then initiated in the 3' direction, as nucleotide triphosphates bind to complementary bases on the template strand, and are joined by phosphate diester linkages. An animation of this process for DNA replication was presented earlier. A characteristic "stop sequence" of nucleotides terminates the RNA synthesis. The messenger molecule (coloured orange above) is released into the cytoplasm to find a ribosome, and the DNA then rewinds to its double helix structure.

In eucaryotic cells the initially transcribed m-RNA molecule is usually modified and shortened by an "editing" process that removes irrelevant material. The DNA of such organisms is often thousands of times larger and more complex than that composing the single chromosome of a procaryotic bacterial cell. This difference is due in part to repetitive nucleotide sequences (ca. 25 per cent in the human genome). rts of genes. The informational DNA segments that make up genes are called exons, and the non-coding segments are called introns. Before the mRNA molecule leaves the nucleus, the non-sense bases that make up the introns are cut out, and the informationally useful exons are joined together in a step known as RNA splicing. In this fashion shorter mRNA molecules carrying the blueprint for a specific protein are sent on their way to the ribosome factories.

The Central Dogma of molecular biology, which at first was formulated as a simple linear progression of information from DNA to RNA to Protein, is summarized in the following illustration. The replication process on the left consists of passing information from a parent DNA molecule to daughter molecules. The middle transcription process copies this information to a mRNA molecule. Finally, this information is used by the chemical machinery of the ribosome to make polypeptides.

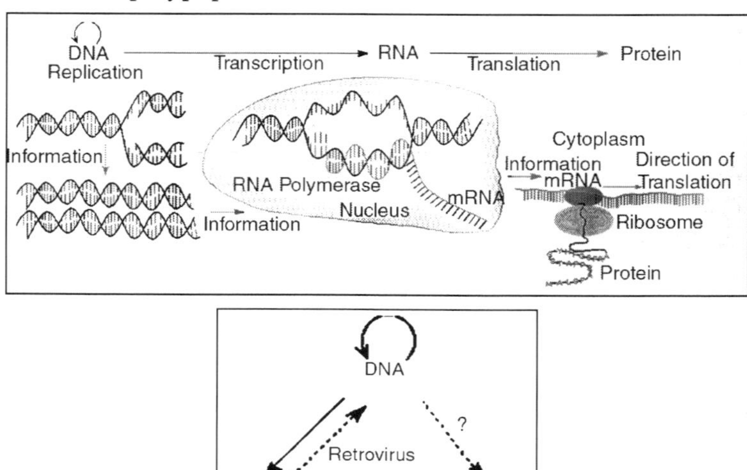

As more has been learned about these relationships, the central dogma has been refined to the representation displayed on the right. The dark blue arrows show the general, well demonstrated, information transfers noted above. It is now known that an RNA-dependent DNA polymerase enzyme, known as a reverse transcriptase, is able to transcribe a single-stranded RNA sequence into double-stranded DNA (magenta arrow). Such enzymes are found in all cells and are an essential component of retroviruses (*e.g.* HIV), which require RNA replication of their genomes (green arrow). Direct translation of DNA information into protein synthesis (orange arrow) has not yet been observed in a living organism. Finally, proteins appear to be an informational dead end, and do not provide a structural blueprint for either RNA or DNA.

In the following section the last fundamental relationship, that of structural information translation from mRNA to protein, will be described

TRANSLATION

Translation is a more complex process than transcription. This would, of course, be expected. After all, the coded messages produced by the German Enigma machine could be copied easily, but required a considerable decoding effort before they could be read with understanding. In a similar sense, DNA replication is simply a complementary base pairing exercise, but the translation of the four letter (bases) alphabet code of RNA to the twenty letter (amino acids) alphabet of protein literature is far from trivial. Clearly, there could not be a direct one-to-one correlation of bases to amino acids, so the nucleotide letters must form short words or codons that define specific amino acids. Many questions pertaining to this genetic code were posed in the late 1950's:

- How many RNA nucleotide bases designate a specific amino acid? If separate groups of nucleotides, called codons, serve this purpose, at least three are needed. There are 43 = 64 different nucleotide triplets, compared with 42 = 16 possible pairs.
- Are the codons linked separately or do they overlap? Sequentially joined triplet codons will result in a nucleotide chain three times longer than the protein it describes. If overlapping codons are used then fewer total nucleotides would be required.
- If triplet segments of mRNA designate specific amino acids in the protein, how are the codons identified? For the sequence ~CUAGGU~ are the codons CUA & GGU or ~C, UAG & GU~ or ~CU, AGG & U~?
- Are all the codon words the same size? In Morse code the most widely used letters are shorter than less common letters. Perhaps nature employs a similar scheme.

Physicists and mathematicians, as well as chemists and microbiologists all contributed to unravelling the genetic code. Although earlier proposals assumed efficient relationships that correlated the nucleotide codons uniquely with the twenty fundamental amino acids, it is now apparent that there is considerable redundancy in the code as it now operates. Furthermore, the code consists exclusively of non-overlapping triplet codons.

Clever experiments provided some of the earliest breaks in deciphering the genetic code. Marshall Nirenberg found that RNA from many different organisms could initiate specific protein synthesis when combined with broken E.coli cells (the enzymes remain active). A synthetic polyuridine RNA induced synthesis of poly-phenylalanine, so the UUU codon designated phenylalanine. Likewise an alternating ~CACA~ RNA led to synthesis of a ~His-Thr-His-Thr~ polypeptide.

The following table presents the present day interpretation of the genetic code. Note that this is the RNA alphabet, and an equivalent DNA codon table would have all the U nucleotides replaced by T. Methionine and tryptophan are uniquely represented by a single codon. At the other extreme, leucine is represented by eight codons. The average redundancy for the twenty amino acids is about three. Also, there are three stop codons that terminate polypeptide synthesis.

PEPTIDE OR PROTEIN

Once a peptide or protein has been synthesized and released from the ribosome it often undergoes further chemical transformation. This post-translational modification may involve the attachment of other moieties such as acyl groups, alkyl groups, phosphates, sulfates, lipids and carbohydrates. Functional changes such as dehydration, amidation, hydrolysis and oxidation (*e.g.* disulfide bond formation) are also common. In this manner the limited array of twenty amino acids designated by the codons may be expanded in a variety of ways to enable proper functioning of the resulting protein. Since these post-translational reactions are generally catalyzed by enzymes, it may be said: "Virtually every molecule in a cell is made by the ribosome or by enzymes made by the ribosome."

Modifications, like phosphorylation and citrullination, are part of common mechanisms for controlling the behaviour of a protein. As shown on the left below, citrullination is the post-translational modification of the amino acid arginine into the amino acid citrulline.

Arginine is positively charged at a neutral pH, whereas citrulline is uncharged, so this change increases the hydrophobicity of a protein. Phosphorylation of serine, threonine or tyrosine residues renders them more hydrophilic, but such changes are usually transient, serving to regulate the biological activity of the protein. Other important functional changes include iodination of tyrosine residues in the peptide thyroglobulin by action of the enzyme thyroperoxidase. The monoiodotyrosine and diiodotyrosine formed in this manner are then linked to form the thyroid hormones T3 and T4, shown on the right below.

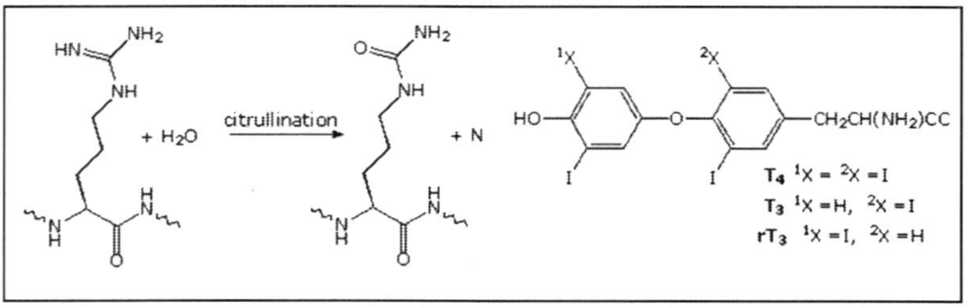

Amino acids may be enzymatically removed from the amino end of the protein. Because the "start" codon on mRNA codes for the amino acid methionine, this amino acid is usually removed from the resulting protein during post-translational modification. Peptide chains may also be cut in the middle to form shorter strands. Thus, insulin is initially synthesized as a 105 residue preprotein. The 24-amino acid signal peptide is removed, yielding a proinsulin peptide. This folds and forms disulfide bonds between cysteines 7 and 67 and between 19 and 80. Such dimeric cysteines, joined by a disulfide bond, are named cystine. A protease then cleaves the peptide at arg31 and arg60, with loss of the 32-60 sequence (chain C). Removal of arg31 yields mature insulin, with the A and B chains held together by disulfide bonds and a third cystine moiety in chain A. The following cartoon illustrates this chain of events.

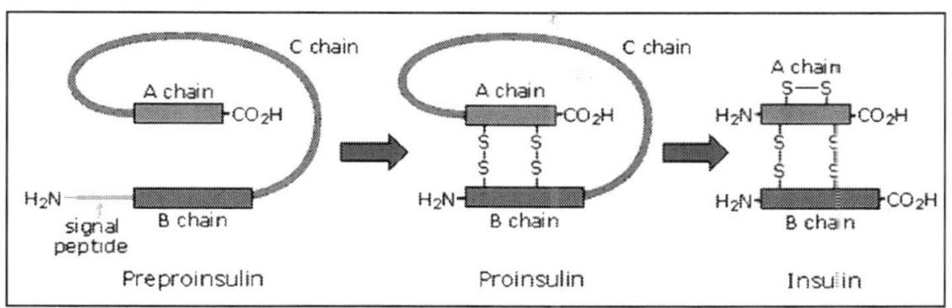

Nisin is a polypeptide (34 amino acids) made by the bacterium Lactococcus lactis. Nisin kills gram positive bacteria by binding to their membranes and targeting lipid II, an essential precursor of cell wall synthesis. Such antimicrobial peptides are a growing family of compounds which have received the name lantibiotics due to the presence of lanthionine, a non-proteinogenic amino acid with the chemical formula HO_2C-CH(NH_2)-CH_2-S-CH_2-CH(NH_2)-CO_2H.

Lanthionine is composed of two alanine residues that are crosslinked on their β-carbon atoms by a thioether linkage (*i.e.* it is the monosulfide analog of the disulfide cystine). Lantibiotics are unique in that they are ribosomally synthesized as prepeptides, followed by post-translational processing of a number of amino acids (*e.g.* serine, threonine and cysteine) into dehydro residues and thioether crossbridges. Nisin is the only bacteriocin that is accepted as a food preservative. Several nisin subtypes that differ in amino acid composition and biological activity are known. A typical structure is drawn below, and a Jmol model will be presented by clicking on the diagram.

Nisin

Ile-Aca-Dal-Ile-Aaa-Leu-Cys-Abu-Pro-Gly-Cys-Lys-Abu-Gly-Ala-Leu-Met-Gly-Cys-Asn-Met-Lys-Abu-Ala-Abu-Cys-Asn-Ser-Ile-His-Val-Aaa-Lys

Dal = D-Alanine; Aaa = 2-Aminoacrylic Acid (or Dha Didehydroalanine); Abu = 2-aminobutyric Acid (or Dbb = Didehydrobutyrine); Aca = 2-aminocrotonic Acid Dhb didehydroaminobutyric Acid

The bacterial cell wall is a cross-linked glycan polymer that surrounds bacterial cells, dictates their cell shape, and prevents them from breaking due to environmental changes in osmotic pressure. This wall consists mainly of peptidoglycan or murein, a three-dimensional polymer of sugars and amino acids located on the exterior of the cytoplasmic membrane.

The monomer units are composed of two amino sugars, N-acetylglucosamine (NAG) and N-acetylmuramic acid (NAM), shown on the right. Transglycosidase enzymes join these units by glycoside bonds, and they are further interlinked to each other via peptide cross-links between the pentapeptide moieties that are attached to the NAM residues.

Peptidoglycan subunits are assembled on the cytoplasmic side of the bacterial membrane from a polyisoprenoid anchor. Lipid II, a membrane-anchored cell-wall precursor that is essential for bacterial cell-wall biosynthesis, is one of the key components in the synthesis of peptidoglycan. Peptidoglycan synthesis via polymerization of Lipid II is illustrated in the following diagram. Cross-linking of the peptide side chains is then effected by transpeptidase enzymes. A model of Lipid II complexed with nisin may be examined as part of the previous Jmol display.

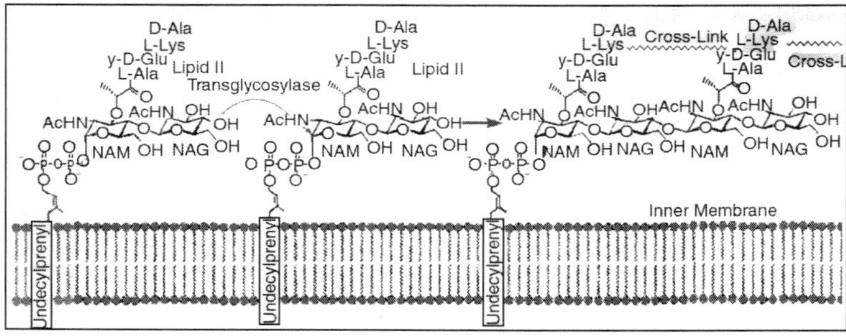

In order for bacteria to divide by binary fission and increase their size following division, links in the peptidoglycan must be broken, new peptidoglycan monomers must be inserted, and the peptide cross links must be resealed. Transglycosidase enzymes catalyze the formation of glycosidic bonds between the NAM and NAG of the peptidoglycan monomers and the NAG and NAM of the existing peptidoglycan. Finally, transpeptidase enzymes reform the peptide cross-links between the rows and layers of peptidoglycan making the wall strong. Many antibiotic drugs, including penicillin, target the chemistry of cell wall formation. The effectiveness of choosing Lipid II for an antibacterial strategy is highlighted by the fact that it is the target for at least four different classes of antibiotic, including the clinically important glycopeptide antibiotic vancomycin. The growing problem of bacterial resistance to many current drugs, including vancomycin, has led to increasing interest in the therapeutic potential of other classes of compound that target Lipid II. Lantibiotics such as nisin are part of this interest.

DNA REPLICATION

In their 1953 announcement of a double helix structure for DNA, Watson and Crick stated, "It has not escaped our notice that the specific pairing we have postulated immediately suggests a possible copying mechanism for the genetic material."

The essence of this suggestion is that, if separated, each strand of the molecule might act as a template on which a new complementary strand might be assembled, leading finally to two identical DNA molecules. Indeed, replication does take place in this fashion when cells divide, but the events leading up to the actual synthesis of complementary DNA strands are sufficiently complex that they will not be described in any detail.

As depicted in the following drawing, the DNA of a cell is tightly packed into chromosomes. First, the DNA is wrapped around small proteins called histones (coloured pink below). These bead-like structures are then further organized and folded into chromatin aggregates that make up the chromosomes. An overall packing efficiency of 7,000 or more is thus achieved. Clearly a sequence of unfolding events must take place before the information encoded in the DNA can be used or replicated.

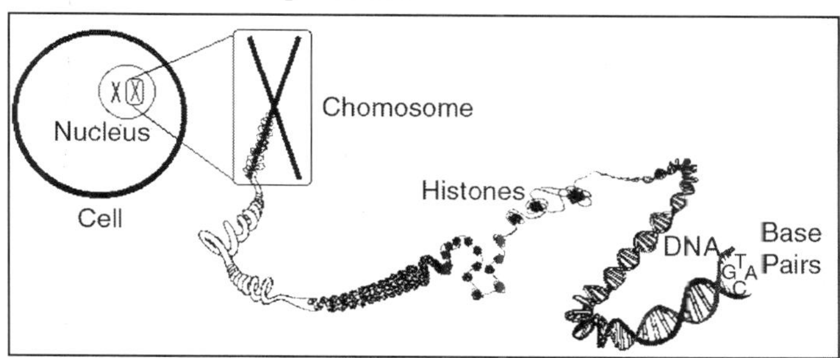

Once the double stranded DNA is exposed, a group of enzymes act to accomplish its replication.

These are described briefly here:

- *Topoisomerase*: This enzyme initiates unwinding of the double helix by cutting one of the strands.
- *Helicase*: This enzyme assists the unwinding. Note that many hydrogen bonds must be broken if the strands are to be separated..
- *SSB*: A single-strand binding-protein stabilizes the separated strands, and prevents them from recombining, so that the polymerization chemistry can function on the individual strands.
- *DNA Polymerase*: This family of enzymes link together nucleotide triphosphate monomers as they hydrogen bond to complementary bases. These enzymes also check for errors (roughly ten per billion), and make corrections.
- *Ligase*: Small unattached DNA segments on a strand are united by this enzyme.

Polymerization of nucleotides takes place by the phosphorylation reaction described by the following equation.

$$\underset{\text{an alkyl triphosphate}}{{}^1R{-}O{-}\overset{O}{\overset{\|}{P}}(O^-){-}O{-}\overset{O}{\overset{\|}{P}}(O^-){-}O{-}\overset{O}{\overset{\|}{P}}(O^-){-}O^-} + {}^2R{-}O{-}H \longrightarrow \underset{\text{a diphosphate ester}}{{}^1R{-}O{-}\overset{O}{\overset{\|}{P}}(O^-){-}O{-}R^2} + \underset{\text{pyrophosphate}}{{}^-O{-}\overset{O}{\overset{\|}{P}}(O^-){-}O{-}\overset{O}{\overset{\|}{P}}(O^-){-}O{-}H}$$

Di- and triphosphate esters have anhydride-like structures and are consequently reactive phosphorylating reagents, just as carboxylic anhydrides areacylating reagents. Since the pyrophosphate anion is a better leaving group than phosphate, triphosphates are more powerful phosphorylating agents than are diphosphates. Formulas for the corresponding 5'-derivatives of adenosine will be displayed by Clicking Here, and similar derivatives exist for the other three common nucleosides. The DNA polymerization process that builds the complementary strands in replication, could in principle take place in two ways. Referring to the general equation above, R1 could represent the next nucleotide unit to be attached to the growing DNA strand, with R2 being this strand. Alternatively, these assignments could be reversed. In practice, the former proves to be the best arrangement. Since triphosphates are very reactive, the lifetime of such derivatives in an aqueous environment is relatively short. However, such derivatives of the individual nucleosides are repeatedly synthesized by the cell for a variety of purposes, providing a steady supply of these reagents. In contrast, the growing DNA segment must maintain its functionality over the entire replication process, and can not afford to be changed by a spontaneous hydrolysis event. As a result, these chemical properties are best accommodated by a polymerization process that proceeds at the 3'-end of the growing strand by 5'-phosphorylation involving a nucleotide triphosphate.

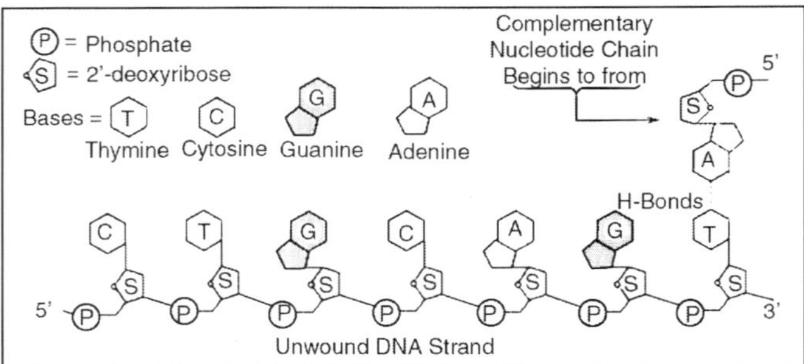

The polymerization mechanism described here is constant. It always extends the developing DNA segment towards the 3'-end (*i.e.* when a nucleotide triphosphate attaches to the free 3'-hydroxyl group of the strand, a new 3'-hydroxyl is generated). There is sometimes confusion on this point, because the original DNA strand that serves as a template is read from the 3'-end towards the 5'-end, and authors may not be completely clear as to which terminology is used.

Because of the directional demand of the polymerization, one of the DNA strands is easily replicated in a continuous fashion, whereas the other strand can only be replicated in short segmental pieces. This is illustrated in the following diagram. Separation of a portion of the double helix takes place at a site called the replication fork. As replication of the separate strands occurs, the replication fork moves away (to the left in the diagram), unwinding additional lengths of DNA. Since the fork in the diagram is moving towards the 5'-end of the red-coloured strand, replication of this strand may take place in a continuous fashion (building the new green strand in a 5' to 3' direction). This continuously formed new strand is called the leading strand. In contrast, the replication fork moves towards the 3'-end of the original green strand, preventing continuous polymerization of a complementary new red strand. Short segments of complementary DNA, called Okazaki fragments, are produced, and these are linked together later by the enzyme ligase. This new DNA strand is called the lagging strand.

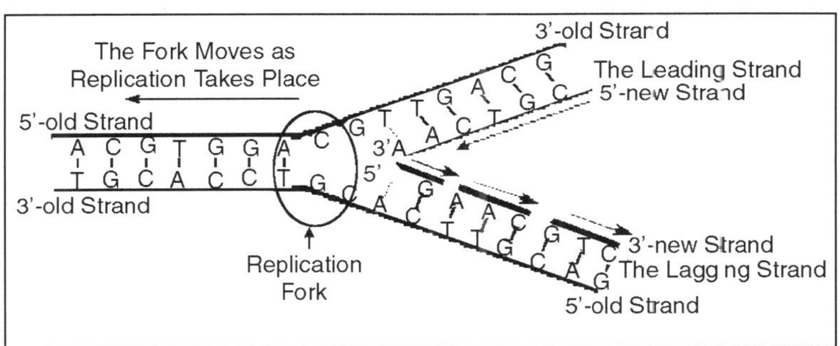

When you consider that a human cell has roughly 109 base pairs in its DNA, and may divide into identical daughter cells in 14 to 24 hours, the efficiency of DNA replication must be extraordinary. The procedure described above will replicate about 50 nucleotides per second, so there must be many thousand such replication sites in action during cell division. A given length of double stranded DNA may undergo strand unwinding at numerous sites in response to promoter actions. The unraveled "bubble" of single stranded DNA has two replication forks, so assembly of new complementary strands may proceed in two directions.

The polymerizations associated with several such bubbles fuse together to achieve full replication of the entire DNA double helix. A cartoon illustrating these concerted replications will appear by clicking on the above diagram. Note that the events shown proceed from top to bottom in the diagram.

RNA CONTAINS RIBOSE AND URACIL

We now turn our attention to RNA, which differs from DNA in three respects. First, the backbone of RNA contains ribose rather than 2'-deoxyribose. That is, ribose has a hydroxyl group at the 2' position. Second, RNA contains uracil in place of thymine.

Uracil has the same single-ringed structure as thymine, except that it lacks the 5' methyl group. Thymine is in effect 5'methyl-uracil. Third, RNA is usually found as a single polynucleotide chain. Except for the case of certain viruses, RNA is not the genetic material and does not need to be capable of serving as a template for its own replication. Rather, RNA functions as the intermediate, the mRNA, between the gene and the protein-synthesizing machinery.

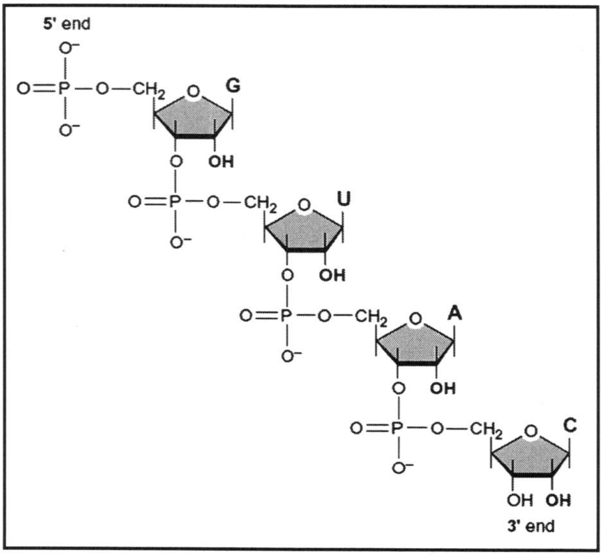

Fig. Structural Features of RNA

Another function of RNA is as an adaptor, the tRNA, between the codons in the mRNA and amino acids. RNA can also play a structural role as in the case of the RNA components of the ribosome.

Yet another role for RNA is as a regulatory molecule, which through sequence complementarity binds to, and interferes with the translation of, certain mRNAs. Finally, some RNAs (including one of the structural RNAs of the ribosome) are enzymes that catalyze essential reactions in the cell. In all of these cases, the RNA is copied as a single strand off only one of the two strands of the DNA template, and its complementary strand does not exist. RNA is capable of forming long double helices, but these are unusual in nature.

RNA CHAINS FOLD BACK ON THEMSELVES TO FORM LOCAL REGIONS

Despite being single-stranded, RNA molecules often exhibit a great deal of double-helical character. This is because RNA chains frequently fold back on themselves to form base-paired segments between short stretches of complementary sequences. If the two stretches of complementary sequence are near each other, the RNA may adopt one of various stem-loop structures in which the intervening RNA is looped out from the end of the double-helical segment as in a hairpin, a bulge, or a simple loop. The stability of such stem-loop structures is in some instances enhanced by the special properties of the loop. For example, a stem-loop with the "tetraloop" sequence UUCG is unexpectedly stable due to special base-stacking interactions in the loop. Base pairing can also take place between sequences that are not contiguous to form complex structures aptly named pseudoknots. The regions of base pairing in RNA can be a regular double helix or they can contain discontinuities, such as noncomplementary nucleotides that bulge out from the helix. A feature of RNA that adds to its propensity to form double-helical structures is an additional, non-Watson-Crick base pair. This is the G:U base pair, which has hydrogen bonds between N3 of uracil and the carbonyl on C6 of guanine and between the carbonyl on C2 of uracil and N1 of guanine. Because G:U base pairs can occur as well as the four conventional, Watson-Crick base pairs, RNA chains have an enhanced capacity for self-complementarity.

Thus, RNA frequently exhibits local regions of base pairing but not the long-range, regular helicity of DNA. The presence of 2'-hydroxyls in the RNA backbone prevents RNA from adopting a B-form helix. Rather, double-helical RNA resembles the A-form structure of DNA. As such, the minor groove is wide and shallow, and hence accessible, but recall that the minor groove offers little sequence-specific information. Meanwhile, the major groove is so narrow and deep that it is not very accessible to amino acid side chains from interacting proteins. Thus, the RNA double helix is quite distinct from the DNA double helix in its detailed atomic structure and less well suited for sequence-specific interactions with proteins (although some proteins do bind to RNA in a sequence-specific manner).

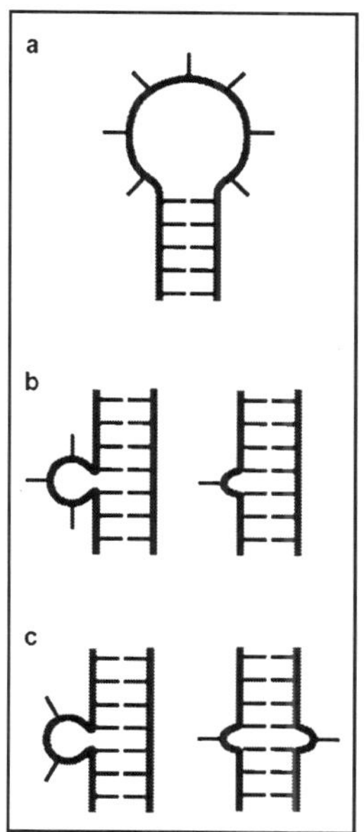

Fig. Double Helical Characteristics of RNA.

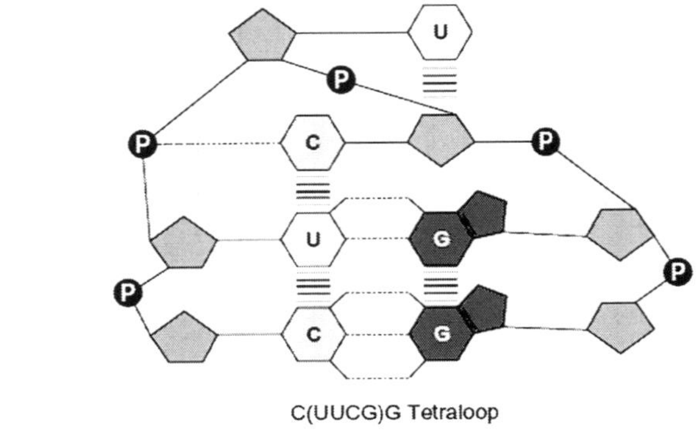

Fig. Tetraloop.

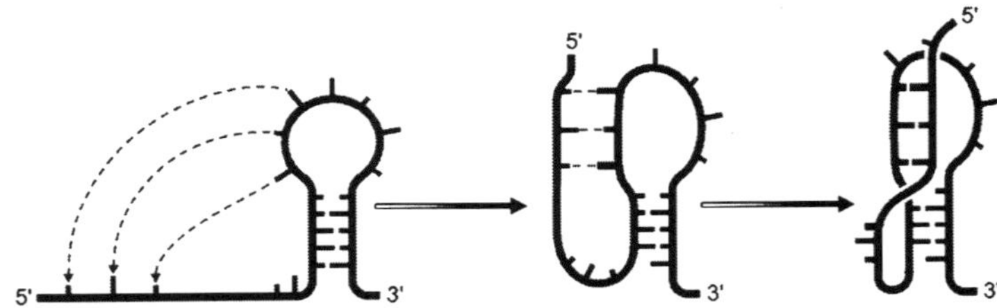

Fig. Pseudoknot

RNA CAN FOLD UP INTO COMPLEX TERTIARY STRUCTURES

Freed of the constraint of forming long-range regular helices, RNA can adopt a wealth of tertiary structures. This is because RNA has enormous rotational freedom in the backbone of its non-base-paired regions. Thus, RNA can fold up into complex tertiary structures frequently involving unconventional base pairing, such as the base triples and base-backbone interactions seen in tRNAs.

Proteins can assist the formation of tertiary structures by large RNA molecules, such as those found in the ribosome. Proteins shield the negative charges of backbone phosphates, whose electrostatic repulsive forces would otherwise destabilize the structure.

Researchers have taken advantage of the potential structural complexity of RNA to generate novel RNA species (not found in nature) that have specific desirable properties. By synthesizing RNA molecules with randomized sequences, it is possible to generate mixtures of oligonucleotides representing enormous sequence diversity.

For example, a mixture of oligoribonucleotides of length 20 and having four possible nucleotides at each position would have a potential complexity of 420 sequences or 1012 sequences! From mixtures of diverse oligoribonucleotides, RNA molecules can be selected biochemically that have particular properties, such as an affinity for a specific small molecule.

SOME RNAS ARE ENZYMES

It was widely believed for many years that only proteins could be enzymes. An enzyme must be able to bind a substrate, carry out a chemical reaction, release the product and repeat this sequence of events many times. Proteins are well suited to this task because they are composed of many different kinds of amino acids and they can fold into complex tertiary structures with binding pockets for the substrate and small molecule cofactors and an active site for catalysis. Now we know that RNAs, which as we have seen can similarly adopt complex tertiary structures, can also be biological catalysts.

Such RNA enzymes are known as ribozymes, and they exhibit many of the features of a classical enzyme, such as an active site, a binding site for a substrate and a binding site for a cofactor, such as a metal ion.

One of the first ribozymes to be discovered was RNase P, a ribonuclease that is involved in generating tRNA molecules from larger, precursor RNAs. RNase P is composed of both RNA and protein; however, the RNA moiety alone is the catalyst.

The protein moiety of RNase P facilitates the reaction by shielding the negative charges on the RNA so that it can bind effectively to its negatively charged substrate. The RNA moiety is able to catalyze cleavage of the tRNA precursor in the absence of the protein if a small, positively charged counter ion, such as the peptide spermidine, is used to shield the repulsive, negative charges. Other ribozymes carry out trans-esterification reactions involved in the removal of intervening sequences known as introns from precursors to certain mRNAs, tRNAs, and ribosomal RNAs in a process known as RNA splicing.

THE HAMMERHEAD RIBOZYME

Before concluding our discussion of RNA, let us look in more detail at the structure and function of one particular ribozyme, the hammerhead. The hammerhead is a sequence-specific ribonuclease that is found in certain infectious RNA agents of plants known as *viroids*, which depend on self-cleavage to propagate. When the viroid replicates, it produces multiple copies of itself in one continuous RNA chain.

Single viroids arise by cleavage, and this cleavage reaction is carried out by the RNA sequence around the junction. One such self-cleaving sequence is called the *hammerhead* because of the shape of its secondary structure, which consists of three base-paired stems surrounding a core of non-complementary nucleotides required for catalysis.

The tertiary structure of the ribozyme, however, looks more like a wishbone. To understand how the hammerhead works, let us first look at how RNA undergoes hydrolysis under alkaline conditions. At high pH, the 2' hydroxyl of the ribose in the RNA backbone can become deprotonated, and the resulting negatively charged oxygen can attack the scissile phosphate at the 3' position of the same ribose.

This reaction breaks the RNA chain, producing a 2', 3' cyclic phosphate and a free 5' hydroxyl. Each ribose in an RNA chain can undergo this reaction, completely cleaving the parent molecule into nucleotides. Many protein ribonucleases also cleave their RNA substrates via the formation of a 2', 3' cyclic phosphate. Working at normal cellular pH, these protein enzymes use a metal ion, bound at their active site, to activate the 2' hydroxyl of the RNA.

The hammerhead is a sequencespecific ribonuclease, but it too cleaves RNA via the formation of a 2', 3' cyclic phosphate. Hammerhead-mediated cleavage involves a ribozyme-bound Mg'' ion that deprotonates the 2' hydroxyl at neutral pH, resulting in nucleophilic attack on the scissile phosphate. Because the normal reaction of the hammerhead is self-cleavage, it is not really a catalyst; each molecule normally promotes a reaction one time only, thus having a turnover number of one.

But the hammerhead can be engineered to function as a true ribozyme by dividing the molecule into two portions—one, the ribozyme, that contains the catalytic core and the other, the substrate, that contains the cleavage site. The substrate binds to the ribozyme at stems I and III. After cleavage, the substrate is released and replaced by a fresh uncut substrate, thereby allowing repeated rounds of cleavage.

LIFE EVOLVE FROM AN RNA

The discovery of ribozymes has profoundly altered our view of how life might have evolved. We can now imagine that there was a primitive form of life based entirely on RNA. In this world, RNA would have functioned as the genetic material and as the enzymatic machines.

This RNA world would have preceded life as we know it today, in which information transfer is based on DNA, RNA, and protein. A hint that the protein world might have arisen from an RNA world is the discovery that the component in the ribosome that is responsible for the formation of the peptide bond, the peptidyl transferase, is an RNA molecule.

Unlike RNase P, the hammerhead, and other previously known ribozymes which act on phosphorous centers, the peptidyl transferase acts on a carbon centre to create the peptide bond. It thus links RNA chemistry to the most fundamental reaction in the protein world, peptide bond formation.

Perhaps then the ribosome ribozyme is a relic of an earlier form of life in which all enzymes were RNAs. DNA is usually in the form of a right-handed double helix. The helix consists of two polydeoxynucleotide chains. Each chain is an alternating polymer of deoxyribose sugars and phosphates that are joined together via phosphodiester linkages.

One of four bases protrudes from each sugar: adenine and guanine, which are purines, and thymine and cytosine, which are pyrimidines. While the sugar phosphate backbone is regular, the order of bases is irregular and this is responsible for the information content of DNA. Each chain has a 5' to 3' polarity, and the two chains of the double helix are oriented in an antiparallel manner—that is, they run in opposite directions.

Pairing between the bases holds the chains together. Pairing is mediated by hydrogen bonds and is specific: Adenine on one chain is always paired with thymine on the other chain, whereas guanine is always paired with cytosine.

This strict base-pairing reflects the fixed locations of hydrogen atoms in the purine and pyrimidine bases in the forms of those bases found in DNA. Adenine and cytosine almost always exist in the amino as opposed to the imino tautomeric forms, whereas guanine and thymine almost always exist in the keto as opposed to enol forms.

The complementarity between the bases on the two strands gives DNA its self-coding character. The two strands of the double helix fall apart (denature) upon exposure to high temperature, extremes of pH, or any agent that causes the breakage of hydrogen bonds. Upon slow return to normal cellular conditions, the denatured single strands can specifically reassociate to biologically active double helices (renature or anneal).

DNA in solution has a helical periodicity of about 10.5 base pairs per turn of the helix. The stacking of base pairs upon each other creates a helix with two grooves. Because the sugars protrude from the bases at an angle of about 120°, the grooves are unequal in size. The edges of each base pair are exposed in the grooves, creating a pattern of hydrogen bond donors and acceptors and of van der Waals surfaces that identifies the base pair.

The wider—or *major*—groove is richer in chemical information than the narrow (*minor*) groove and is more important for recognition by nucleotide sequence-specific binding proteins. Almost all cellular DNAs are extremely long molecules, with only one DNA molecule within a given chromosome. Eukaryotic cells accommodate this extreme length in part by wrapping the DNA around protein particles known as nucleosomes.

Most DNA molecules are linear but some DNAs are circles, as is often the case for the chromosomes of prokaryotes and for certain viruses. DNA is flexible. Unless the molecule is topologically constrained, it can freely rotate to accommodate changes in the number of times the two strands twist about each other. DNA is topologically constrained when it is in the form of a covalently closed circle, or when it is entrained in chromatin.

The linking number is an invariant topological property of covalently closed circular DNA. It is the number of times one strand would have to be passed through the other strand in order to separate the two circular strands. The linking number is the sum of two interconvertible geometric properties: twist, which is the number of times the two strands are wrapped around each other; and the writhing number, which is the number of times the long axis of the DNA crosses over itself in space.

DNA is relaxed under physiological conditions when it has about 10.5 base pairs per turn and is free of writhe. If the linking number is decreased, then the DNA becomes torsionally stressed, and it is said to be negatively supercoiled. DNA in cells is usually negatively supercoiled by about 6%.

The left-handed wrapping of DNA around nucleosomes introduces negative supercoiling in eukaryotes. In prokaryotes, which lack histones, the enzyme DNA gyrase is responsible for generating negative supercoils. DNA gyrase is a member of the type II family of topoisomerases. These enzymes change the linking number of DNA in steps of two by making a transient break in the double helix and passing a region of duplex DNA through the break.

Some type II topoisomerases relax supercoiled DNA, whereas DNA gyrase generates negative supercoils. Type I topoisomerases also relax supercoiled DNAs but do so in steps of one in which one DNA strand is passed through a transient nick in the other strand. RNA differs from DNA in the following ways: its backbone contains ribose rather than 2'-deoxyribose; it contains the pyrimidine uracil in place of thymine; and it usually exists as a single polynucleotide chain, without a complementary chain. As a consequence of being a single strand, RNA can fold back on itself to form short stretches of double helix between regions that are complementary to each other. RNA allows a greater range of base pairing than does DNA.

Thus, as well as A:U and C:G pairing, U can also pair with G. This capacity to form a non-Watson-Crick base pair adds to the propensity of RNA to form doublehelical segments. Freed of the constraint of forming longrange regular helices, RNA can form complex tertiary structures, which are often based on unconventional interactions between bases and between bases and the sugarphosphate backbone. Some RNAs act as enzymes—they catalyze chemical reactions in the cell and in vitro.

These RNA enzymes are known as ribozymes. Most ribozymes act on phosphorous centers, as in the case of the ribonuclease RNase P. RNase P is composed of protein and RNA, but it is the RNA moiety that is the catalyst. The hammerhead is a self-cleaving RNA, which cuts the RNA backbone via the formation of a 2', 3' cyclic phosphate in a reaction that involves an RNA-bound Mg^{+} ion. Peptidyl transferase is an example of a ribozyme that acts on a carbon centre. This ribozyme, which is responsible for the formation of the peptide bond, is one of the RNA components of the ribosome.

The discovery of RNA enzymes that can act on phosphorous or carbon centers suggests that life might have evolved from a primitive form in which RNA functioned both as the genetic material and as the enzymatic machinery.

2

Biochemical and Immunocytological Characterizations of Plant Physiology

In order to understand the mechanism of a plant in its entirety many separate phases of the dynamic activity of plants have been recognized and studied as individual processes. In his efforts to interpret these individual processes and their interrelationships the plant physiologist is confronted with many problems: What is the mechanism by which water, gases, and solutes enter a plant from its environment? How do such substances pass out of a plant into its surroundings? How are foods and other complex organic compounds synthesized in the plant?

How are they utilized in the development and mainte-nance of plants as living systems? What transform-ations of energy occur within a plant, and what exchanges of energy take place between a plant and its environment? How are water and solutes transported from one part of a plant to another? How are new tissues constructed? How is the development of one organ or tissue coordinated with the development of other organs or tissues? Why does a plant produce only vegetative organs at certain stages in its life cycle and reproductive organs only at other stages?

How are individual plant processes and the development of a plant as a whole influenced by environmental conditions? All of these problems and many other related or subsidiary ones lie within the province of the branch of science which is known as plant physiology.

For convenience, the study of plant life is subdivided into various branches such as physiology, morphology, anatomy, ecology, pathology, genetics, etc. Such a classification is necessarily more or less arbitrary. Neither physiology nor any other phase of plant life can be singled out for study without some consideration of plants from other viewpoints. A particularly intimate interrelationship exists between the structures and processes of plants.

Every physiological process is conditioned by the anatomical arrangement of the tissues, and by the size, configuration, and other structural features of the cells in which it occurs. Furthermore, the coordinated development of cells and tissues, *i.e.*, of the plant itself, is a complex of physiological processes. Thus the sciences of plant physiology and plant anatomy merge in the study of plant growth.

Just as different species of plants differ in outward configuration and internal anatomy, so do they also differ in physiology. The world of plants includes a great number of very diverse kinds of organisms that range in size from simple bacterial cells a few microns long to the enormous redwoods of our Pacific coast and mountain forests. Their difference in size, although striking enough, is not as fundamental as another distinction which exists between redwoods and bacteria. The redwoods and all other green plants are able to manufacture their own food, while the bacteria (with a very few exceptions) and all other non-green plants must obtain their nourishment from some outside source.

The physiology of the green and the non-green plants is therefore basically unlike. This book is essentially a discussion of the physiology of the chlorophyllous (green) plants with the emphasis on the vascular green plants. The physiological processes of the bacteria and fungi have been considered only when they have a direct bearing on the physiology of the green plants.

With few exceptions all vascular green plants carry on the same fundamental physiological processes, but there are many differences in subsidiary processes. Most green plants, for example, synthesize starch, but many don't. Numerous similar variations occur in the kinds of metabolic products built up by various species of plants. Most of the physiological differences among the species of green plants, however, are quantitative rather than qualitative. All chlorophyllous plants carry on photosynthesis, but when different species are exposed to the same environmental conditions the rate at which the process takes place may differ greatly from species to species.

A similar situation holds with respect to practically all of the other processes occurring in green plants. Even varieties of the same species often exhibit marked differences in physiological behaviour when exposed to a given environ-mental complex. Some varieties of wheat, for example, are markedly cold resistant while others are not.

RELATION BETWEEN PHYSIOLOGY AND PHYSICAL SCIENCES

Formerly the opinion was almost universal that living organisms owe their distinctive properties to the possession of subtle and unknown forces which are peculiar to living matter. At the present time such vitalistic theories find very few advocates. The contrary and now widely held assumption is that living organisms operate in accordance with the same physico-chemical principles that hold in the inanimate world.

The complexity and elusiveness of living processes are not assumed to be due to intangible unknown varieties of energy but to the interplay of recognizable physico-chemical forces in the complex organized system of the protoplasm.

Adoption of this latter point of view has led to a widespread use of the tools of physics and chemistry in experimental work on plants, and to the interpretation of plant processes in terms of these two sciences. This has led to notable progress in our understanding of the physiology of plants and has permitted the analysis and expression of many physiological relations in quantitative terms.

A knowledge of certain fundamental principles of physics and chemistry is therefore essential to the understanding of physiological processes. For this reason several of the earlier and parts of some of the later chapters in this book are devoted to a brief exposition of those underlying principles of the physical sciences with which a student of plant physiology should be familiar.

RELATION BETWEEN PLANT PHYSIOLOGY AND AGRICULTURAL SCIENCES

Green plants are not only the ultimate source of all food but supply the raw materials for many of our basic industries. With the rise of modern industrial civilization both the quantities and kinds of plant products which we utilize have increased rather than decreased. In addition to foods some of the more important raw products obtained from plants are wood, textile fibres, pulp, rubber, vegetable oils, gums and drugs. Even most of the so-called synthetic products of the chemist are not synthetic in the sense that they have been built up from simple inorganic compounds, but only insofar as they represent modifications of naturally occurring plant products.

An industrial civilization not only requires a wide variety of plant products but insists that these products meet certain standards of quality. The successful cultivation of plants has, therefore, become a highly skilled occupation, and the agricultural sciences are rapidly becoming a domain of the specialist. Success in controlling the activities of living plants can not be achieved without some understanding of the processes which occur within them, and of the effects of environmental conditions upon these processes. The problems of the forester, the fruit grower, the cotton planter, the floriculturist, the grain farmer, and of all others who cultivate plants differ in detail, yet they all have a fundamental similarity in requiring an application of the principles of plant physiology for their solution.

Fundamental investigations in plant physiology have contributed in many ways to improved methods of propagating, cultivating and harvesting economically important plants, and to methods of handling and storing many plant products. Furthermore, control of the fungous diseases and insect predators of plants often requires applic-ation of the principles of plant physiology.

Much of the investigational work carried on by scientific agronomists, horticulturists, floriculturists, and foresters actually lies in the field of pure or applied plant physiology, although often it is not formally classed as such.

PLANT PHYSIOLOGY AS A SCIENCE

The origin of mans first conscious interest in plants long antedates recorded history. Agriculture had already become a highly developed art thousands of years before any experimental study of plant processes began. Consequently, there had grown up a vast body of traditional plant lore which passed orally from father to son, generation after generation. This practical knowledge of plants developed over many centuries as a result of countless, mostly involuntary trial and error experiences and innumerable observations of plant behaviour under all kinds of circumstances.

Which of this mass of mingled facts and beliefs regarding plants is alive in the consciousness of the common man today. The closer he lives to the soil and the more familiar he is with traditional plant lore, the more it influences his thoughts and actions. Many of these customarily accepted beliefs are essentially sound and most of them contain elements of truth. Others, however, are entirely erroneous, and not a few are tempered with superstitions, so me of which have an unbroken lineage back to the days of witch-doctors and savagery. No reputable botanist has held for generations, for example, that plants obtain their food from the soil, yet this and many other fallacious beliefs are still widely entertained among the general population.

The value of practical information about plants should not be underrated, since its perpetuation in the mind of man permitted the development of agriculture to a high plane as a practical art before any widespread investigation of plants from a scientific point of view was undertaken. Nevertheless, traditional plant lore is not only often inadequate, but is riddled with misconceptions, and often suffers from points of view which are inherently stultifying to the acquisition of further knowledge. The layman, for example, often personifies plants in an attempt to explain their behaviour. Man has desires and foresight and it is often assumed, either consciously or tacitly, that plants are similarly endowed. To many, for example, the statement that roots grow downwards in search of water, or that stems grow upwards in order to reach the light are accepted as adequate explanations of plant behaviour.

Mans knowledge that water and light are essential to plants is not evidence that plants are similarly aware of these facts. To assume that plants realise their needs and are able to act in conformity with their requirements is equivalent to crediting them with a high order of intelligence. Explanations of plant behaviour are commonly encountered in which purposeful action on the part of plants is tacitly or deliberately implied, although there is no justification for the adoption of such a point of view. Furthermore, the layman seldom pursues his quest for information about plants beyond the stage of observation, while the scientist frequently does.

Observation has suggested, for example, that light is necessary for the continued existence of plants. To one who is scientifically minded, either by instinct or training, the obvious next step is to test this postulate experimentally. If the suggested hypothesis is substantiated by experiment it is tentatively accepted as a theory. Theories such as those proposed in explanation of the processes or reactions of plants, together with the experimental results which are considered to support them, are usually published in a scientific journal or monograph and thus exposed to evaluation and further experimental testing by other scientists.

Continued experimentation may lead to substantiation, rejection, or modification of the theory as originally proposed. Modification would undoubtedly be the fate of the hypothesis used as an example, since sooner or later some investigator would find that non-green plants can thrive in the absence of light. Experiments often raise more questions than they answer. New approaches to the problem under consideration as well as desirable new lines of inquiry are constantly opening up to the alert investigator. In this way experimentation leads to more experimentation, more facts accumulate, and more theories are proposed. Some of the suggested hypotheses are confirmed, others are rejected, and still others are modified. Most of them, sound or fallacious, in turn suggest further observation and experimentation.

As a result of such endless and painstaking labours there is slowly built up that vast, complex, and ever changing body of knowledge which we refer to as a science. The system of subjecting all hypothetical explanations of natural phenomena to experimentation is the essence of the scientific method. It is the characteristic methodology of all sciences. Progressive modification of accepted concepts in the light of new experimental findings continually increases the soundness of scientific generalizations. Thus there are incorporated into any science theories and generalizations in various stages of acceptance.

Some stand upon such a firm substructure of facts that they are accepted by all authorities in the field. Others, less securely supported by experimental results, are subscribed to by some but rejected by other workers. Finally, in any science there are always some theories which are so dubious that they find only a few advocates. Furthermore, some of the theories now widely held sooner or later will be discarded as a result of new findings or as a consequence of different interpretations of facts already known.

However, not all scientists are in agreement regarding the interpretation of the same sets of facts. Although this state of affairs is entirely consistent with the spirit of scientific research, it is frequently puzzling to students and laymen. Differences of opinion regarding the hypotheses which suitably explain scientific phenomena are most likely to arise when experimental data are inadequate. Disagreements regarding the interpretation of experiments and observations are often inevitable steps in the clarification of scientific generalizations.

Controversies usually focus attention upon gaps in our factual information. Frequently, therefore, they are stimulating to research, and often lead to a further enrichment of human knowledge. Without the knowledge of plant processes which has been slowly accumulated through more than two centuries of observation, experimentation, and critical evaluation by numerous workers in all parts of the world, this book could never have been written. In spite of the patient labours of these many workers vast gaps still exist in our understanding of the physiology of plants gaps which are reflected in the necessarily inadequate treatment of many topics in this book. The future of this science and all others lies in the hands of the front-line investigators who wage a continual struggle for the extension of human knowledge.

POSSESSIONS OF SOLUTIONS

Water is the most abundant compound present in all physiologically active plant cells. The water which occurs in a liquid state in plant cells invariably contains other substances dissolved in it and usually also contains dispersed particles which are not in true solution. When the particles dispersed throughout the water are within a certain range of sizes the system of water plus particles falls into the category of colloidal systems. The complicated dynamics of living systems can be largely interpreted in terms of the physicochemical properties of solutions and colloidal systems, one component of which is water, although it should not be inferred that non-aqueous solutions and colloidal systems are entirely absent in living organisms.

Similarly liquid water never occurs in the pure state in the natural environment of living organisms. The water of streams, lakes, and oceans invariably contains various substances in solution and usually in the form of dispersed particles as well. This is likewise true of the soil water. Even raindrops, the products of natural distillation, contain gases which have dissolved in them from the atmosphere.

COMMON NATURE OF SOLUTIONS

Simple solutions are systems in which one component (the solute) is dispersed throughout the other (the solvent) in the form of molecules or molecules and ions. Theoretically the solvent may be a gas, a solid, or a liquid, but solutions in which the solvent is a liquid are by far the most important in living organisms. Except in extremely concentrated solutions the average distance between the solute particles is usually very great relative to their size. Naturally occurring solutions, whether in living organisms or their environment, usually contain a number of different solutes and are often exceedingly complex. Water is the commonest and most important of all solvents both in the inorganic world and in the realm of living organisms. The further discussion will be principally in terms of aqueous solutions.

SOLUTIONS OF A GAS IN A LIQUID

The water present in living organisms usually contains dissolved gases. Those most commonly present are carbon dioxide, oxygen, and nitrogen. Practically all the water in the environment of organisms in oceans, lakes, streams, soils, and raindrops also contains these gases in solution, and sometimes others as well.

A given volume of water or any other liquid will hold only a limited quantity of gas in solution at a given temperature. When no more of a certain gas can be dissolved in a liquid it is said to be saturated with respect to that gas. Gases vary widely in their solubility in water, but in general fall in to two groups: those which are sparingly soluble and those which are highly soluble.

When only a small fraction of a unit volume of a gas will dissolve in a unit volume of water the gas is classified in the sparingly soluble group. When from one to many unit volumes of a gas will dissolve in one unit volume of water it is classified in the highly soluble group. Oxygen, hydrogen, and nitrogen are familiar examples of gases belonging to the first group, while carbon dioxide, ammonia, and hydrogen chloride are examples of the second. When gases are highly soluble in water it is usually evidence that a chemical reaction takes place between the gas and water.

There actions between water and carbon dioxide and water and ammonia are indicated in the following equations:

$$CO_2 + H_20 \rightarrow H_2C0_3NH_3 + H_20 + 1NH_4OH$$

In a solution of such gases not only are molecules of the gas present, but also molecules of a compound formed by the reaction of the gas with water. The apparently great solubility of gases such as these is due to the formation of relatively soluble compounds by the reaction of the gas with the water.

Increase in pressure of a gas above a liquid increases the solubility, *i.e.* the concentration of that gas in the liquid. For sparingly soluble gases the increase in solubility is directly proportional to the increase in pressure (Law of Henry). This principle also holds qualitatively, but not quantitatively, for the very soluble gases.

When a mixture of several gases is maintained over water, each dissolves independently and in accordance with the gaseous pressure (partial pressure) which it exerts against the surface of the water (Law of Dalton). This principle holds strictly only for those gases which are slightly soluble in water. For example about one-fifth of the atmospheric pressure is due to the oxygen present.

Assuming an atmospheric pressure of 760 mm. Hg (the value at standard conditions), the pressure due to the oxygen is equivalent to about 152 mm. Hg. The quantity of oxygen which dissolves in water exposed to the air is the same as that which would dissolve if oxygen only at a pressure of 152 mm. Hg occupied the space over the water.

SOLUTIONS OF A LIQUID IN A LIQUID

In general solutions of a liquid in a liquid fall into two classes: those in which the liquids are freely miscible with each other in all proportions and those in which each liquid reaches a definite point of saturation in the other. Alcohol and water, for example, mix with each other in all proportions; such a solution is an example of the first mentioned type. Many oily liquids also are miscible with each other in all proportions. A familiar example is the solution of lubricating oil in gasoline. In solutions of this kind the liquid present in excess is usually considered to be the solvent. In a 50 per cent solution of alcohol and water, either liquid could be considered the solvent, or either could be considered the solute. Ether, chloroform, and many other liquids are sparingly soluble in water. After water and ether, for example, are shaken together in a flask, two distinct layers of liquid separate upon standing. The upper layer consists of the lighter ether, saturated with water, while the lower consists of water saturated with ether. In both layers, however, the concentration of solute present at saturation is small. In the upper layer ether is the solvent, water the solute; the converse is true of the lower layer.

SOLUTIONS OF A SOLID IN A LIQUID

This is by far the most familiar type of solution and in many respects the most important. Substances in the solid state vary greatly in their solubility in water, ranging all the way from those which are virtually insoluble to those which are extremely soluble. There is usually a limit to the amount of any solute which can be dissolved in a given volume of water at a given temperature. When this limiting concentration is reached the solution is said to be saturated, following the same terminology used with solutions of gases and liquids in liquids. It is almost impossible to prepare a true saturated solution of any solute unless some of it is also present in the solid state.

Under certain conditions, and only when none of the undissolved solid is present, a super-saturated solution can be prepared; that is, a solution in which the concentration of the solute is greater than that in the saturated solution. If a fragment of the solid solute be added to such a solution the excess solute usually crystallizes out immediately, and the concentration of the solution decreases to that usually present at saturation. Usually increase in temperature increases the solubility of a solid in a liquid but there are some exceptions to this principle. The solubility of common salt (NaCl) in water, for example, is only slightly influenced by temperature, while the solubility of many calcium compounds is decreased by an increase in the temperature of water.

METHODS OF THE COMPOSITION OF SOLUTIONS

If a mol of any soluble compound be dissolved in just enough water to make exactly one litre of solution the result is a volume molar solution. Since the volume of water changes with temperature it is usual to specify that the solution is to be made up to a litre volume at 200°C. Whenever the word molar is used without qualification this type of solution is meant. Similarly, if half the molar weight, or one-tenth the molar weight of a substance be dissolved in enough water to make exactly one litre of solution the result is a half molar (0.5 IVI) or tenth molar (o.i IVl) solution, respectively, etc. Gram molar weights of all substances contain the same number of molecules.

Estimates made by various methods show this number to be about 6.o6 × 1023 molecules. Hence one litre of any volume molar solution will contain this number of solute molecules, one millilitre (0.001 litre) will contain one thousandth of this number, and so on. Equal volumes of all solutions of the same volume molarity contain the same number of solute molecules but different numbers of solvent molecules. If a given volume of a volume molar solution be diluted with an equal volume of water, the result is a 0.5M solution; if a given volume be diluted with nine volumes of water the result is a 0.1M solution etc. Therefore a volume molar solution of any strength may be diluted with water and the resulting more dilute solution will have a volume molarity in proportion as it has been diluted.

If a mol of any soluble substance be completely dissolved in 1000g. ofwater the result is a weight molar solution. Such solutions are often called molal solutions. They are used principally in experimental work upon various osmotic phenonema.The addition of a mol of most solids to a litre of water will increase the volume of the resulting solution to more than one litre. This increase in volume is called the solution volume of the solute. The solution A mol is the molecular weight of a compound in grams.

There is a shrinkage in volume when the solute is added to the solvent. On the other hand the solution volume of some compounds, especially the sugars, is considerable. When a mol of sucrose is added to 1000g. of water the resulting solution will have a volume of 1207cc. at 0^0C. Hence the solution volume of sucrose is 207cc. The solution volume of a mol of sodium chloride, on the other hand, is only about 18cc. Since every solute has a different solution volume, it follows that equal volumes of weight molar solutions do not contain the same number of either solvent or solute molecules.Neither will the dilution of a given volume of a weight molar solution with an equal volume of water result in a 0.5 molal solution. In other words the concentration of a weight molar solution does not change in proportion to the amount by which it has been diluted. This is a fact which is sometimes overlooked in experimental work.

In physiological work it is often convenient to make up solutions on a percentage basis. Such solutions are made up either on the basis of percentage by weight, or percentage by volume. Solutions of solids in water or other solvents are made up on a weight percentage basis. A 10 per cent sodium chloride solution, for example, is made by dissolving 10 g. of sodium chloride in 90 g. of water. Solutions of liquids in water or other solvents can also be prepared on a weight percentage basis. It is simpler, however, to make up such solutions on a volume percentage basis. On this basis a 10 per cent solution of alcohol is prepared by adding 10cc. of alcohol to 90cc. of water. The percentage system of solutions has no direct relation to either the volume molar or weight molar systems.

RELATIONSHIP BETWEEN ELECTROLYTES AND NON-ELECTROLYTES

Some aqueous solutions readily conduct an electric current; others do not. The former are called electrolytes; the latter non-electrolytes.The solutions of all acids, bases, and salts are electrolytes. Solutions of most organic compounds such as the sugars, alcohols, ketones, and ethers are non-electrolytes.

This process is called electrolysis. If hydrochloric acid is the electrolyte, for example, hydrogen gas will be liberated at the negative pole (cathode) and chlorine gas will be liberated at the positive pole (anode).If electrolysis is continued long enough eventually all of the hydrochloric acid present in the system will be decomposed into hydrogen gas and chlorine gas. Strictly speaking the term electrolyte refers only to a solution of an ionized substance, but it is also often applied to any compound which, when dissolved in water, produces ions.A similar dual usage of the term non-electrolyte also prevails.

The occurrence of electrolysis, as well as the unusually large effects of electrolytes on the osmotic pressures of solutions, have led to an explanation of the behaviour of electrolytes in terms of the theory of electrolytic dissociation. This theory was first advanced by the Swedish chemist Arrhenius who introduced it in essentially its present form in 1887. According to the Arrhenius theory when an electrolyte is dissolved in water some of the molecules dissociate into two kinds of particles, one positively charged, the other negatively charged. Each of these particles is called an ion. Ions are supposed to be present in a solution whether any electrolysis occurs or not.Two, three, four or even more ions may be formed from a single molecule.The conduction of an electrical current by an electrolyte is due to the presence of these ions.

The positively charged ions, which in electrolysis migrate towards the cathode, are called cations; the negatively charged ions which migrate towards the anode are called anions. A cation may carry one, two, three, or even four positive charges; an anion may carry from one to several negative charges.When an electrolyte dissociates the number of positive charges carried on thecations is always equal to the number of negative charges carried by the anions.

The Arrhenius theory does not assume complete ionization of all of the solute molecules in an electrolyte but that the molecules are continuously dissociating into ions, while free ions in the solution are continuously reuniting and forming molecules, both processes proceeding at an equal rate whenever an equilibrium condition prevails.An equilibrium of this sort, which is maintained by two opposing processes proceeding at equal rates is termed adynamic equilibrium. Electrolytes vary greatly in degree of dissociation. Those in which a large proportion of the molecules are maintained in a dissociated condition are termed strong electrolytes; those of which the contrary is true, weak electrolytes.Strong electrolyte solutions conduct electric currents better than weak electrolyte solutions of equal concentration. In general the more dilute a given electrolyte solution the larger the proportion of dissociated molecules present. In extremely dilute solutions ionization is practically complete. Increase in temperature reduces the dissociation of an electrolyte.

It is probable that many of the so-called non-electrolytes also dissociate very slightly in solution, but the degree of dissociation cf such compounds is so small that it can be detected only by very refined methods, if at all. Even water, as the subsequent discussion will show, dissociates slightly, producing hydrogen and hydroxyl ions.

ACIDS, BASES, AND SALTS

An acid may be defined as a substance which produces hydrogen ions (H^+) when dissolved in water. The characteristic properties of acids are due to the hydrogen ions produced. The most common in Organic acids are hydrochloric (HCl), nitric(HNO_3), and sulfuric($H_{25}0_4$). In living organisms a large group of more complex, but much weaker acids, known as organic acids, play important roles. The strength of an acid depends upon its degree of ionization; the greater the proportion of hydrogen ions an acid produces in solution at a given concentration, the stronger it is.

A base may be defined as a substance which produces hydroxyl ions (OH^-) when dissolved in water. The characteristic properties of bases are due to the hydroxyl ions produced. Some of the commonest bases are sodium hydroxide (NaOH), calcium hydroxide ($Ca(OH)_2$), and potassium hydroxide (KOH). The strength of a base, like that of an acid, depends upon its degree of ionization. The greater the proportion of hydroxyl ions a base produces in solution at a given concentration, the stronger it is.

The theory of electrolytic dissociation accounts much more satisfactorily for the behaviour of weak electrolytes than of strong electrolytes. For this and other reasons Debye and Hiickel in 1923, and other modern investigators, have postulatedthat strong electrolytes, at least, are completely dissociated. The fact that they behave as if only partly dissociated is accounted for in terms of inter ionic attractions. These attractions are supposed to exert what may be roughly described as a braking effect upon the movement of the individual ions.This prevents them from operating at their maximum effectiveness in conducting an electric current or in other phenomena depending upon ionic mobilities. The greater the concentration of an electrolyte, the closer together are the ions, and, according to this view, the less their apparent dissociation.

A salt may be defined as a compound which has been produced by the union of an anion or anions of an acid with the cation or cations of a base. Salts are produced when an acid and a base are brought together in a solution as a result of a chemical union between the hydrogen ion (s) of the acid and the hydroxyl ion (s) of the base, forming water. This reaction is called neutralization.

Since water is practically undissociated, neutralization reactions go rapidly to completion. The reverse reaction, indicated in the above equations by the arrows pointing to the left, is called hydrolysis. Under certain conditions hydrolytic reactions may proceed at a rapid rate, although in the examples presented above the speed of the reaction towards the left (hydrolysis) is negligible compared with the speed of the reaction towards the right (neutralization).

Normal Solutions

The concentrations of acids and bases are most commonly expressed in terms of normal solutions. A normal solution of an electrolyte contains in a dissolved state per litre of solution at 200 C. a weight of the compound in grams equal to its molar weight divided by its hydrogen equivalent. The hydrogen equivalent of a compound is defined as the number of replaceable hydrogen atoms in one molecule, or the number of atoms of hydrogen with which one molecule could react. Thus a normal solution of an acid contains 1.008 g. of replaceable hydrogen per litre; a normal solution of a base 17.008 g. of replaceable hydroxyl per litre. By this system the concentration of any acid, base, or salt can be designated as 0.1 N, 0.5 N, 2 N etc., as the case may be. The normality of an acid solution is a measure of its total acidity, i. e.of its concentration in terms of replaceable hydrogen ions. Similarly the normality of a base solution is a measure of its total basicity. Since 1.008 g.of replaceable hydrogen represents the same number of ions as 17.008 g. of replaceable hydroxyl (why?) it is evident that equal volumes of acid and base solutions of equal normality will exactly neutralize each other. Hydrogen Ion Concentration. The effects of acids upon chemical reactions and upon physico-chemical conditions generally in both inorganic systems and in living organisms are due principally to the hydrogen ions which they produce when in solution.

Some of the most fundamental of physiological phenomena are markedly influenced by the concentration of hydrogen ions in the medium in which they occur. For many purposes therefore it is more important to have some sort of a measuring stick of the concentration of hydrogen ions present in a solution than of the total acidity of the solution.Total acidity, as already noted, is customarily expressed in terms of normality. It is also entirely possible to use the normal system for expressing the concentration of hydrogen ions. In a normal solution of hydrochloric acid, for example, about 78 per cent of the molecules are dissociated.

The term normal as used in the preceding sentence refers to total acidity, that is, to all the ionizable hydrogen whether actually present as ions or combined with anions in the form of molecules. In terms of the hydrogen ions present, however, such a solution is only 0.78 N. The term normal as used in this latter sense refers only to the ionized hydrogen. We may speak therefore of a normal solution of hydrogen ions as well as of a normal solution of an acid.Since no acid is ever completely dissociated, a normal solution of any acid will always be less than normal when its concentration is expressed in terms of the hydrogen ions present.

In order to prepare a normal solution of hydrogen ions it is necessary to make up a solution which is more than normal in terms of total acidity. Such a solution must be of the precise strength and degree of dissociation that exactly i.008 g. of the ionizable hydrogen present are actually in the dissociated form as ions per litre of the solution. Although hydrogen ion concentration can be readily expressed in terms of normalities, in actual practice this system is not generally used because it is apt to prove cumbersome, especially when it is necessary to refer to the very small concentrations of hydrogen ions usually dealt with in biological problems.

The hydrogen ion concentration of a solution is now quite generally defined in terms of its pH value. The pH value bears a simple mathematical relation to the hydrogen ion concentration of a solution in terms of its normality. Because of the practically universal acceptance of this system it is necessary to understand the significance of the term pH and its relation to hydrogen ion concentration expressed in terms of normality. The relation between pH and hydrogen ion concentration in terms of normality is a logarithmic one. The pH of a normal solution of hydrogen ions is 0, of a 0.1 N solution i, of a 0.01 N solution 2, etc. Zero is the logarithm of i, i is the logarithm of 10, 2 is the logarithm of 100, etc. The pH value for any solution is the negative of the logarithm of the hydrogen ion concentration in terms of normality. It may also be defined as the logarithm of the number of litres of solution that contains one gram atomic weight of hydrogen ions.

All aqueous solutions as well as pure water contain hydrogen ions in some concentration. Corresponding to the concentration of hydrogen ions is a certain definite concentration of hydroxyl ions. It can be shown by the principle of mass action that the mathematical product of the concentration of hydroxyl ions and the concentration of hydrogen ions in a solution is a constant.

This may be expressed as follows:

$$(H+) \times (0H)K.Kio14 \text{ at } 2O \text{ C}.$$

Hence as the pH of a solution is increased the pOH decreases and vice versa. For example, if the pH increases from 5 to 6 the poll decreases from 9 to 8.

Furthermore, only at pH 7 can the concentration of hydrogen ions equal the concentration of hydroxyl ions. This is therefore the neutral point on the pH scale and corresponds to the dissociation of pure water. This pH value represents a dissociation of only one water molecule in approximately every 555,000,000.

Values below 7 on the pH scale represent the acid range, those above 7 the alkaline range of the scale. An acid solution is one with a larger concentration of hydrogen ions than hydroxyl ions, while in an alkaline solution the reverse is true. The lower the pH value the greater the hydrogen ion concentration of a solution. A pH value of 5 represents ten times the hydrogen ion concentration of a solution with a p11 of 6 and one hundred times the hydrogen ion concentration of one with a ph value of 7, etc. This is due to the fact, previously emphasized, that the numbers on the pH scale are related to each other as logarithms and not as ordinary arithmetic numbers.

The hydroxyl ion concentration of a solution could be expressed in p011 units instead of in pH units. For alkaline solutions especially this would seemto be a logical practice. But because of the definite mathematical relationship between the pH and pOH the p11 value alone also defines the p011 value. Hence both the acidity of a solution in terms of H + ions, and its alkalinity in terms of OH ions may be, and customarily are, expressed in terms of pH units. It is possible for aqueous solutions to exist which have a pH value of less than 0 (*i.e.* a minus pH value) or more than 14.

An solution of hydrochloric acid, for example, would have a pH value of a little less than o; correspondingly a N solution of sodium hydroxide would have a pH value a little greater than 14. In actual experience, however, solutions of such extreme concentrations are seldom encountered that the sodium acetate solution retards in some way changes in pH value upon the addition of an acid while a solution of sodium chloride does not.

Solutions of such compounds as sodium acetate which are relatively resistant to changes in p11 due to the addition or loss of hydrogen or hydroxyl ions are known as buffer solutions, and this property of solutions is called buffer action. Solutions such as sodium chloride which show no buffer effect are called unbuffered solutions.If 1 cc. of a normal sodium hydroxide solution be added to 10 cc. of a normal solution of sodium chloride, a marked increase in p1-f will occur. Ifi cc. of the same solution be added to 10 cc. of a normal solution of sodium acetate, its pH will also increase markedly. In other words the sodium acetate solution is buffered against the addition of an acid, but not against the addition of a base. If, on the other hand, 10 cc. of a normal solution of acetic acid be substituted for the sodium acetate it will be found that this solutionis strongly buffered against the introduction of hydroxyl ions into the solution.

A mixture of equal volumes of molar sodium acetate solution and molar acetic acid solution will exhibit buffer action against both acids and bases over a considerable range of the pH scale.Two important points regarding buffer solutions have been brought out by the foregoing discussion. No one solute as a rule will act as a buffer against both acids and bases. Furthermore no buffer solution will exert a buffer effect over the entire range of the p11 scale.Different buffer solutions vary greatly in their effectiveness in maintaining pH stability. Some are strongly buffered; others weakly. Some are strongly buffered against acids and weakly buffered against bases; of others the converse is true. The commonest types of buffer systems are those composed of a weak acid plus one of its salts.

The sodium acetate-acetic acid buffer system already described is such a system. Practically all of the buffer solutions of importance in living organisms belong to this group. Buffer action consists essentially in the tying up of free hydrogen or hydroxyl ions nearly as rapidly as they are introduced into the solution in the formation of compounds which are only slightly dissociated. The ensuing change of pH is therefore relatively small in proportion to the volume of acid or base added. As an illustration let us consider once more a solution consisting of both sodium acetate and acetic acid dissolved in water. The sodium acetate is strongly dissociated, but the acetic acid will dissociate only slightly, being a weak acid.Suppose now that a little HCl be added to this solution. This is equivalent to adding H_1 and Cl ions and HCl molecules; the latter, however will dissociate, forming additional ions as rapidly as the H_1 and Cl ions already present are bound up in chemical combination.

H+ and CH_3COO ions cannot exist side by side in the same solution in appreciable concentrations since CH_3COOH is a poorly dissociated compound. Hence the added H^+ ions are almost all tied up in the formation of CH_3COOH. The Cl ions form NaCl with the Na ions which dissociates in the usual way. The result is that there is only a slight increase in the concentration of hydrogen ions in the solution, and hence only a very slight reduction in pH value. Suppose now that instead of HCl, a little NaOH be added to this solution.

This is equivalent to adding Na^+ and OH ions and NaOH molecules; the latter, however, will produce additional ions by dissociating as rapidly as the Na^+ and 0H ions already present are tied up in chemical combination. But 0H and H+ ions cannot exist side by side in the same solution in appreciable concentrations since H_20 is only slightly dissociated. Most of the added 0H ions therefore combine with the H1 ions produced by the CH_3COOH and form H_20.

More of the CH_3COOH hydrolyzes producing more H+ ions, which in turn unite with more of the OH ions. This continues until practically all of the added OH ions are tied up. The Na1 ions form CH_3COONa with the CH_3COO ions which dissociates in the usual way. The final result is that there is only a very slight decrease in the concentration of H^+ ions in the system, and hence only a very slight increase in its pH value.

Any mechanism which will act in such a way as to remove hydrogen or hydroxyl ions from a solution may operate as a buffer system. Other types of buffering are known but they are relatively of much less importance in living organisms than the type of chemical mechanism which has just been considered. Hydration of Solutes. Molecules of water adhere to the particles of many solutes. Water thus associated with the particles of a solute is called water of hydration. For example, each molecule of dissolved sucrose apparently has six molecules of water associated with it. This means that if a mol of sucrose (342.2 g.) is dissolved in water there will be bound to the sucrose molecules a total of six mols of water (108.096 g.). Ions also are hydrated. Different species of ions appear to have different numbers of water molecules associated with them.

Although exact values have been assigned by several experimenters for the number of molecules of water of hydration for certain ions these values are open to question as the results of different investigations do not agree. It seems certain, however, that both anions and cations are hydrated, and that for some kinds of ions, at least, as many as a hundred or more molecules of water of hydration may be associated with a single ion.

ROLE OF INTERFACIAL PHENOMENA IN PLANT PHYSIOLOGY

There are a number of familiar observations which indicate that the surface layer of water possesses certain distinctive properties which are not exhibited within the body of the liquid. For example, if a perfectly dry, clean, steel needle be carefully laid on the surface of some still water it will float, in spite of the fact that steel is heavier than water. Careful observation shows that the water surface beneath the needle is depressed, forming a tiny liquid cradle, within which the needle is supported. This phenomenon is clearly due to the properties of the surface film of the water, because if the needle is laid on the water at a slight angle, so that this film is punctured, it will immediately sink to the bottom of the vessel.

A number of other familiar phenomena are possible principally because of the distinctive properties of the surface film of water. The ability of many kinds of insects to walk on water, the formation of drops by water under certain conditions the rise of water in capillary tubes, soil, blotting paper, etc., the upward bulging of the water surface in a vessel when it is filled to the very last drop, and many other aspects of the behaviour of water, are due largely or entirely to the properties of its surface layer. The surface layers of other liquids also exhibit distinctive properties, but in no common liquid except mercury are such properties as well marked as in water.

SURFACE TENSION

Every molecule within the body of a liquid, although in rapid motion as a result of its kinetic energy attracts and is attracted by other molecules. The attractive forces which any molecule exerts upon its neighbours may be considered as acting along lines of force which radiate from it in all directions. The pull exerted by any molecule upon surrounding molecules, however, diminishes very rapidly with increasing distance from that particular molecule, reaching a maximum at distances not exceeding one or two molecular diameters. Liquids possess a definite volume and distinct boundaries because the molecules are close enough together to attract each other with forces of considerable magnitude.

The important property of cohesion in liquids is due to these forces. The magnitude of the cohesive force between the molecules is greater in water than in most liquids. The molecules of a solid are so closely bound together by intermolecular attractions that it has lost ability to change its form, and possesses not only definite, but fixed boundaries. The molecules of both liquids and solids are kinetically active, but their movements are limited more or less rigidly by cohesive forces between the molecules. This limitation upon the freedom of movement of molecules is much greater in solids than in liquids. In gases the distance between the molecules is relatively so great that the intermolecular attractive forces are of negligible magnitude.

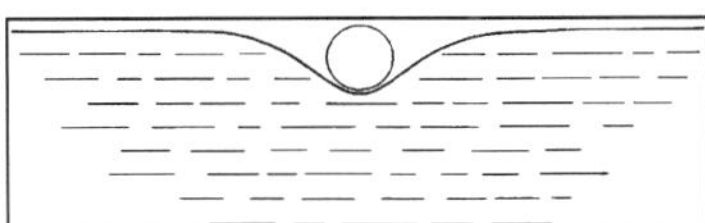

Fig. Needle Floating on water

Molecules at the surface of a liquid or a solid are affected differently bythe cohesive forces between molecules than those within the interior. The further more detailed discussion of this point will be in terms of water. In a vessel of water each of the individual molecules in the body of the liquid is attracted in all directions by the cohesive forces exerted upon it by the surrounding molecules.

At the surface, however, the molecules are not surrounded on all sides by other water molecules. Every molecule in the surface layer is pulled strongly towards the interior by the water molecules beneath it. It is for this reason that drops of water falling freely through space acquire approximately a spherical shape. Since there are no outwardly directed attractions which balance these internally directed forces the surface layer of molecules in any liquid is constantly under a tension. The strained condition of the surface layer of liquid molecules at liquid-gas boundaries is known as surface tension.

Solid surfaces also manifest surface tensions, but gases, having enormous intermolecular distances and no surface boundaries, do not. In addition to the unbalanced molecular attractions which the molecules in the surface layer of a liquid sustain, such molecules are usually definitely oriented. This results in a greater concentration of molecules in the surface layer than in the body of the liquid, and contributes towards maintenance of rigidity of the surface film. Surface tension values are generally stated in terms of dynes per centimeter, that is, as the force in dynes which is exerted by the liquid along a line 1 cm. in length.

ISSUES AFFECTING SURFACE TENSION

Temperature

Since the phenomenon of surface tension is one manifestation of the cohesive forces existing among molecules, the greater these cohesive forces, the greater the surface tension. Temperature is a measure of the kinetic energy of the molecules, hence an increase in temperature means an increase in their kinetic energy and vice versa.

As the kinetic activity of water molecules increases,the effectiveness of the cohesive forces existing between them decreases. Hence as the temperature of water increases, its surface tension decreases. The decrease in surface tension of any liquid is, in fact, directly proportional to increase in temperature, except when a liquid approaches its critical temperature. The variation in the surface tension of water within the temperature range at which most physiological processes occur is not sufficient, however to be of any very great significance.

Solutes

The surface tension of a liquid is influenced by the presence of the molecules or ions of a solute. Some solute increase the surface tension of water; others decrease it. If alcohol, for example, be added to water the surface tension of the resulting solution will be less than that of pure water. The attraction between an alcohol molecule and a water molecule is less than that between two water molecules. Since the alcohol molecules are distributed among the water molecules, the mean cohesive force in the solution is less than that in pure water.

The reduction in surface tension occasioned by the addition of alcohol to the water is one aspect of the reduction of the cohesive forces between the molecules in the solution. Most organic compounds have a similar and often very marked effect on the surface tension of water when dissolved in it. However a few such as sucrose have a slight increasing effect on surface tension. Most inorganic salts when dissolved in water increase its surface tensions. The presence of such solutes increases the cohesion within the liquid by attracting the water molecules more strongly than the water molecules attract each other.

The mean cohesive force of the molecules in the solution is thus increased. The result of this effect is a greater surface tension in the solution than in pure water. The magnitude of the effect of inorganic salts upon surface tension is never very great, however.

As a rule, increasing the concentration of a solute in water increases the magnitude of its effect, whether positive or negative, upon surface tension. The influence of solutes upon surface tension is not proportional to their concentration, however, as low concentrations of any solute produce relatively greater effects upon surface tension than high concentrations.

Interfacial Tension

The condition of tension or strain in a limiting layer of molecules is not confined to boundaries between liquids and gases.At a boundary between two immiscible liquids or between a solid and a liquid each of the abutting layers of molecules is subject to similar tensions. Boun-1 In general compounds which decrease the surface tension of water are of the type known as non-polar, while those which increase it are of the type known as polar.

The term surface tension is usually restricted to tensions developed at the surface of a liquid when in contact with a gas, *i.e.* at the interface between a liquid and a gas. Surface tension is therefore merely one variety of interfacial tension. The magnitude of interfacial tensions is influenced by the factors of temperature and solutes in the same general way as surface tensions.

Adsorption

The molecules of most organic compounds, as we have already seen, have a lesser attraction for water molecules than the molecules of water have for each other. This relatively greater attraction between water molecules than between the molecules of water and those of such non-polar solutes results in the displacement into the surface or interfacial layers of a disproportionately large number of the solute molecules. Hence the concentration of most organic solutes is greater in the surface or interfacial layers than in the body of the liquid. The concentrating of solute molecules in the surface or interface of a liquid is called positive adsorption.

The molecules of inorganic salts, on the other hand, are more strongly attracted to water molecules than the water molecules are attracted to each other. Because of this stronger attraction of the water molecules for such polar solutes relatively more solute molecules are pulled from the surface films into the body of the liquid than water molecules. Molecules of polar solutes, therefore, generally occur in greater concentration in the body of the liquid than in the surface layers. This phenomenon is called negative adsorption. When negative adsorption occurs the solvent molecules become relatively more concentrated in the surface layer than in the body of the liquid.For reasons which will become clear later, negative adsorption rarely occurs at interfaces.

The difference in concentration between the surface layer and interior of a liquid are in general much less in negative adsorption than in positive adsorption. Adsorbed molecules are not static, but exhibit a continuous although reduced kinetic activity. Molecules adsorbed at surfaces or interfaces are in dynamic equilibrium with the molecules of the same species in the body of the liquid. An adsorption equilibrium is attained when the number of adsorbed molecules passing out of the interface in a unit of time is equal to the number passing into the interface.

When adsorption occurs from a liquid containing many solutes, as is true of most biological liquids, all solutes are adsorbed to a greater or lesser degree, depending upon their specific properties in relation to the adsorbing surface. In general, however, under such conditions no one solute will be adsorbed as completely as if it alone were present. If a substance is introduced into an aqueous solution which will lower the surface tension of water more than another substance, already adsorbed, the first compound will largely displace the second compound in the interfaces of that solution.

Interfacial Adsorption

Adsorption occurs at all types of interfaces.Many solids adsorb gas molecules very powerfully at solid-air interfaces.Solutes may be adsorbed at interfaces between immiscible liquids, and at interfaces between liquids and solids, as well as at liquid-gas interfaces (surfaces). Water molecules are strongly adsorbed at many solid-liquid interfaces. Adsorption is, therefore, a phenomenon of a very general occurrence. Adsorption at liquid-gas interfaces involves only the forces of attraction between the solvent and solute molecules, since the molecules of a gas are too far apart to have any appreciable attraction for the solute molecules. Similarly adsorption at solid-gas interfaces involves only the forces of attraction between the molecules of the solid and those of the adsorbed gas. At liquid-liquid, or solid-liquid interfaces, however, adsorbed molecules are subject to the attraction of the molecules of the two abutting substances at the interface.

At an interface between carbon particles and water, molecules are attracted by both the carbon and the water. If the carbon exerts a greater attraction for the solute than the water as is usually the case a greater accumulation of solute molecules will occur in the interface, than if the opposite is true. The concentration of adsorbed molecules which will be attained at any solid-liquid or liquid-liquid interface is therefore controlled by the relative magnitude of the attractive forces of the two phases for the molecules of the adsorbate (substance adsorbed). Inorganic salts which are negatively adsorbed at surfaces (water-air interfaces) are usually positively adsorbed at water-liquid interfaces, and almost always so at water-solid interfaces.

This is due to a greater attraction for the solute particles by the non-waterphase of the interface than the water molecules themselves exert. Specially treated (activated) charcoal is one of the best adsorbents known both for gases and for solutes. Many important applications are made of this property of charcoal in chemistry and technology. Gas masks for protection against poison gases in warfare owe their efficiency to the adsorptive capacity of charcoal. *Colour*ed or other soluble impurities are often removed from liquids by passing them through layers of charcoal.

Solids such as charcoal which are powerful adsorbents usually possess an enormous surface area in proportion to their mass, which accounts for their effectiveness as adsorbing agents. The surface of g. of charcoal has been variously estimated at from about 50 to about 6oo square metres. Much of this surface is in the form of internal submicroscopic. Adsorption in which the fundamental controlling forces are the cohesive and adhesive forces between molecules is called mechanical adsorption in order to distinguish it from certain more complex kinds of adsorption phenomena. Of these the most important is electrical adsorption. Most surfaces in contact with water bear an electrical charge. The presence of such charges markedly influences the adsorption of ions or other charged particles.

Cellulose, in common with many other substances, acquires a negative charge when immersed in water. If a strip of cellulose filter paper be arranged so that its lower end dips in a solution of eosin (a red dye) it will be observed that the eosin will rise through the filter paper almost as rapidly as the water from the solution rises by capillarity. If another strip of filter paper be similarly arranged to dip in a solution of methylene blue (a blue dye) the water will rise up the filter paper as rapidly as it does from the eosin solution, but the dye will scarcely rise at all. The difference in behaviour of these two dyes is due principally to the electrical charge which their coloured ions carry.

The coloured ions of eosin are negatively charged; those of methylene blue positively charged. The negatively charged particles of eosin are repelled from the negatively charged surface of the cellulose fibres. The eosin particles are therefore driven towards the centre of the capillary columns of water in between the fibres, and move up the filter paper almost as rapidly as the water. As soon as the positively charged particles of methylene blue come in contact with the negative charges of the cellulose, however, they are held to the surface of the cellulose by the forces of electrical attraction. In other words they are electrically adsorbed. While the water rises by capillarity at about the same rate as it does from the eosin solution, the rise of methylene blue is relatively very slow.

Certain examples of adsorption have also been recognized which are designated by the name of chemical adsorption. Chemical reactions are involved in adsorption phenomena of this type. The familiar reaction of 12K1 with starch in which the starch stains purple is often considered to be an example of chemical adsorption. Actually there exists no clean cut distinction between mechanical, electrical, and chemical adsorption. The several phenomena intergrade, and elements of more than one of these mechanisms are usually operating whenever adsorption occurs. Since the surface or interface of even pure water bears an electric charge, at least minor electrical effects are involved in even the simplest adsorption phenomena.

Adsorption of Water

Water itself is strongly adsorbed at certain types of interfaces. The adsorption of water is a phenomenon which can be regarded as more nearly similar to electrical adsorption than to mechanical adsorption. Water molecules, while not charged in the sense that ions are strongly polar and behave somewhat as charged bodies. Water-vapour molecules become adsorbed on many different kinds of surfaces.

The water-vapour adsorbed upon glass weighing bottles often becomes an appreciable source of error in accurate quantitative work. The particles in certain colloidal systems have the property of adsorbing large quantities ofwater, as shown in the next two chapters. A large portion of the water which moves into various types of substances in the process of imbibition becomes adsorbed upon the internal surfaces of the imbibing substance.

Biological impact of Adsorption

Adsorption phenomena are known to play a manifold role in living organisms, probably being involved in practically all cell activities. Protoplasm and many other constituents of plant cells are essentially colloidal, and adsorption phenomena are of general occurrence in colloidal systems. Within plant cells many interfaces occur, as at the boundaries between the proto-plasm and vacuole, protoplasm and cell wall, and nucleus and cytoplasm, at all of which interfacial concentration of solutes undoubtedly occurs. The adsorption of certain compounds at the cytoplasmic interfaces is generally believed to exert a marked influence on the permeability of the cytoplasm. It is quite possible that interfacial tensions play an important role in the process of cell division. Imbibitional phenomena, of basic importance in the water relations of plant cells, involve the adsorption of water. The action of enzymes as well as of other catalysts is generally believed to involve adsorption phenomena. Much information regarding the structure of cells has been gained by the use of dyes which are differentially adsorbed by various constituents of cells. Chromosomes, for example, are so named because they strongly adsorb certain stains.

COLLATERAL READING

The relation of adsorption to many of the processes and phenomena mentioned above will receive further attention in subsequent chapters of this book. There are probably few if any processes occurring in living organisms in which adsorption phenomena are not at least indirectly involved. Adsorption and other interfacial phenomena are responsible for many of the so-called vital activities of living cells.

UNDERSTANDING THE COLLOIDAL SYSTEMS

AN OVERVIEW

Most biological problems lead ultimately to a consid-eration of one phase or another of the more fundamental problems of the structure, constitution, and physico-chemical properties of protoplasm. Certain characteristic physicoc-hemical properties of protoplasm have been recognized and studied. Protoplasm is more or less elastic. It may vary in viscosity from values not much greater than that of pure water to those of jellies. It possesses a pronounced capacity for the imbibition of water.

Protoplasm is usually high in water content, but nevertheless often behaves as if immiscible with water. Mechanical disturbances may markedly influence its physical state. When exposed to very high or very low temperatures, to high concentrations of salts or to certain other factors, protoplasm may be irreversibly coagulated. Under certain conditions it may lose its fluidity and assume a jelly-like condition, a change in physical state which is usually reversible. Protoplasm is predominantly composed of substances in the colloidal state and it is to these colloidal systems that it owes most of its characteristic physico-chemical properties, some of the better known of which have been listed above. The world of living organisms has, in fact, been largely molded on a colloidal pattern. Most physiological processes, reduced to their ultimate tangible mechanism, either occur in a colloidal matrix, or under such conditions as to be strongly influenced by the colloidal organization of the cells in which they take place.

Many physiological processes occur only under the influence of the organic catalysts known as enzymes, which are supposed by most authorities to be in the colloidal condition and to owe many of their properties to that fact. It is impossible, therefore, to obtain any adequate comprehension of the properties of protoplasm without a background of facts and principles regarding colloidal systems. The more fundamental properties of such systems will be discussed in this and the following chapter.

Some of the most important components of the material environment of plants and animals are also essentially colloidal. Most soils contain a considerable proportion of matter in the colloidal or near-colloidal condition, and owe many of their most distinctive properties to this fact. Few streams or bodies of water are entirely free from matter in the colloidal condition. Clouds, fogs, mists, and smoke also represent matter in the colloidal state.

General Nature of Colloidal Systems

If a little sucrose or sodium chloride be shaken up in water the solid will soon disappear, and the resulting system will be a solution. Although there is abundant evidence which indicates that the solute particles are dispersed throughout the solvent, molecules and ions are so small that it is impossible to detect their presence by direct observation, even with the most powerful optical systems available. If, instead of sucrose or sodium chloride, we stir some line river bottomsilt into water, a different sort of a system will result. The silt particles don't dissolve but simply become dispersed throughout the liquid. Such a system is called a suspension. The particles in it are of such size that they can readily be detected under a microscope.

Suspensions are not stable, however, as the particles gradually settle out, and the system becomes separated into its two original components within a relatively short period of time. Similar systems can be prepared by vigorously shaking together two immiscible liquids such as oil and water. Such systems are called emulsions. Emulsions are not stable unless there is also present in the system a third component called an emulsifier.

Still another type of system consisting of particles dispersed through water can be prepared. If a trace of sulphur be dissolved in a small volume of alcohol which is then poured into a some what larger volume of water, a cloudy opalescent liquid will result which is composed of sulphur particles dispersed through water. This type of system is intermediate in its properties between solutions and suspensions. The dispersed particles are not molecules, but, as in suspensions, aggregates of molecules.

Unlike suspensions, however, such systems are relatively stable, as the particles will remain dispersed throughout the liquid indefinitely. The dispersed particles of such systems cannot be seen under a microscope, showing that they are smaller than the particles in suspensions or emulsions. They can, however, be detected with the aid of an ultramicroscope. The sulphur-in-water system which has just been described is a simple example of a colloidal system.

Colloidal systems, as the preceding discussion has indicated, are two phased systems like solutions. Unlike solutions, however, the particles of the dispersed phase are not in the molecular or ionic condition, but with certain exceptions to be noted shortly are molecular aggregates. One colloidal particle is often composed of hundreds or even thousands of molecules lumped together. The molecular aggregates must not be so large however, that the particles readily settle out of the system, as stability is one of the essential attributes of colloidal systems. Hence molecular dispersions of such substances are simultaneously both solutions and colloidal systems.

The limits generally accepted for the range of sizes of colloidal particles have been somewhat arbitrarily set and actually there is no sharp boundary between colloidal systems and suspensions on the one hand, nor between colloidal systems and solutions on the other.There is a perfect gradation in properties from one type of system to the next. The properties of suspensions or emulsions in which the suspended particles are of small dimensions approach those of colloidal systems, while the smaller the dispersed particles in a colloidal system the more closely it approaches a solution in its properties. The term to designate a certain group of substances, which seemed to be set apart from other substances by several distinctive properties. When dispersed in water these substances had a slow rate of diffusion, and failed to diffuse through membranes of parchment paper. Furthermore they did not form crystals. He applied the alternate term crystalloid to those crystal-forming substances which, when in solution, diffused relatively rapidly and passed readily through parchment membranes. We now know that, strictly speaking,no such distinction can be made; the word colloid properly refers to a distinctive state of matter, and cannot be applied with accuracy to any one class of substances.

Theoretically any substance can, by proper manipulation, be brought into the colloidal state, and actually this experience fall naturally into two groups generally known as oil-in-water emulsions, and water-in-oil emulsions. In the former type an oil or some other liquid insoluble in water, or practically so, is dispersed throughout a water dispersion medium. In the latter type the converse is true, the oil or other liquid immiscible with water constitutes the dispersion medium, while small aggregations of water molecules are dispersed through it. The proportions of the components of most emulsions can be varied between wide limits.Emulsions are not generally considered to be true colloidal systems, but, like suspensions, approach them in properties. Emulsions occur commonly in the cells of plants and animals, and are generally believed to be essential components of the protoplasmic matrix. Both water-in-oil and oil-in-water emulsions are known to exist in living cells, but the latter type is commoner. When examined under high magnification with a microscope protoplasm in its grosser aspects often presents the appearance of an emulsion of fats and fat-like substances dispersed through the body of the protoplasm.

Some common examples of oil-in-water emulsions are milk, cream, emulsions of olive oil in water, and the latex of the rubber tree, milk weed, etc. Perhaps the only generally familiar water-in-oil emulsion is butter. Many other similar systems can be prepared, however. Water in olive oil is one example of such a system. Water in kerosene is another. The word oil as employed in this discussion is not restricted to those substances usually classified chemically as oils, but is here somewhat generalized to include other liquids which are immiscible with water. Kerosene, for example, is not chemically an oil.

Emulsions with the exception of some very dilute ones, lack stability unless there is also present in the system an emulsifier. In the absence of an emulsifier, the two components of an emulsion rapidly separate, the oil, being the component of lower specific gravity rising to the top. The group of substances classified as emulsiflers is chemically a very heterogeneous one. Some of the best known emulsifiers are the soaps, saponins, various substances which form hydrophilic colloids when dispersed in water (gums, gelatin, etc.), andfine suspensions of certain rather inert materials such as sulphur, carbon, silica and resin. The last group is probably of little importance in living organisms. Oil-in-water emulsions may be stabilized by such substances as soaps of the alkali metals (Na, K, Li, etc.), gum acacia, and proteins; while soaps ofthe alkaline earth metals (Ca, Ba, etc.), and gum dammar are examples of stabilizers of water-in-oil emulsions. In general, emulsifying agents which are soluble in water or form hydrophilic sols in water, stabilize oil-in-water emulsions while those which are insoluble in water but soluble in oil stabilize water-in-oil emulsions.Many pharmaceutical preparations are stabilized by means of gum acacia or gum dammar.

Emulsions found in nature are usually stabilized by proteins. This is true of milk and cream, in which the droplets of fat are dispersed throughout a water medium. The latex of plants, one of the best examples of a naturally occurring emulsion, is also stabilized by proteins. Mayonnaise dressing is essentially an emulsion of olive oil in water, stabilized by the proteins of the egg which is the third essential component of the system. Protoplasmic emulsions are probably also stabilized by proteins, although soaps may also act a emulsifiers in plant cells.

The mechanism of the stabilizing effect of emulsifiers upon emulsions is not well understood, and it is probable that it varies with different types of emulsifiers. It is well known, however, that any substance which greatly lowers the interfacial tension between water and another liquid will usually stabilize an emulsion composed of those two liquids. The usual interfacial tension between water and benzene, to cite one example, is about 35 dynes per centimeter. Addition of a little sodium oleate (a soap) to water will lower this interfacial tension to about 2 dynes per centimeter. Sodium oleate is an excellent stabilizing agent for an emulsion of benzene in water.

PLANT PHYSIOLOGY AND SOLS AND GELS

The discussion in this chapter will deal only with sols in which water is the dispersion medium. Such systems are classified into the two groups of hydrophilic sols and hydrophobic sols. The dispersed particles in all such colloidal systems fall within the size range of 0.001 to 0.1, but their size may vary greatly with different sols. Even in a given sol the particles may exhibit a wide range of dimensions, although in some they are quite homogeneous in size. Most of the important differences between hydrophilic and hydrophobic sols are due to the hydration of the dispersed particles in systems of the former type. There is no general agreement regarding the exact physico-chemical relationship between the micelle and its water of hydration; but only two conceptions have any wide currency at the present time.

It is presumed that the first layer of oriented water molecules is so firmly attached to the micelle that it is essentially an integral part of it. The successively enveloping layers of water molecules are also oriented, but with increasing distance from the periphery of the micelle the forces of attraction and orientation progressively decrease. In this zone there is a gradual transition from water molecules which are practically all oriented to those which are completely unoriented. Oriented molecules fit together more closely than unoriented molecules. They are packed more closely, just as more bricks can be stacked in a given space if arranged regularly, than if tossed in indiscriminately. As a result of this packing the water in these oriented shells has a greater density than that in the bulk of the liquid;in other words there is an actual contraction in the volume of the liquid associated with the micelles.

The fact that micelles may, under certain conditions, lose their water of hydration very rapidly would seem to favour the latter theory. It is possible of course, that in some hydrophilic systems the water is actually dissolved in the micelles, that in others it is present only as a shell of oriented molecules while in still others both of these two suggested modes of hydration may exist.The important properties of sols will now be summarized, with special attention to differences between the properties of sols of the hydrophilic type and sols of the hydrophobic type.

DILUTION

Most sols are extremely dilute. In other words the actual mass of substance dispersed throughout the dispersion medium is extremely small in proportion to the total volume of the system. The well-known colloidal gold sols, for example, rarely contain more than i g. of gold dispersed per litre of sol. Other hydrophobic sols are correspondingly dilute.

Most hydrophilic sols are also extremely dilute, although there are some exceptions to this statement, as very viscous sols of such substances as starch and gum acacia can be prepared which contain parts or more of dispersed material per 100 parts of sol.

SLOW RATE OF DIFFUSION OF THE DISPERSED PARTICLES

In general the particles which make up the disperse phase of a sol diffuse much more slowly than substances in true solution, *i.e.* in the molecular or ionic state. The slow rate of diffusion of micelles as compared with most ions and molecules is clearly correlated with the relatively large size of the colloidal particles.There is, however, no sharp line of demarcation in terms of diffusion rates between colloidal micelles and solutes. The diffusion rates of some solutes with large molecules are not perceptibly faster than those of the dispersed particles of colloidal systems in which the micelles are relatively small in size.

FILTERABILITY

Sols are usually filterable; that is they pass through ordinary filter papers without any appreciable separation of the disperse phase from the dispersion medium by the filter. Usually there is some loss of the disperse phase due to an initial adsorption when the sol first comes in contact with the filter. Sometimes this may be very considerable. Since the pores in ordinary filter papers are about 2–5 L in diameter, and even porcelain filters such as those widely used in bacteriological work, have pores 0.2–0.6 in diameter, it is easy to understand why micelles with diameters in the size range 0.001–0.1 pass through.

Special filters have been devised, however, with pores of such a small diameter that the disperse phase can be separated from the dispersion medium by filtering a sol through them. Such filters are known as ultra filters. The process of filtering through such a filter is known as ultrafiltration. The most commonly used types of ultrafilters are those made of collodion or gelatin.The size of the pores in such filters can be controlled by the length of time allowed for drying and in other ways. It is possible to prepare ultra filters with pores of such dimensions that solutes can pass through them, but colloidal micelles cannot.

TYNDALL PHENOMENON

Suppose that a clean glass vessel, preferably one with flat, parallel sides, be filled with pure water and held so that a strong pencil of light passes laterally through the vessel. If an observation be made laterally and at right angles to the path of the beam of light, no distinctive trace of its path through the water can be detected. Such a liquid is said to be optically empty. The same will be true if the water in the vessel be replaced by a sugar or salt solution, or in fact, by any true solution.

Suppose however, that the vessel be filled with a hydrophobic sol and observed in the manner just described. The results are strikingly different. The path of the light will be clearly delineated as a cloudy, often opalescent track through the sol. Even a colloidal system which seems perfectly transparent to the unaided eye will usually show some turbidity when submitted to this test. The intensity of this effect varies greatly with the specific colloidal system, and with the concentration of the disperse phase, but it is universally shown by hydrophobic sols.

The phenomenon just described is called the Tyndall phenomenon, and is due to the scattering or diffraction of light. The difference in the index of refraction between the two phases of the colloidal system is also a factor determining the intensity of the Tyndall effect. The greater this difference the stronger the effect. Since in the diffraction of light, the short wave lengths (blue end of the spectrum) are bent more than the longer wave lengths, a partial separation of the spectrum results., For this reason a sol with a colourless disperse phase often appears to be pale blue when viewed in the path of a strong beam of light.

Similar Tyndall phenomena are exhibited by hydrophilic sols, but usually the effect is less striking when sols of this type are viewed in the path of a beam of light than the effect observed when hydrophobic sols are employed. Actually a trace of the light track will usually be perceived even in water or true solutions, due to the presence of contaminating dust particles. In order to prepare a truly optically empty liquid, provision must be made for the removal of such dust particles.

The instrument known as the ultramicroscope is based on the principle of the Tyndall phenomenon. It is not possible to observe colloidal particles directly in the ultramicroscope; only the light diffracted from their surfaces can be seen. Neither can any definite image of particles in this small range of sizes be obtained.The ultramicroscope is a microscope which is so arranged that the colloidal system or other material to be examined can be illuminated laterally (*i.e.* at right angles to the tube of the microscope). This lateral illumination is usually provided by a powerful source of light and a suitable series of condensing and focussing lenses, so arranged that the light is focussed to a point within the mount. Under the ultramic-roscope the dispersed particles of a hydrophobic sol appear as bright spots of light varying in size and brilliancy.Very little concerning the actual size or shape of the micelles can be determined since each bright spot represents merely the light diffracted by a single particle. It is possible, however, to determine the number of particles in a given volume of solution by means of the ultramicroscope. The use of the ultramicroscope consists essentially in an observation of the Tyndall phenomenon in a small volume of a sol under the microscope.

BROWNIAN MOVEMENT

In 1828, the botanist Robert Brown observed through a microscope that pollen grains which were suspended in water showed a rapid oscillatory motion. Brown at first was inclined to attribute this motion to the fact that the pollen grains were alive, but examination of preparations of dead pollen grains and spores showed that they likewise exhibited such a motion. It became evident therefore that this movement was in no way connected with living processes.

This phenomenon is termed Brownian movement, after its discoverer.Many suspensions in which the particles are within the range of microscopic visibility exhibit Brownian movement. It is clearly shown by many of the smaller species of bacteria when suspended in water. In solid-in-gas colloids, such as tobacco smoke, the dispersed particles show a very vigourous Brownian movement. Particles in the protoplasm of slime molds and certain other species frequently exhibit a Brownian movement which is clearly discernible under the microscope.

For particles of a given mass, the smaller their volume the greater the amplitude of their Brownian movement. For particles of equal volume, the less their mass the more vigorously they will exhibit Brownian movement. In general this phenomenon is exhibited more clearly by the micelles of hydrophobic sols than by those of hydrophilic sols. The viscosity of the liquid phase is also an important factor governing the rapidity with which the dispersed particles move. The more viscous the liquid, the more sluggish the movement of the particles.Brownian movement is caused by the kinetic activity of the molecules of the solvent. Even the smallest particles in which Brownian movement can be observed are very large in proportion to the size of the solvent molecules which impinge upon them.

A particle suspended in a liquid such as water suffers a continual bombardment by the molecules of the liquid. If the particle be relatively large, at any given moment it is bombarded on every side by numerous molecules, moving in all possible directions, and at different speeds. The effects of the individual impacts largely counteract each other however, and there is little or no movement of the particle.

If the particle be smaller, however, the results are quite different. At any given moment a much smaller number of water molecules impinge upon the particle. The resulting forces cease to be balanced and the sum total effect of the blows which the particle sustains on some one side are greater than the effect of the blows sustained on any other side. Hence the particle moves. The next moment a greater impetus may be given to the particle from some other direction and the course of its movement is changed.

In this way the highly erratic movements of suspended particles known as Brownian movement originate. Increase in temperature increases the rate of Brownian movement because of an increase in the kinetic energy of the solvent molecules. This phenomenon is the nearest approach we have to actual visible evidence of the validity of the kinetic theory of matter. It almost brings before our eyes the veritable dance of the molecules.

OSMOTIC PRESSURE

The osmotic pressure of solutions is a colligative property; that is one which depends on the proportion of solute particles (molecules or ions or both) to solvent molecules regardless of the kind of solute particles. A sol should also possess an osmotic pressure since theoretically a micelle has the same effect upon the magnitude of the osmotic pressure as a molecule or an ion. A brief discussion should aid in making this point clear. Large molecules have the same theoretical effect upon osmotic pressure as small molecules. The individual molecules of some compounds are large enough to fall within the colloidal range of sizes.

They are both molecules and micelles at one and the same time. Such a micelle should have just as much of an effect on osmotic pressure as a molecule.

Actually the osmotic pressures of sols never exceed a small fraction of an atmosphere and are therefore much less than the osmotic pressure of any but the most extremely dilute solutions. The osmotic pressure of colloidal systems depends upon the concentration of the particles of the disperse phase. In most sols the total mass of the disperse phase is not only relatively small; but it is present in the form of particles which are much larger than molecules or ions. The negligible osmotic pressures of sols area necessary corollary of the relatively small concentration of dispersed particles.

VISCOSITY

The viscosity of a fluid is its resistance to flow. The more viscous a liquid the less readily it will flow. Glycerine, for example, is much more viscous than water. The viscosity of hydrophobic sols never varies appreciably from that of the dispersion medium water. Unlike hydrophobic sols the viscosity of hydrophilic sols is usually greater than that of the dispersion medium.

The viscosity of hydrophilic sols increases appreciably with increase in the concentration of the sol, but the relation is not a linear one to the concentration of the viscosity of all liquids, including sols, is disperse phase.influenced by temperature. In general, increase in temperature decreases viscosity. In hydrophilic systems this reduction in viscosity with increase in temperature is probably due to two factors: the decrease in viscosity of the medium itself, and the decrease in the hydration of the micelles.

ELECTRICAL PROPERTIES

The dispersed particles of all hydrophobic sols carry electrical charges. A colloidal system, however, is electrically neutral because for every charge carried on a micelle an equal charge of opposite value is carried by ions in the dispersion medium. The situation is similar to that in a solution of an electrolyte. Although the individual ions are charged for every negative charge carried by an anion an equal positive charge is carried by a cation. In some colloidal systems the dispersed particles are negatively charged, in others positively charged, but normally all the dispersed particles in any one system bear a charge of the same sign.

Whatever the origin of the micellar charges they invariably are produced in such a way as to involve the release of ions into the dispersion medium which may therefore be regarded as also being charged. When the micelles are negatively charged, the dispersion medium is positively charged and vice versa. Electrostatic attraction therefore exists between the surface charges of a colloidal particle and the ions of opposite charge in the dispersion medium.The result is that surrounding each colloidal particle with its charged surface is a shell of ions of opposite charge.

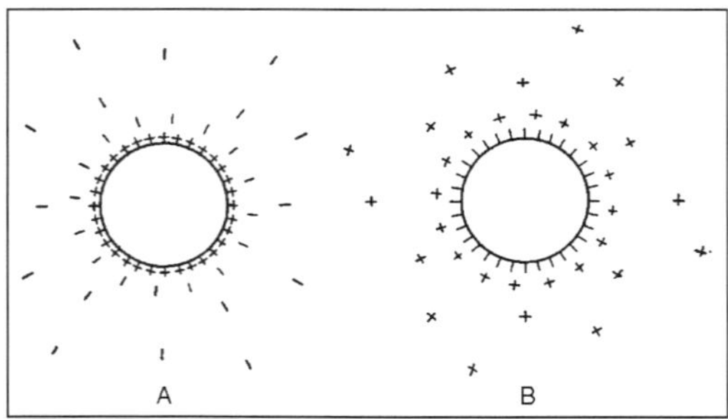

Fig. Diagrammatic Representation of the Electric Charge Around Micelles: (A) Positively Charged (B) Negative Charged

This arrangement of charges at the surface of a micelle is called an electric double layer. Similar electric double layers exist also at boundaries between solid surfaces and liquids, as for example along the walls of a capillary tube. One positively charged, one negatively charged. The innermost layer of ions in the dispersion medium is probably compactly oriented around the oppositely charged micelle, which is in turn surrounded by progressively more diffuse layers. That is, while most of the ions are close to the charged surface of the micelle, some are farther away, although with increasing distance from the surface of a micelle, the number of ions associated with that micelle decreases rapidly.

The ions of the double layer are in dynamic equilibrium with undissociated molecules at the periphery of the micelle proper. Anions and cations of the double layer are continually uniting and forming uncharged molecules which become part of the micelle. Contrariwise molecules at the surface of the micelle are continually dissociating into cations and anions. If the micelle is negatively charged the anions adhere to its surface; if positively charged, the cations. The ions of the opposite charge become part of the outer shell around the micelle.

The presence of this electrical double layer results in a difference of electrical potential between the micelle and the intermicellar liquid. The magnitude of this difference of potential varies with different sols and with the same sol under different conditions. The usual method of determining the sign of the charge upon the dispersed particles of sols is to observe the direction of their migration in an electrical field. If two electrodes are so arranged as to dip into a hydrophobic sol contained in a suitable vessel, and the electrodes connected with a direct current of proper potential, it will be found that the dispersed particles will move towards one of the poles.

By suitable arrangements, using a microscope or ultramicroscope (depending upon the size of the particles), the migration of the particles may be actually watched. In a positively charged ferric hydroxide sol, for example, the micelles will move towards the negatively charged electrode, while in a negatively charged arsenious sulfide sol they will move towards the positively charged pole. This phenomenon is known as cataphoresis or electrophoresis. As a micelle moves under the influence of an electric current only the lnnermost layer of the electric double layer the one which determines its electrical charge clings to the micelle and flow with it. This inner layerslides past the oppositely charged ions of the outer shell of one micelle after another as it moves towards the anode if negatively charged, or towards the cathode if positively charged. Simultaneously ions of the outer layer move towards the opposite pole from the one towards which the micelles migrate.There is a close analogy between this phenomenon and the behaviour of the ions of an electrolyte during electrolysis.

The sign of the charge on the dispersed particles of any sol can be determined by cataphoresis. Particles considerably above the colloidal range of dimensions frequently can be shown to exhibit cataphoretic migration. The phenomenon can be demonstrated with bacteria, unicellular algae, spores, etc.Practically all such small living organisms are negatively charged.

The dispersed particles of hydrophobic sols apparently acquire their charges either by electrolytic dissociation or by adsorption. In some such systems the charges seem to arise as a result of ionization of some of the molecules composing the micelle. The ions released into the dispersion medium become the outer envelope of the double layer, leaving the micelle with a residual and compensating charge of ions of opposite sign. Individual molecules of some substances are large enough to fall within the colloidal range of sizes. The dye Congo Red, which is the sodium salt of a complex organic acid, is an example of such a substance. Dispersed in water this compound produces sodium ions and a colloidal anion, the latter, of course, being negatively charged.

In other systems the electrical charge on the dispersed particles is apparently acquired by adsorption of either the positive or negative ions of an electrolyte, the ion of opposite charge becoming the outer shell of the double layer. Ferric hydroxide sols 2 are normally positively charged. The charge on the micelles of these sols is often ascribed to the adsorption of+ 1 ions of the $FeCl_3$ from which ferric hydroxide sols are usually prepared, the Cl ions forming the outer layer around the micelles. Similarly the negative charge of the micelles of arsenious sulfide sols is often ascribed to the adsorption of H_25 which is used in the preparation of this sol. Dissociation of H_25 results in the release of H^+ ions into the dispersion medium leaving the dispersed particles with a residual and compensating negative charge.

Similarly it is believed that some substances acquire an electrical charge by adsorbing hydrogen or hydroxyl ions more frequently the latter from their water dispersion medium. Certain inert substances such as cellulose,carbon, quartz, collodion, etc., are believed to become charged in this way.All of these substances acquire a negative charge when in contact with water,indicating that the hydroxyl ions are adsorbed, the hydrogen ions becoming the outer shell of the double layer. The micelles of some Izydrophilic sols are charged; those of others are not.As in hydrophobic systems the dispersed particles of different hydrophilic sols acquire their charges in different ways. In some the charges on the particles 2 Most investigators believe that the so-called ferric hydroxide sol is actually a sol of hydrated ferric oxide ($Fe_2O_3(H_2O)$).

The electrical charges on the micelles of protein sols arise in this way. In othersystems the charges may originate from traces of electrolytes which are present as impurities. This is probably true of agar sols. Such charges may be regarded as similar in their origin to those acquired by adsorption on the micelles of hydrophobic sols, in that the charge is due to some compound associated with the substance of which the micelle is composed, and not to that substance itself. Most of the better known hydrophilic sols are negatively charged.

FLOCCULATION

Since the dispersed particles of any sol are in rapid motion it would seem that repeated collisions would result in a progressive agglomeration of the particles into larger and larger masses which eventually would settle out of the system. Sols, however, are relatively stable systems, and it is important to understand the mechanism by which the stability of such colloidal systems is maintained.The stability of hydrophobic sols is maintained almost entirely by the charge which each micelle carries. Although by Brownian movement the dispersed particles are repeatedly brought close together actual collision seldom occurs because the shells of ions around the micelles are mutually repellent.

Reduction of the charge on the micelles of any hydrophobic sol to the point where there is no difference of electrical potential across the double layer results in agglomeration of the individual particles into flakes of a size which rapidly settle out of the surrounding liquid. This phenomenon is called flocculation, coagulation, or precipitation. The point at which there is no difference of electrical potential across the double layer is known as the iso electric point of the sol. At the isoelectric point the micelles of a sol are,relative to the surrounding medium, completely uncharged.

As, by Brownian movement, two such micelles are brought into contiguity they no longer repel each other with sufficient intensity to prevent their agglomeration. By the addition of other micelles such particles rapidly increase in size, soon resulting in flocculation of the sol. Merely reducing the electric charge to a value approaching that of the isoelectric point is sufficient to induce instability and slow flocculation in many sols.Flocculation is most commonly initiated by the introduction of electrolytes into the system. Very small quantities of an electrolyte are often sufficient to cause the flocculation of a relatively large volume of sol.

The important principles regarding the flocculation of hydrophobic sols by electrolytes canbe most easily elucidated by a discussion of some of the data in. The flocculating effect of an electrolyte is due primarily to the added ion of opposite charge from that borne by the colloidal particle. Thus the As_2S_3sol may be flocculated by cations such as Na1, Ca^{+1}, or Al^{+11}, while theFe(OH)3 sol may be flocculated by anions such as C1, NO_3, or SO_4^-.

Furthermore, the flocculating effect increases with an increase in the valency of the effective ion. However, the influence of the valency of an ion upon its flocculating effectiveness does not follow a simple 1 2 3 arithmetical ratio. In the flocculation of hydrophobic sols the ion of the added electrolyte with a charge of the same sign as that of the micelle is not entirely without effect. The influence of such ions is usually to increase the stability of the system. Precisely stated, therefore, the influence of an electrolyte upon the micelles of a hydrophobic sol is a differential effect between the anions and cations but the influence of the ion of opposite charge predominates.

The mechanism of the flocculation of a hydrophobic sol is too complex a process to be considered in more than a general way in an introductory discussion. In general flocculation results from destruction of the double layers around the micelles. There is a close analogy between the flocculation of a hydrophobic sol by an electrolyte and a precipitation reaction between one electrolyte and another. The micelles of hydrophobic sols may be regarded as giant ions bearing numerous charges.

When $CaCl_2$ is added to a solution of Na_2SO_4 slightly dissociated $CaSO_4$ is precipitated, leaving paired Na1 and Cl ions in the solution. An analogous phenomenon results when $CaCl_2$ is added to a negatively charged As_2S_3 sol in which the outer shells of the double layers are composed of hydrogen ions derived from the dissociation of adsorbed H_25. The resulting flocculant consists of uncharged particles of As_2S_3 + Ca, while the H^+ and Cl ions left behind in the solution pair off in the usual manner, forming hydrochloric acid.

In other words $Ca^{+\ +}$ ions have replaced the H + ions in the outer shells of the micelles. The result, however, is an unstable particle, since Ca1 1 ions do not remain in the dissociated state when associated with As_2S_3 micelles which owe their negative charge to the dissociaion of adsorbed H_2S molecules. They combine with such micelles, thus dissipating their electrical charges. Flocculation of the uncharged micelles then ensues. Flocculation may also be initiated by introducing into a hydrophobic sol another hydrophobic sol with micelles bearing a charge of opposite sign. If $Fe(OH)_3$ sol be slowly added to an As_2S_3 sol a point will be reached at which complete flocculation will occur. The particles of the two sols will settle out as an intimate mixture. The same phenomenon occurs when any negatively charged hydrophobic sol is added to any positively charged hydrophobic sol in sufficient quantity, or vice versa. This process is called mutual flocculation. When this phenomenon occurs the ions of the outer zone of one kind of particle pair off with the oppositely charged ions of the outer shell of the other kind.

The addition of a small amount of a hydrophilic sol, such as a gelatin or gum arabic sol, to a hydrophobic sol makes flocculation of the latter by electrolytes or micelles of opposite charge difficult or impossible. This effect of a hydrophilic on a hydrophobic sol is termed protective action. Protective action is apparently due to the adsorption of the micelles of the hydrophilic sol around the micelles of the hydrophobic sol. The properties of the sol,therefore, become essentially those of the hydrophilic system.

As will be seen shortly hydrophilic sols are much less easily flocculated by electrolytes than hydrophobic sols and this is undoubtedly the basis for protective action.We turn now to the question of the mechanism of the flocculation of hydrophilic sols. The micelles of such sols may or may not carry an electrical charge, but whether charged or not such colloidal systems are stable. Hydrophobic sols, as already shown, are stable only when the micelles bear an electrical charge. One of the most important differences between hydrophobic and hydrophilic sols is the possession by the latter of a second stability factor,which in itself is effective in keeping such sols stable for long periods of time.

In our previous discussion we have seen that uncharged micelles of hydrophobic sols soon agglomerate and settle out of the dispersion medium. Why do not the uncharged micelles of a hydrophilic sol behave in the same way? This is probably due to the effect of hydration upon the properties of the dispersed particles. The micelles of all hydrophilic sols, it will be remembered, are highly hydrated. The first layer of molecules of water of hydration is probably so firmly bound to the particle as to virtually constitute an integral part of the micelle itself. Surrounding this are other layers of water molecules more or less completely oriented depending on their distance from the surface of the particles. Each such particle is cushioned against impacts with other particles by its enveloping shell of oriented water molecules.Agglomeration of the micelles is thus prevented, and hence even uncharged hydrophilic sols are stable.

The possession of two stability factors by hydrophilic sols complicates the mechanism of flocculation in such systems. The manner in which flocculation of hydrophilic sols may occur can be illustrated by reviewing a simple experiment. The experimental material is a dilute (about o.1 per cent) sol of agar-agar. Such a sol is perhaps the most typical example of a simple hydrophilic system. If a relatively large volume of alcohol be added to a small portion of such a sol, the micelles lose their water of hydration, and the sol acquires the cloudy, bluish, opalescent appearance typical of many lyophobic sols. In fact it now is a lyophobic sol, and shows all the typical properties of such systems.

The alcohol, which is a powerful dehydrating agent, has robbed the micelles of their shells of oriented water molecules. Nevertheless, the sol is still stable, since the micelles retain their original negative charges. Finally, let a drop of a solution of an electrolyte such as $AlCl_3$ be added to the sol. The sol now flocculates almost immediately since the only remaining stability factor the electrical charge has been destroyed by the addition of the electrolyte, and the opalescent cast of the sol disappears.The stability factors of a hydrophilic sol can also be dissipated in the opposite order. Suppose that the $AlCl_3$ solution first be added to the agar sol until the charges on the micelles are neutralized. Although now at its isoelectric point, unlike hydrophobic systems, the sol does not flocculate.

If however, alcohol now be added to the system, immediate flocculation occurs because of a dehydration of the micelles, resulting in an elimination of the only remaining stability factor in the system. Briefly then, in order to flocculate a hydrophilic sol, its micelles must be both dehydrated and electrically discharged, except in the occasional systems in which micelles are uncharged, in which dehydration alone will suffice. Dehydration alone of a hydrophilic sol with charged micelles results in its conversion into a lyophobic sol. The addition of relatively large quantities of certain electrolytes to hydrophilic sols will result in their flocculation without any previous removal of the water of hydration of the micelles by means of alcohol or any other dehydrating agent. Only very soluble salts are effective in this way. It is evident that this phenomenon, usually called salting out, involves a more complicated reaction than that taking place in simple flocculation and must be distinguished from the latter phenomenon. Three salts which are especially suitable for salting out hydrophilic sols are $(NH_4)_2SO_4$, $MgSO_4$, and Na_2SO_4. In these three salts the ions, especially the anions, acquire relatively large quantities of water of hydration.

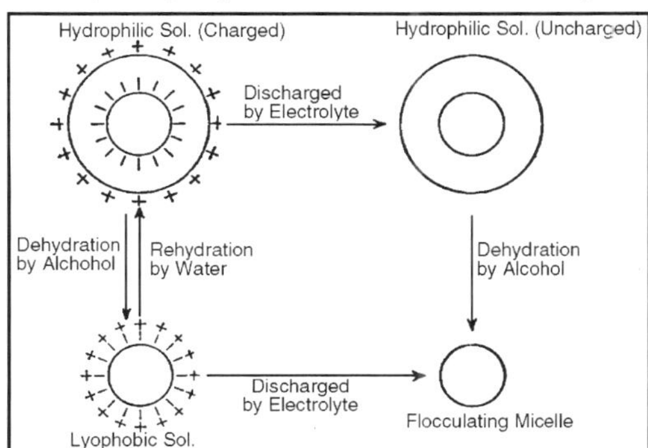

Fig. Diagrammatic Representation of the Flocculation of a Micelle of a Hydrophilic Sol.

The result of the addition of a very strong solution of such a salt to a hydrophilic sol is a two-fold one. A small initial amount of the added electrolyte discharges the micelles. Addition of further increments of an electrolyte eventually brings about dehydration of the micelles due to the great attraction of the added ions for water, or to the effect of the solute in reducing the diffusion pressure of the water or to both, and the resultant separation of the dispersed phase out of the system. Salting out, therefore, consists in an electrical discharge of the micelles, followed by their dehydration. It results in the destruction of both of the stability factors of the system.

AMPHOTERIC PROPERTIES OF PROTEIN SOLS

Protein sols differ from most others in that the micelles are amphoteric, *i.e.* they may act either as an acid or as a base. The acid properties of proteins depend upon their COOH groups; their basic properties upon their NH_2 groups.Whether the proteins will combine with acids or bases depends principally upon the pH of the dispersion medium. In a gelatin sol, for example, in which the pH of the medium is above the isoelectric point the COOH groups of the molecules react with a base such as sodium hydroxide, forming sodium gelatinate. This compound then dissociates into sodium ions and negatively charged gelatin micelles. If the pH of the medium is below the isoelectric point the NH_2 groups of the gelatin molecules may combine with the molecules of an acid such as HCl forming gelatin hydrochloride.

Dissociation of this compound produces chloride ions and positively charged gelatin micelles. Like the micelles of other sols, those of proteins are uncharged with respect to the medium at the isoelectric point. Therefore no migration of the micelles occurs if an electric current is passed through a protein sol at its isoelectric point. At pH values higher than the isoelectric point protein micelles migrate towards the anode, while at values below the isoelectric point they migrate towards the cathode.

The principles governing the stability and flocculation of a protein solare similar to those which hold for other hydrophilic sols with the one further complication that protein micelles may be either positively or negatively charged. Protein sols are stable at their isoelectric point because, although uncharged, they possess, like all hydrophilic sols, micelles which are highly hydrated. On either side of the isoelectric point the micelles of a protein sol have the additional stability factor of an electrical charge.

Addition of a sufficient quantity of a dehydrating agent such as alcohol to a gelatin sol at its isoelectric point will result, as it does with an uncharged agar sol, in immediate flocculation. If the micelles are charged, however, addition of alcohol will result not in flocculation, but in the conversion of the system into a lyophobic sol. Addition of a suitable electrolyte to this sol will result in a discharge ofthe particles and their cons-equent flocculation.

PROPERTIES OF GELS

Under certain conditions most hydrophilic sols change into gels. Any gelatin sol which is not too dilute, for example, will set upon standing and form a gel.

Everyone is familiar with such gelatin gels, often as they appear on the table under the guise of desserts. Other familiar gels are the agar gels widely used as a medium for the culturing of bacteria, fungi, algae, etc., ordinary household jellies which are basically pectin gels, and starch gels.

The latter also sometimes come before our eyesby the dessert route, under the name of corn starch puddings. Some hydrophilic sols, however, do not ordinarily form gels. This is true of sols of gumacacia and of some protein sols. The gel-forming capacity of some substances is very remarkable. A gelatin sol containing as low a proportion as one part of gelatin to one hundred ofwater will usually gelate. Agar gels containing only 0.15 per cent of agar can be prepared. In such a gel one part of agar has the property of removing the liquidity from nearly 700 parts of water.

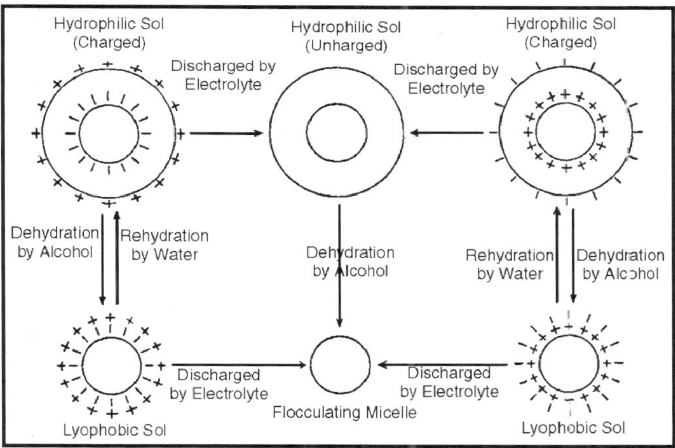

Fig. Diagrammatic Representation of the Flocculation of a Micelle of a Protein Sol.

Since gels form from two-phase colloidal systems, it is generally assumed that they also are two-phase systems. In all of the gels which we shall have occasion to consider, the liquid phase is water, although gels in which one component is some other liquid are well known. Gels are usually rigid enough to maintain their shape under the stress of their own weight. This means that they will be molded to the shape of the vessel in which the gelation has occurred, and will retain the shape of that vessel after being removed from it.

Solutes diffuse through gels almost as rapidly as through pure water unless the gels are very concentrated. The diffusion of a solute through a. gel is easily demonstrated by a simple experiment. A gelatin sol is allowed to solidify in a test tube which is then inverted in a shallow dish containing a solution of a dye such as methylene blue. Within 24 hours diffusion of the dye into the gelatin gel can be detected. Because of the fact that there are no convection currents in a gel to complicate the results, this is perhaps the best visual method of demonstrating the diffusion of solutes.

Two equal quantities of an electrolyte, one dispersed in water, the other in a gel of equal volume, will conduct an electric current almost equally well. In other words ionic mobility is apparently as great when the ions are dispersed in a gel as when they are dispersed in pure water.The velocities of chemical reactions occurring in a gel medium are not appreciably different from the velocities of the same reactions occurring under the same conditions of solute concentration, temperature, etc., in a water medium. Neither of the last two general statements are strictly true for very concentrated gels. The above three properties of gels distinguish them clearly from the solid or amorphous states of matter. These properties must be reconciled with any acceptable theories of the structure of gels.

ELASTIC GELS

Two general types of gels are usually recognized, the elastic type, and the non-elastic type. The best known example of the latter is the silica gel. Elastic gels are the important type biologically. Gelatin-water and agar-water gels are probably the best-known examples of this type of colloidal system. When such a gel dries a gradual and consistent shrinkage in its volume ensues until desiccation is complete. After desiccation the dry matter of the gel will imbibe water, but no other liquid. Gels in which the liquid phase was other than water will imbibe, after desiccation, only the liquid originally present. Elastic gels are generally heat reversible. When heated such gels are converted into sols (solation) and when cooled such sols resume their gelcondition (gelation).

Usually this reversal of state can occur a number of times to a colloidal system without greatly affecting its physical properties when in either the sol or gel condition. The temperatures at which the solation and gelation of a given colloidal system occur are not identical. The processes differ in this respect from the melting and freezing of a solid. For example, a 4 per cent gelatin sol gelates at about 280 C, but the resulting gel must be raised to about 31 0 C. before solation will occur. For agar gels the temperature spread between gelation and solation is much greater. The viscosity of a heat reversible sol increases steadily with a decrease in temperature and shows no sudden change as the sol passes into the gel condition.

THE STRUCTURE OF GELS

Numerous theories purporting to explain the structire of gels have been advanced, and there is no doubt that there are elements of truth in most of them. Since it seems improbable that the molecular architecture of all gels is the same, it is unlikely that any one of the proposed theories will apply equally well to all gels. The structure of gelsis too fine to be resolved by the microscope, and only rarely has the ultramicroscope revealed anything when it has been used as an instrument for studying gel structure.

Necessarily, therefore, most evidence of the structure of gels is indirect. There seems to be little doubt that in non-elastic gels of the silicic acid type that the solid phase is crystalline, and forms a sort of a rigid framework. The liquid phase is held in the interstices of this solid framework.In some respects this may be regarded as the simplest type of gel structure.One of the older theories of the structure of elastic gels is the so-called honeycomb theory.

According to this theory the fluid phase of the gel is discontinuous, being separated into minute chambers or compartments,bounded on all sides by films of the solid, or at least of a more solid phase.The analogy between such a cellular structure and a honeycomb is obvious.The solid phase is not necessarily supposed to be composed of the chemically pure, originally dispersed material; it may represent simply a phase which isrelatively rich in the dispersed substance as compared with the fluid phase.In a gelatin gel, for instance, it might be supposed that the solid phase is relatively rich in gelatin but poor in water while the converse is true for the fluid phase. At the present time an hypothesis more generally favoured is that both the solid and liquid phases of a gel are continuous. The solid phase is usually visualized as a meshwork of long, tangled fibrillae of ultramicroscopic, or sub-ultramicroscopic dimensions; the spaces within this interwoven mesh being occupied by the fluid phase.

This theory is often called the brushpile theory in allusion to the supposed jumble of intermeshing threads of the solid phase.The remarks made in the preceding paragraph concerning the constitution ofthe two phases are also valid for this theory. This last theory appears to be most acceptable in the light of the known facts regarding the very slight effect of gels upon diffusion, conductivity, and the velocity of chemical reactions. Ultramicroscopic examination of certain gels has also yielded evidence which appears to support this theory.Since elastic gels can be readily transformed into hydrophilic sols, and hydrophilic sols into gels, there is good reason for believing that hydrophilic sols may also possess a fibrillar structure. Such a structure is also postulated by many authorities for protoplasm, as later discussion will show.

HYSTERESIS

The statement is sometimes encountered that gels possess the faculty of memory. This statement is not, of course, to be accepted literally. It is merely a way of saying that the previous treatment to which a gel has been subjected have an influence, often marked, upon its behaviour, so that in an allegorical sense the gel may be said to remember those treatments. This phenomenon of the influence of the previous treatment of a gel upon its behaviour is known as hysteresis. Three gelatin gels were prepared containing respectively 10, 20, and 40 g.of gelatin per 100 cc. of water. Strips of these gels of equal rectilinear dimensions and thickness were then dried in a current of warm air until all of them were reduced to a moisture content of about 3.5 per cent.

In spite of the fact that all three of these gels possessed the same water content when immersed in water their swelling behaviour was quite different,depending in each sample on the previous history of the gel. The gel which was originally prepared in the proportion of io parts of gelatin to 100 of water swelled the most, followed in order by the one originally prepared in aproportion of 20 parts of gelatin to 100 of water, and finally by the gel originally prepared in a proportion of 40 parts of gelatin to ioo parts of water.The example cited is just one of the many hysteresis effects which have been recognized in gels.

This use of the term hysteresis should not be confused with its commonuse in another sense in physics and engineering and even the factor of time may induce such effects in gels. It follows that in experimental work with gels, if results are to be comparable, all the gels used in a given experiment or set of experiments must have had identical previous histories.

SYNERESIS

Syneresis may be defined as the spontaneous separation of aportion of the liquid component of a gel. The liquid which separates is not quite pure, however, being essentially a very dilute sol. Syneresis is a widely observed phenomenon. It may be observed in both gelatin and agar gels.The so-called bleeding of agar culture media is familiar to all who work with them. The liquid which usually forms around gelatin gels which have stood for some time is often a result of syneresis. It has been suggested that several important biological phenomena involve the process of syneresis.These include the formation of vacuoles in plant and animal cells, the separation of serum from a blood clot, the exudation of serum into a blister, glandular secretion, and the contraction of muscles.

THIXOTROPY

If a trace of sodium chloride is added to a test tube full of 1o per cent bentonite (a colloidal clay), and vigorously shaken, a colloidal sol will result which will set to a gel after standing a few minutes. By shaking, this gel can be converted to a sol which will again set upon brief standing. The process may be repeated an indefinite number of times. This phenomenon is called thixotropy. Protoplasm also exhibits thixotropic reactions. Stirring of the protoplasm or subjection of a cell to pressure can beshown to greatly reduce its viscosity and probably induces gel to sol changes.Thixotropic phenomena may therefore play important roles in cellular physiology.

3

Tracer Techniques in Plant Pathology

Globally, enormous losses of the crops are caused by the plant diseases. The loss can occur from the time of seed sowing in the field to harvesting and storage. Important historical evidences of plant disease epidemics are Irish Famine due to late blight of potato, Bengal famine due to brown spot of rice and Coffee rust. Such epidemics had left their effect on the economy of the affected countries.

OBJECTIVES OF PLANT PATHOLOGY

Plant Pathology (Phytopathology) deals with the cause, etiology, resulting losses and control or management of the plant diseases. The objectives of the Plant Pathology are the study on:

- The living entities that cause diseases in plants;
- The non-living entities and the environmental conditions that cause disorders in plants; iii. the mechanisms by which the disease causing agents produce diseases;
- The interactions between the disease causing agents and host plant in relation to overall environment; and
- The method of preventing or management the diseases and reducing the losses/damages caused by diseases.

SCOPE OF PLANT PATHOLOGY

Plant pathology comprises with the basic knowledge and technologies of Botany, Plant Anatomy, Plant Physiology, Mycology, Bacteriology, Virology, Nematology, Genetics, Molecular Biology, Genetic Engineering, Biochemistry, Horticulture, Tissue Culture, Soil Science, Forestry, Physics, Chemistry, Meteorology, Statistics and many other branches of applied science.

CONCEPT OF PLANT DISEASE

The normal physiological functions of plants are disturbed when they are affected by pathogenic living organisms or by some environmental factors. Initially plants react to the disease causing agents, particularly in the site of infection. Later, the reaction becomes more widespread and histological changes take place. Such changes are expressed as different types of symptoms of the disease which can be visualized macroscopically. As a result of the disease, plant growth in reduced, deformed or even the plant dies.

CAUSES OF PLANT DISEASES

Plant diseases are caused by pathogens. Hence a pathogen is always associated with a disease. In other way, disease is a symptom caused by the invasion of a pathogen that is able to survive, perpetuate and spread. Further, the word "pathogen" can be broadly defined as any agent or factor that incites 'pathos or disease in an organism or host. In strict sense, the causes of plant diseases are grouped under following categories:

- *Animate or biotic causes*: Pathogens of living nature are categorized into the following groups.
  - Fungi
  - Bacteria
  - Phytoplasma
  - Rickettsia-like organisms
  - Algae
  - Phanerogams
  - Protozoa
  - Nematodes
- *Mesobiotic causes* : These disease incitants are neither living or non-living, e.g.
  - Viruses
  - Viroides
- Inanimate or abiotic causes: In true sense these factors cause damages (any reduction in the quality or quantity of yield or loss of revenue) to the plants rather than causing disease. The causes are:
  - Deficiencies or excess of nutrients (e.g. 'Khaira' disease of rice due to Zn deficiency)
  - Light
  - Moisture
  - Temperature
  - Air pollutants (e.g. black tip of mango)
  - Lack of oxygen (e.g. hollow and black heart of potato)
  - Toxicity of pesticides
  - Improper cultural practices
  - Abnormality in soil conditions (acidity, alkalinity)

When a plant is suffering, we call it diseased, i.e. it is at 'dis-ease'. Disease is a condition that occurs in consequence of abnormal changes in the form, physiology, integrity or behaviour of the plant. According to American

Phytopathological Society, disease is a deviation from normal functioning of physiological processes of sufficient duration or intensity to cause disturbance or cessation of vital activities. The British Mycological Society defined the disease as a harmful deviation from the normal functioning of process. A plant is diseased when it is continuously disturbed by some causal agent that results in abnormal physiological process that disrupts the plants normal structure, growth, function or other activities. This interference with one or more plant's essential physiological or biochemical systems elicites characteristic pathological conditions or symptoms.

CLASSIFICATION OF PLANT DISEASE

To facilitate the study of plant diseases they are needed to be grouped in some orderly fashion. Plant diseases can be grouped in various ways based on the symptoms or signs (rust, smut, blight etc.), nature of infection (systemic or localized), habitat of the pathogens, mode of perpetuation and spread (soil-, seed- and air-borne etc.), affected parts of the host (aerial, root disease etc.), types of the plants (cereals, pulses, oilseed, ornamental, vegetable, forest diseases etc.).

But the most useful classification has been made based on the type of pathogens that cause plant diseases. Since this type of classification indicates not only the cause of the disease, but also the knowledge and information that suggest the probable development and spread of disease alongwith their possible control measures. The classification is as follows:

- Infectious plant diseases:
  - Disease caused by parasitic organisms: The organisms included in animate or biotic causes can incite diseases in plants.
  - Diseases caused by viruses and viroids.
- Non-infectious or non-parasitic or physiological diseases: The factors included in inanimate or abiotic causes can incite such diseases in plants under a set of suitable environmental conditions.

PARASITISM AND PATHOGENESIS

An organism which lives in or on other living organisms and derives its nutrients from the latter is called parasite. The relationship between a parasite and its hosts is known as parasitism. Many fungi and most bacteria grow on a non-living substrate within a living plant. The organism of this type of mode of nutrition is called saprophyte.

Based on the different types of modes or nature of nutrition, the relationship between the host and parasite or saprophyte is termed in many ways viz., obligate parasite (biotroph), obligate saprophyte, facultative parasite, facultative saprophyte, hemibiotroph and necrotroph (perthotrophs or perthophyte).

Parasitism in cultivated crops is common phenomenon. Any agent that can cause suffering or damage or disease is called a pathogen. In plant pathology, the term 'pathogen' is usually used to the living or infectious organisms. The ability of a pathogen or parasite to cause disease is known as pathogenicity.

It is obvious that a plant becomes diseased when it is attacked by a pathogen or parasite. The ultimate condition i.e. disease occurs by passing through some distinct events. Thus, the genesis or chain of events or stages of disease development are called pathogenesis.

This is also called as disease cycle. The events that occur in specific order are namely inoculation, penetration, establishment of infection, invasion or colonization, growth and reproduction, dissemination and survival of the pathogen (over wintering or over summering in absence of the host).

The events will continue to repeat in the same order in presence of both the host and pathogen/parasite that may lead to severe disease condition.

Koch's Postulates

Robert Koch forwarded four essential procedural steps called postulates for correct diagnosis of a disease and its actual causal agent. The postulates are:

- *Recognition:* The pathogens must be found associated with the disease in the diseased plant. The symptom of the disease should be recorded.
- *Isolation:* The pathogen should be isolated, grown in pure culture in artificial media. The cultural characteristics of the pathogen should be noted.
- *Inoculation:* The pathogen of pure culture must be inoculated on healthy plant of same species/variety. It must be able to reproduce disease symptoms on the inoculated plant identical to step 1.
- *Re-isolation:* The pathogen must be isolated form the inoculated plant in culture media. Its cultural characteristics should be similar to those noted in step 2 (This step was added by E.F. Smith).

If all the postulates are proved true, then the isolated pathogen is identified as the actual causal organism responsible for the disease.

EFFECT OF PATHOGEN ON THE PLANTS

During the course of pathogenesis, normal activities of the infected host plant undergo malfunction. Consequently, morphological and physiological changes occur.

- Morphological or structural changes: Physiological malfunctioning of the host cells causes disturbances in chemical reaction which ultimately lead to some structural changes viz., overgrowth, phyllody, sterile flowers, hairy roots, witches broom, bunchy top, crown gall, root knot, leaf curling, rolling, puckering etc.

- Physiological changes:
  - Disintegration of the tissues by the enzymes of the pathogen.
  - Effect on the growth of the host plant due to growth regulators produced by the pathogen or by the host under the influence of the pathogen.
  - Effect on uptake and translocation of water and nutrients.
  - Abnormality in respiration of the host tissues due to disturbed permeability of cell membrane and enzyme system associated with respiration.
  - Impairing the phenomenon of photosynthesis due to loss of chlorophyll and destruction of leaf tissue.
  - Effect on the process of translation and transcription.
  - Overall reproduction system of the host.

SYMPTOMS OF PLANT DISEASES

A visible or detectable abnormality expressed on the plant as a result of disease or disorder is called symptom. The totality of symptoms is collectively called as syndrome while the pathogen or its parts or products seen on the affected parts of a host plant is called sign. Different types of disease symptoms are cited below:

Necrosis: It indicates the death of cells, tissues and organs resulting from infection by pathogen. Necrotic symptoms include spots, blights, burn, canker, streaks, stripes, damping-off, rot etc.

Wilt: Withering and drooping of a plant starting from some leaves to growing tip occurs suddenly or gradually. Wilting takes place due to blockage in the translocation system caused by the pathogen.

Die-back: Drying of plant organs such as stem or branches which starts from the tip and progresses gradually towards the main stem or trunk is called die-back or wither tip.

Mildew: White, grey or brown coloured superficial growth of the pathogen on the host surface is called mildew.

Rusts: Numerous small pustules growing out through host epidermis which gives rusty (rust formation on iron) appearance of the affected parts.

Smuts: Charcoal-like and black or purplish-black dust like masses developed on the affected plant parts, mostly on floral organs and inflorescens are called smut.

Blotch: A large area of discolouration of a leaf, fruit etc. giving a blotchy appearance. White blisters: Numerous white coloured blister-like ruptures are surfaced on the host epidermis that forms powdery masses of spores of fungi. They are called white blisters or white rust.

Colour change: It denotes conversion of green pigment of leaves into other colours mostly to yellow colour, in patches or covering the entire leaves.

- *Etioliation*: Yellowing due to lack of light,
- *Chlorosis*: Yellowing due to infection viruses, bacteria, fungi, low temperature lack of iron etc.
- *Albino*: Lack of any pigment and turned into white or bleached
- *Chromosis*: Red, purple or orange pigmentation due to physiological orders etc.

Exudation: Such symptom is commonly found in bacterial diseases when masses of bacterial cells ooze out to the surface of affected plant parts and form some drops or smear, it is called exudation. This exudation forms a crust on the host surface after drying.

Overgrowth: Excessive growth of the plant parts due to infection by pathogens. Overgrowth takes place by two processes (i) Hyperplasia: abnormal increase in size due to excessively more cell division (ii) Hypertrophy: abnormal increase in size or shape due to excessive enlargement of the size of cell of a particular tissue.

Atrophy: It is known as hypoplasia or dwarfing which is resulted from the inhibition of growth due to reduction in cell division or cell size.

Sclerotia: These are dark and hard structures of various shaped composed of dormant mycelia of some fungi. Sometimes, sclerotia are developed on the affected parts of the plant. Presence of sclerotia on the host surface is specifically called a sign of disease rather than symptom.

DEVELOPMENT OF EPIDEMICS

Sudden outbreak of a disease within a relatively short period covering a large area and affecting many individuals in a population is called epidemic. Although, this term was originally designated to the human diseases, now applied in the diseases of animals, poultry, plants etc. Epidemic form of plant diseases is called as epiphytotics. For a disease to occur, coincidence of three parameters of disease triangle is essential, namely, the vulnerable host, virulent pathogen and favourable environment. Under such circumstances, the pathogen not only completes its life cycle but also undergoes repeated generations. Then an epidemic develops only when few repeated generations are completed by the pathogen on the same host. As each generation or cycle of the pathogen takes a few days for completion, the fourth parameter i.e. time factor (forms a disease tetrahedron or disease pyramid) is also involved in epidemic build up. In other words, epidemic growth is both temporal (pertaining to time) and a spatial (relating to space or area) process.

The initial stages of an epidemic growth curve have a lag phase, when the incubation period is longer, inoculum load is weak and prevalent environmental conditions are unfavourable. Subsequently, when the conducive conditions occur, the growth of the disease is rapid and the severity of the epidemic explodes like a time bomb.

Later, severity declines either due to unfavourable weather or crop maturity or both.

PLANT DISEASE MANAGEMENT

The word 'control' is a complete term where permanent 'control' of a disease is rarely achieved whereas, 'management' of a disease is a continuous process and is more practical in influencing adverse affect caused by a disease. Disease management requires a detail understanding of all aspects of crop production, economics, environmental, cultural, genetics and epidemiological information upon which the management decisions are made.

Principles of Plant Disease Management

There is six basic concept or principles or objectives lying under plant disease management.

- *Avoidance of the pathogen*: Occurrence of a disease can be avoided by planting/sowing a crop at times when, or in areas where, inoculum remain ineffective/inactive due to environmental conditions, or is rare or absent.
- *Exclusion of the pathogen*: This can be achieved by preventing the inoculum from entering or establishing in a field or area when it does not exist. Legislative measures like quarantine regulations are needed to be strictly applied to prevent spread of a disease.
- *Eradication of the pathogen*: It includes reducing, inactivating, eliminating or destroying inoculum at the source, either form a region or from an individual plant (rouging) in which it is already established.
- *Protection of the host*: Host plants can be protected by creating a toxin barrier on the host surface by the application of chemicals.
- *Disease resistance*: Preventing infection or reducing the effect of infection of the pathogen through the use of resistance host which is developed by genetic manipulation or by chemotherapy.
- *Therapy*: Reducing severity of a disease in an infected individual.The first five principles are prophylactic (preventive) procedure and the last one is curative.

Methods of Plant Disease Management

Avoidance of the pathogen:

- Choice of geographical area
- Selection of a field
- Adjustment of time of sowing
- Use of disease escaping varieties
- Use of pathogen-free seed and planting material
- Modification of cultural practices

Exclusion of inoculum of the pathogen

- Treatment of seed and plating materials
- Inspection and certification
- Quarantine regulations
- Eradication of insect vector

Eradication of the pathogen

- Biological control of plant pathogens
- Eradication of alternate and collateral hosts
- Cultural methods:
  - Crop rotation
  - Sanitation of field by destroying/burning crop debris
  - Removal and destruction of diseased plants or plant parts
  - Rouging
- Heat and chemical treatment of diseased plants
- Soil treatment: by use of chemicals, heat energy, flooding and fallowing

Protection of the host

- Chemical control: application of chemicals treatment, dusting and spraying
- Chemical control of insect vectors
- Modifications of environment
- Modification of host nutrition

Disease resistance

Use of resistant varieties: Development of resistance in host is done by

- Selection and hybridization for disease resistance.
- Chemotherapy
- Host nutrition
- Genetic Engineering, tissue culture

Therapy

Therapy of diseased plants can be

- Chemotherapy
- Heat therapy
- Tree Surgery

FUNGI

The term fungus (plural fungi) includes eukaryotic, spore-bearing, achlorophyllus, organisms that generally reproduce sexually and asexually, and whose usually filamentous, branched somatic structures are typically surrounded by cell walls containing chitin or cellulose, or both of these substances, together with many other complex organic molecules.

General Morphology, Characters and Somatic Structures of Fungi

The thallus: Thallus is a growth form lacking differentiation into root, stem and leaves. In fungi, it is known as somatic (soma= body) phase. It may be plasmodial, unicellular, pseudoplasmodial or mycelial. A filamentous structure of fungal body composed of multicells is known as mycelium (pl. mycelia) and a fragment (unit) of mycelium is called hypha (pl. hyphae i.e. web). Hyphae or mycelia may be septate (having cross wall in the filament) or aseptate (without cross wall or septum).

Branching habit of mycelium: Dichotomous, sympodial, lateral, opposite, verticilliate, monopodial etc.

Other somatic structures: Rhizoides (rootlike), appressorium (pl. appressoria), haustorium (pl. haustoria), hyphopodium (pl. hyphopodia).

Hyphal aggregations and tissues: During certain stages of life cycle, fungal mycelia become organized loosely or compactly that form some structures called plectenchyma (i.e. woven tissue). Its two general types are known as prosenchyma (i.e. approaching a tissue) and pseudoparenchyma (a type of plant tissue). These two types compose various other somatic and reproductive structures like stroma (mattress), sclerotium (hard structure) and rhizomorph (root shaped).

Reproduction in fungi: Fungi reproduce by three processes viz., (A) Vegetative, (B) Asexual and (C) Sexual reproduction.

Vegetative Reproduction

- Fragmentation
- Fission
- Budding
- Sclerotium
- Rhizomorph
- Chlamydospores
- Oidia (small egg)

A Sexual Reproduction

- *Exogenous:* The spores (reproductive units) borne at the tip or outside the vegetative structure called conidia (sing. conidium i.e. dust).

  Two types of conidia are thallospores and conidiospores. The letter has got three types viz., Blastospores, Aleuriospores, and Phialospores. The bearing structure is called conidiophore (phore=bearer). Generally, conidia are developed on a simple (without branch) tubular conidiophore. Some other types of conidia bearing structures are phialids (small bottle type), synnema, coremia, acervulus (heap), sporodochia, pycnidia and sori (sing. sorus) or pustule.
- *Endogenous:* The spores produces in sporangia (sing. sporangium; vessel or container) and hence called sporangiospores.

Sporangiospores are of two types viz.

- Plasmospores or zoospores or swarm spores which are motile due to having flagella and
- Aplanospores which are non-motile due to lacking flagella.

Sexual Reproduction

The sexual reproduction takes place by fusion of two compatible haploid nuclei, usually the gametes.

There are two distinct fungal species

- *Monoecious or hermaphroditic:* They are bisexual having both sex organs on one thallus (homothallic).
- *Dioecious:* They are unisexual having either male or female sex organs on one thallus (heterothallic)

Four distinct phases of sexual reproduction are: somatogamy, plasmogamy, karyogamy and meiosis. These phases occur by any one of the following five general methods of sexual reproduction,

- Gametic copulation - (a) Isogamy and (b) Anisogamy
- Gametangial contact
- Gametangial copulation
- Spermatization
- Somatogamy (Anastomosis)

Classification of Fungi, Taxonomy and Nomenclature

A. Taxonomy and Units of Classification:

Super kingdom	-Eukaryonta
Kingdom	-Protista (now Fungi)
Sub-kingdom	-Mycota
Division	-Mycota (suffix)
Sub-division	-Mycotina (suffix)
Class	-Mycetes (suffix)
Sub-class	-Mycetidae (suffix)
Order	-Ales (suffix)
Family	-Aceae (suffix)
Genus	-
Species	-

Suffix are used or added to the scientific names of a taxon. A taxon (pl. taxa) is a category in the classification system.Whittaker (1969) provided five kingdom system viz., Monera, Protista, Plantae, Animalia and Fungi, and thus Fungi is separated from Protista on the basis of nutrition pattern.

B. Various Classifications

Classification of fungi was given by various authors viz. Gwynne-Vaughan and Barnes, Martin, E.A. Bessey. C.J. Alexopoulos etc. The classification forwarded by Ainsworth is most widely accepted that has been given below:

Nomenclature

Naming of fungi and their classification fall under the rule of *International Code for Botanical Nomenclature*. The scientific name of an organism follows the pattern of binomial (bi = two+ nomen = name) nomenclature system, which is composed of two words.

The first word designates the genus and the genus name is always capitalized. The second word designates the species and its name is not capitalized.

Binomials when written are underlined and when printed italicized. Modification or updating, if any, in the nomenclature is done by a Committee for Fungal Nomenclature at each International Botanical Congress held at every four years interval.

Study of Selected Genera

- *Plasmodiophora* Woron.

  Kingdom: Protista

  Sub-kingdom : Mycota

  Division: Myxomycota

  Sub-division : Myxomycotina

  Class : Plasmodiophoromycetes

  Order: Plasmodiophorales

  Family : Plasmodiophoraceae

  Genus : *Plasmodiophora*

The genus is an obligate parasite and causes important diseases such as club-rot of brassicas.

Resting spore: Hyaline, spherical upto 4 ì in diameter, germinate to produce single anteriorly biflagellate primary zoospores.

Zoospores : A naked uninucleate protoplast, active, moving by irregular jerks and then comes in contact with root hair and become amoeboid, penetrate the cell wall and form a thallus in the host cell lumen.

Plasmodium: Zygote and young plasmodia unite to form larger plasmodia found in cortical cells and later in various root and stem tissue, always intracellular, number of nuclei increases as it enlarges and occupies the entire lumen of an abnormally large host cell, in the late stage, cleavage takes place around each nucleus in the plasmodium and develop into resting spores, spores free from each other, held together by the host cell wall, until decomposed in the soil by secondary organisms.

- *Spongospora* Lagerheim
  Kingdom : Protista
  Sub-kingdom: Mycota
  Division :Myxomycota
  Sub-division :Myxomycotina
  Class : Plasmodiophoromycetes
  Order : Plasmodiophorales
  Family: Plasmodiophoraceae
  Genus: *Spongospora*

The genus has got limited host range of crop plants. It causes important disease like powdery scab in potato and acts as a vector of potato-mop-top virus disease.

Resting spore: Germinate by release of biflagellate zoospores, which penetrates root hairs, thallus enlarge, become multinucleate and on germination form zoosporangia a thick walled, round to oval body.

Zoospores: Protoplasm of zoosporangium divides to form upto 50 secondary zoospores, discharge through an opening common to zoosporangial wall and host cell wall, biflagellate of unequal length, zoospores from zoosporangia may reinfect and produce another zoosporangium, this stage is repeated as long as young roots are available and conditions are favourable.

Plasmodium: Amoeboid stage of fused secondary zoospores form intracellularly plasmodium, invaded cells enlarge and divide into several infected cells, wart like structure is formed by abnormal cell growth and cell division, plasmodium gives rise to spore balls (19-85 ì) which consist of spongy mass with irregular internal channels, form yellow brown dust in mature sori, spore balls contain many individual cells, each constituting an individual uninucleate resting spore.

- *Synchytrium* deBary & Woron.
  Kingdom: Protista
  Sub-kingdom : Mycota
  Division : Eumycota
  Sub-division: Mastigomycotina
  Class : Chytridiomycetes
  Order: Chytridiales
  Family : Synchytriaceae
  Genus: *Synchytrium*

The genus causes important wart disease of potato. The pathogen is widespread in many countries and on many hosts.

Thallus: Developing into either a group of zoosporangia or a resting spores, initial thallus functioning as sorus and segmenting directly into a number of zoosporangia or gametangia, or as an evanescent prosorus from which the contents emerge through a pore and form an attached vesicular sporangial or gametangial sorus outside of the initial thallus but within the infected host cell, or developing into a resting spore.

Zoosporangia and Gametangia: Variously shaped but predominantly polyhedral with hyaline wall and red, orange, yellow, reddish yellow, lemon coloured, grey or hyaline granular contents, dehiscing by a tear, cleft or papilla.

Zoospores and Gametes: Ovoid, ellipsoid, solid oblong, spherical or pyriform, usually with 1 and sometimes 2 conspicuous, refringent globules, and a single posterior, whiplash flagellum.

Resting spores: Formed by fusion of isogametes, with the resulting zygotes infecting the host and developing into diploid resting spores; expospores thick, smooth, rough or ridged, hyaline, amber coloured, reddish brown or light or dark brown; endospores relatively thin and hyaline, content variously coloured with a few to numerous refringent globules, functioning as a sporangaium or a prosorus in germination.

- *Physoderma* Wallroth

  Kingdom: Protista

  Sub-kingdom: Mycota

  Division : Eumycota

  Sub-division : Mastigomycotina

  Class : Chytridiomycetes

  Order : Chytridiales

  Family : Physodermataceae

  Genus: *Physoderma*

Among al the chytridiomycete, *P. maydis* only causes a disease of some importance in overground parts of plants, e.g. brown spot disease of corn.

Rhizomycelium: Endobiotic, tenuous portions filamentous, delicate, fine or comparatively coarse, intercalary, enlargements, oval, elliptical, subspherical, broadly spindle shaped, turbinate and occasionally irregular, usually septate, bi or multicellular.

Rhizoides: Haustoria arising from most portions of the rhizomycelium, often reduced, digitate.

Ephemeral zoosporangia: Usually epibiotic, gregarious, sessile, pyriform, oval elongate, slipper shaped, irregular, sunken and deeply lobed, sometimes star shaped and angular, dehiscing by delinquescence of apical papilla, proliferating, subtented by a small endobiotic apophysis, richly branched and bushy, or reduced and digiate rhizoids.

Zoospores: From ephemeral sporangia usually smaller than those from resting sporangia, ellipsoidal oval, almost spherical, usually tapering posteriorelly, with a conspicuous, eccentric, refractive globule.

Resting sporangia: Formed on endobiotic rhizomycelium, terminal or intercalary, ellipsoidal, oval truncate, sub-hemispherical and flattened on one surface, sometimes slightly irregular, smooth or grooved; amber, light to dark brown, usually with thick epispore and thin endospore, contents coarsely granular or with one to several, large refractive globules, with or without an encircling ring of blunt, digitate, haustoria, germinating by an endosporangium which pushes up an oval or a circular, saucer shaped lid, or irregularly cracks the epispores; endosporium emerging partly, to emit zoospores by deliquescence of an apical papilla.

Sexual organ: Unknown, doubtful or lacking.

- *Pythium* Pringsheim

  Kingdom: Protista

  Sub-kingdom: Mycota

  Division: Eumycota

  Sub-division: Mastigomycotina

  Class: Oomycetes

  Order :Peronosporales

  Family :Pythiaceae

  Genus: *Pythium*

Pythium spp. Are common as soil inhabitants with long survival rates. They generally parasites on wide host range and manily causing both pre and post-emergence damping-off; in general root rots of young plants.

Mycelium: Well developed, much branched or unbranched hyphae, occasionally bearing appressoria and chlamydospores.

Zoosporangium: Either entirely filamentous and undifferentiated from the vegetative hyphae, simple or branched, acrogenous or intercalary, or consisting of a series of basal, complex lobulations and a filamentous discharge tube or a well defined sphaeroidal structure sharply dinstinct from its supporting hypha, an acrogenous, intercalary or laterally sessile, with a emission tube of variable length, sometimes internally proliferous.

Zoospores: Somewhat reniform, each containing a single vacuole and with two oppositely directed flagella of equal length, emerging from a shallow, longitudinal groove, expelled from the sporangium as an undifferentiated mass into a delicate vesicle produced by the tip of the discharge tube and where cleavage and maturation take place, capable of repeated emergence before finally encysting and germinating; probably always monoecious.

Oogonia: Terminal or intercalary, spherical or sub-spherical when terminal, ellipsoidal to limoniform when intercalary, smooth walled or variously enchinulated, for the most part forming a single oospore with or without conspicuous periplasm.

Antheridia: None or one to several, hyphogynous, monoclinous or dichlinous, allantoid, clavate, globose, sub or trumpet shaped, terminal or intercalary, borne on a short or long stalk, or sessile, usually one to four to an oogonium, forming a distinct fertilization tube.

Oospores: Usually borne singly within the oogonium, pleuratic or apleurotic, wall smooth or reticulate, thin or inspissate, the granular protoplasm usually bearing a conspicuous reserve, globule and a lateral refringement body; upon germination forming one or several germtubes or zoospores.

- *Phytophthora* deBary

  Kingdom: Protista

  Sub-kingdom: Mycota

  Division: Eumycota

  Sub-division : Mastigomycotina

  Class : Oomycetes

  Order: Peronosporales

  Family : Pythiaceae

  Genus: *Phytophthora*

Phytophthora: Is one of the most important plant pathogenic genera. Members of this genus frequently cause root rot and pre and post seedling diseases and are more often specialized and destructive plant pathogens.

Mycelium: White in mass, in host often with haustoria.

Hyphae: 3-8ì, irregularly swollen undulate, sometimes with characteristic swellings, initial branching at right angles to parent hyphae and often swollen for a short distance.

Chlamydospores: Thick walled secondary spores, usually spherical, intercalary, sometimes terminal, wall smooth, upto 2 ì thick, hyaline.

Sporangiosphores: Usually undifferentiated, branching sympodial or irregular and from below the sporangium or from within an empty one.

Sporangia: Usually terminal, single on long hyphae in sympodia or within an evacuated sporangium; ellipsoid, ovoid, obpyriform, apex differentiated by an internal hyaline, thickening of the inner wall and sometimes protruding to form a papilla, wall smooth upto 2 ì thick, non caducous or shed with a pedical, germination by zoospores emerging individually through apex or by germtube.

Zoospores: Hyaline, ovoid to phaseoliform, biflagellate, interiorly directed tinsel (shorter) and posteriorly directed whiplash (longer), when mortality ceases the spherical cyst may show repetetional emergence but not diplanetism.

Oogonium: Usually terminal, spherical or tapering to the stalk, delimited by a thick septum, wall hyaline, thin, becoming thicker and often yellow to brown, mostly smooth occasionally tuberculate or reticulate.

Antheridium: Usually single, monoclinous or diclinous, spherical oval, clavate or short cylindrical, often angular, amphigynous, or paragynous.

Oospores: Single more or less filling oogonium; spherical, smooth, hyaline, outer wall very thin, inner wall 0.5-6 ì thick, when mature with large central globule.

- *Sclerophthora* Thirum., Shaw and Naras.

  Kingdom: Protista

  Sub-kingdom: Mycota

  Division: Eumycota

  Sub-division : Mastigomycotina

  Class: Oomycetes

  Order : Peronosporales

  Family: Pythiaceae

  Genus : *Sclerophthora*

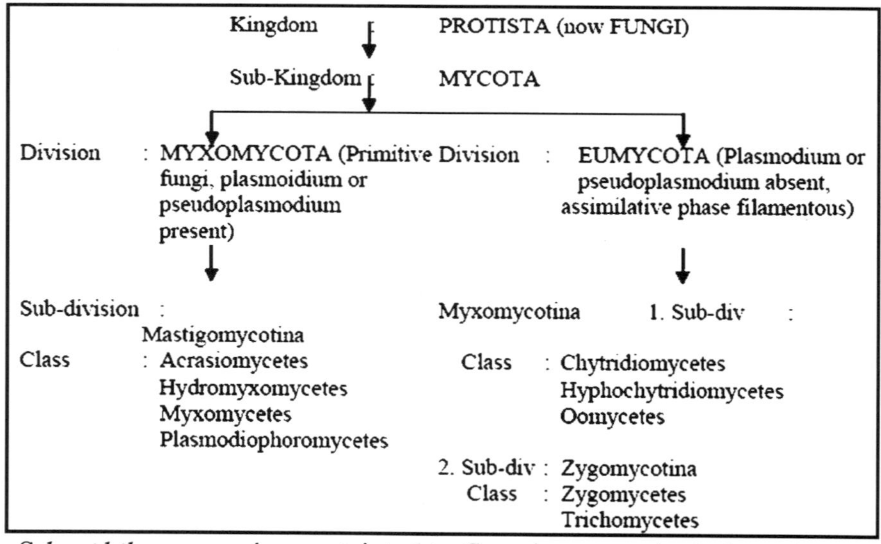

Sclerophthora spp: Are restricted to Gramineae particularly on leaves and inflorescences and cause the distinctive symptoms of chlorotic streaking, stunting, shedding, proliferation and hypertrophy.

Mycelium: Hyaline, coenocytic, sporangial stage like *Phytophthora.*

Sporangiophores: Hyphoid, very little differentiated from the hyphae within the host, simple or sympodially and successively branched.

Sporangia: Large, limoniform or obpyriform, apically poroid, borne singly at the apices of the sporangiophores, germinating in water by division of the cell contents into bicillate zoospores.

Oogonia: Wall thickened and confluent with the wall of oospores.

Oospores: Germination indirect by sporangial formation.

- *Sclerospora* Schroet.

  Kingdom : Protista

  Sub-kingdom : Mycota

  Division: Eumycota

  Sub-division: Mastigomycotina

  Class : Oomycetes

  Order: Peronosporales

  Family: Peronosporaceae

  Genus: *Sclerospora*

The genus has many species which are important obligate pathogen of cereals attacking leaves, stems and inflorescence, germination of the oospores is direct by means of germ tube.

Mycelium: Intercelluar, bearing small, usually knob-like unbranched haustoria.

Sporangio-conidiophores: Typically stout composed of a main trunk and a compact group of rather short apical branches which are one to several times divided, branching dichotomous to indefinite.

Sporangio-conidia: Germinate by one or more germtubes or by zoospores.

Oospores: Wall being confluent with that of the oogonium and by the stout conidiophore with heavy branches clustered at the apex.

- *Peronospora* Corda.

  Kingdom : Protista

  Sub-kingdom : Mycota

  Division : Eumycota

  Sub-division: Mastigomycotina

  Class: Oomycetes

  Order: Peronosporales

  Family: Peronosporaceae

  Genus : *Peronospora*

The genus is an obligate parasite and causes downy mildew disease in crucifers.

Mycelium: Thick, coenocytic, intercellular, branched, lobed with coarse haustoria.

Conidiophores: Emerge through stomata, singly or in clusters, slender, sometimes with bulbous or twisted base, unbranched upto 2/3 of its length, branching dichotomously 2-7 times, at acute angles, sterigmata taper to thin point, swell into sac and form single conidium at the tip.

Conidia: Oval, ellipsoidal, hyaline, 24-27 x 15-20 ìm, readily fall-off and germinate by lateral germtube.

Sex organs: Oogonium spherical, hyaline, both oogonium and antheridia first multinucleate but soon the other nuclei pass to periplasm and degenerate, one male and one female nucleus remain functional.

Oospores: Produced in host tissue in late stage; globose 26-42 ìm, pale yellow, enclosed in crest like folds formed by the periplasm.

- *Peronosclerospora* (Ito) Shirai and K. Hara

  Kingdom : Protista

  Sub-kingdom: Mycota

  Division: Eumycota

  Sub-division : Mastigomycotina

  Class : Oomycetes

  Order: Peronosporales

  Family: Peronosporaceae

  Genus: *Peronosclerospora*

Shaw erected *Peronosclerospora* as new genus attacking graminaceous hosts in the subtropic and tropics causing downy mildew disease particularly in maize, millets, sorghum and sugarcane. Infection takes place in newly germinated plants, sometimes directly from oospores present in the soil.

The secondary infection occurs by air borne spores mostly before sunrise. The spread of disease is generally short distanced as conidiophores are extremely prone to desiccation.

Sporangiosphore-sporangia (Vegetative phase): Primary aerial sporophore, comes out from host surface, 10 ìm or more broad, usually 15-25 x 2-3 ìm, dichotomously branched in the upper part, sporangia usually non-papillate, forming zoospores, oogonia

Oogonia: Wall thick, rough or ornamented,

Oospore: Plerotic

- *Albugo*

  Kingdom: Protista

  Sub-kingdom: Mycota

  Division: Eumycota

  Sub-division: Mastigomycotina

  Class : Oomycetes

  Order: Peronosporales

  Family : Albuginaceae

  Genus: *Albugo*

The genus causes white rust disease in many crops

Mycelium: Intercellular, widespread, bearing simple, globose, intracellular haustoria.

Sori: Subepidermal, later erumpent, forming white to cream, patches on all aerial parts of host.

Sporangiophores: Crowded, clavate, hyaline; forming basipetal chains of sporangia, joined by colourless connective links.

Sporangia (Zoosporangia): All alike or terminal sporangium larger than the others, globose, elliptical, oblong cubical or truncated obovate, hyaline, wall smooth, equally thickened.

Sexual organs: Anthredia and oogonia borne singly on short terminal or lateral mucelial branches within host tissue.

Oogonia: Globose, differentiated into periplasm and oosphere.

Anthredia: Small club shaped, arising near oogonia.

Oospores: Thick walled, globose, epispores dark coloured, ornamented with warts, ridges or reticulations, germination by zoospores.

- *Taphrina* Fr.

  Kingdom: Protista

  Sub-kingdom : Mycota

  Division : Eumycota

  Sub-division : Ascomycotina

  Class : Hemiascomycetes

  Order: Taphrinales

  Family: Taphrinaceae

  Genus : *Taphrina*

The spp. causes gall formation on leaves, stem and fruits. Hypertroph of infected tissues and deformed leaves and fruits occurs, blister-like leaf lesions, and witches brooms, occurs on temperate broad-leaved tree and stone fruits.

Mycelium: Intercellular, sub-cuticular or within the epidermal wall, forming asci in subcuticular layer or a wall locule; overwintering in the form of blastospores derived from ascospores by budding, infection is by blastospores.

Asci: Arise from rounded ascogenous cells, either by elongation of the ascogenous cell or by bursting out from the ascogenous cell wall.

Budding of the ascospores to form blastospores may also occur within the ascus and continue after spore expulsion.

- *Erysiphae* Hedwig ex. Merat

  Kingdom: Protista

  Sub-kingdom: Mycota

  Division: Eumycota

  Sub-division: Ascomycotina

  Class: Pyrenomycetes

Order: Erysiphales

Family: Erysiphaceae

Genus: *Erysiphe*

The fungi belonging to this group are obligate biotrophs that cause a major group of plant diseases commonly known as powdery mildews.

Ascospores (Cleistothecia): With simple, myceloid appendages but with several asci.

Conidiophores: On superficial hyphae with abundant conidia.

Conidia: Borne in chains, base of the conidiophores straight or swollen.

Mycelium: Superficial

- *Claviceps* Tul.

  Kingdom: Protista

  Sub-kingdom: Mycota

  Division : Eumycota

  Sub-division: Ascomycotina

  Class : Pyrenomycetes

  Order : Clavicipitales

  Family: Clavicipitaceae

  Genus: *Claviceps*

The various species of genus *Claviceps* causes disease in plants known as ergot which is name for sclerotia.

The sclerotia, which may contain toxic alkaloids produce disease in humans/animals known as ergotism.

Stromata: Borne on a elongated black sclerotium.

Sclerotium: Club shaped with globose heads, round to elliptical.

Perithecia: Completely, sunk in the head.

Asci: Unitunicate, cylindrical bearing ascospores.

Ascospores: Thread like

- *Sclerotinia* Fuckel

  Kingdom: Protista

  Sub-kingdom: Mycota

  Division: Eumycota

  Sub-division: Ascomycotina

  Class: Pyrenomycetes

  Order: Helotiales

  Family :Sclerotiniceae

  Genus: *Sclerotinia*

Sclerotinia spp. are mostly soil inhabitant and cause many important diseases in crop plants.

Apothecia: Arising from well defined tuberoid sclerotia.

Sclerotia: Freed from the substrate at maturity or only loosely enclosed within it, rind black, medulla usually white.

Microconidia: Borne in sporodochia, superficial or in cavities in the host.

Apothecia: Brown, disk saucer shaped or flat, receptacle downy or scurfy.

Asci: 8 spored.

Ascospores: Elliptical, inequilateral or very slightly reniform, hyaline, aseptate, often binucleate, usually uniseriate.

- *Puccinia*

Kingdom: Protista

Sub-kingdom : Mycota

Division: Eumycota

Sub-division: Basidiomycotina

Class: Teliomycetes

Order: Uredinales

Family :Pucciniaceae

Genus: *Puccinia*

Puccinia spp. are causing rust diseases in many economically important crops and substantial losses in family Grammineae. This is the largest genus of rust fungi with 3-4 thousand spp.

Spermogonia: Subepidermal.

Aecidia: Subepidermal in origin, aecidioid with a membranous peridium and catenulate spores or uredinoid with aecidiospores borne single or endophylloid with aecidioid sori but the spore germinating to form basidia.

Uredia: Subepidermal, paraphysate or a paraphysate.

Urediospores: Borne singly wall usually echinulate.

Teliosori: Subepidermal, erumpent.

Teliospores: Typically 2 celled, less commonly 1, or 3-4 celled, one pore in each cell, pedicels long or short, persistent or deciduous, wall coloured, basidia external, typically 4 celled.

- *Uromyces* (Link) Unger

  Kingdom: Protista

  Sub-kingdom: Mycota

  Division: Eumycota

  Sub-division : Basidiomycotina

  Class: Teliomycetes

  Order: Uredinales

  Family : Pucciniaceae

  Genus: *Uromyces*

Responsible for causing rust disease in many crops. Largest genus after *Puccinia*.

Spermogonia: Deeply embedded in the tissues of the host, flask shaped with conical mouth and ostiolar filaments and flexuous hyphae.

Aecidia: Usually with an evident, generally cup shaped peridium

Aecidiospores: With indistinct pores.

Urediospores: Formed singly on their pedicels, with several distinct pores, rarely accompanied by paraphyses.

Teliospores : Aseptate on distinct pedicels, always with an apical, hyaline epiculus.

Basidiospores: Flattened on one side or kidney shaped, autoecious or heteroecious in nature.

- *Melampsora* Castagne

  Kingdom: Protista

  Sub-kingdom: Mycota

  Division: Eumycota

  Sub-division: Basidiomycotina

  Class: Teliomycetes

  Order: Uredinales

  Family : Melampsoraceae

  Genus : *Melampsora*

Responsible for causing rust in some commercial crops like flax (linseed) and castor.

Spermogonia: Subcuticular or subepidermal, conical or hemispherical, without paraphyses but sometimes with flexuous hyphae.

Aecidia: Coemoid usually foliicolous, without peridium or sometimes with peripherial hyphae like paraphyses which may unite to form a rudimentary peridium.

Aecidiospores: Catenulate, globoid, or elliposoid, with verrucose walls.

Uredia: Subepidermal, pulverulent, with a thin, evanescent peridium, stalked.

Urediospores: Borne single on pedicels, globoid or ellipsoid, with indistinct pores; with capitate or clavate paraphyses.

Tellia: Subcuticular or subepidermal, forming crusts consisting of a single layer of spores.

Teliospores: Single celled, adhering laterlly with coloured walls forming a crust with one indistinct apical pore, germinating in spring.

Basidia: Typically 4 celled, producing globoid, colourless or yellowish basidiospores.

- *Ustilago* (Pers.) Roussel

  Kingdom: Protista

Sub-kingdom: Mycota
Division: Eumycota
Sub-division: Basidiomycotina
Class: Teliomycetes
Order: Ustilaginales
Family: Ustilaginaceae
Genus: *Ustilago*

This is the largest genus causing smut diseases in family Gramineae.

Sori: Borne in various parts of the host, forming dusty, dark, spore masses at maturity.

Spores: Single, produced irregularly in fertile mycelial threads which later disappear through gelatinisation. Germination by septate promycelium producing infection threads or sporidia formed terminally and laterlly near the septa, sporidia germinate in water to form infection hyphae or budding indefinitely in nutrient solutions.

- *Sporisorium* (=*Sphacelotheca*) de Bary

Kingdom: Protista
Sub-kingdom: Mycota
Division: Eumycota
Sub-division: Basidiomycotina
Class: Teliomycetes
Order: Ustilaginales
Family: Ustilaginaceae
Genus: *Sphacelotheca*

Sporisorium spp. cause smut diseases in Gramineae family.

Sori: In the inflorescence, predominantly in the ovaries, covered with a definite pseudomembrane enclosing a dusty spore mass and a central columella; pseudomembrane later flaking away, composed largely or entirely of definite sterile cells; hyaline or slightly tinted in nature, sometimes more or less firmly bound together.

Spores: Free, single, developed in a somewhat centripetal manner, small to medium size, germination as in *Ustilago*.

- *Tolyposporium* Woron.

Kingdom: Protista
Sub-kingdom: Mycota
Division: Eumycota
Sub-division: Basidiomycotina
Class: Teliomycetes
Order: Ustilaginales

Family: Ustilaginaceae

Genus: *Tolyposporium*

The genus is responsible for causing smut in cereals and grasses.

Sori: In the inflorescence, usually in ovaries, forming a granular to agglutinated spore mass.

Spore balls: Composed of numerous spores bound together by ridged folds or thickening of their outer walls known as sterile cells.

Spores: Small to medium size, often with a reticulate aspect due to the characteristic folds or thickenings which bind them together in the balls. Germination by 3-4 celled promycelium with sporidia borne at the septa.

- *Tilletia* Tul.

  Kingdom: Protista

  Sub-kingdom: Mycota

  Division: Eumycota

  Sub-division: Basidiomycotina

  Class: Teliomycetes

  Order: Ustilaginales

  Family: Tilletiaceae

  Genus: *Tilletia*

The species mostly occurs on Gramineae and are important pathogens on temperate cereals and grasses.

Sori: Mostly in the ovaries, occasionally in vegetative tissues, forming powdery to agglutinated spore masses, often foetid, emit foul odour due to trimethylamine. *Spores:* Single, formed from intercalary cells of a sporogenous mycelium or terminally on sporogenous hyphae, commonly encased in a hyaline or thin, gelantinoid sheath, comparatively large and regular, exhibiting a wide range in size, variously sculptured (commonly reticulate, cerebriform, verrucose, spiny, tuberculate or rarely smooth) pallid to opaque, germinating by a continuous promycelium, bearing terminal sporidia that usually fuse *in situ*, giving rise to secondary sporidia; unmature spores few to copious, similar in morphology to the mature spores but pigmented to a lesser degree; sterile cells, single, present in varying numbers, hyaline to tinted, smooth to granular, naked or sheathed, variously shaped.

- *Neovossia* (Mitra) Mundkur

  Kingdom: Protista

  Sub-kingdom: Mycota

  Division: Eumycota

  Sub-division: Basidiomycotina

  Class: Teliomycetes

Order: Ustilaginales
Family: Tilletiaceae
Genus: *Neovossia*

Sori: Mostly in grains and partially infected, containing black powdery mass of teliospores, in severely infected grains most of the seed tissues is converted into bunt spores, infected portion of seed is grey, later turns black, emit foul odour due to trimethylamine.

Spores: Globose to subglobose, 22-49 ìm in size, black, some spores bear hyphal appdendages (apiculus), with curved spines and reticulation on the epispore, spines covered with thin hyaline membrane, spores intermixed with globose to elongate, yellow to yellowish-brown, smooth, thick walled sterile cells.

- *Alternaria* Nees ex Fr.

  Kingdom: Protista
  Sub-kingdom: Mycota
  Division: Eumycota
  Sub-division: Deuteromycotina
  Class: Hyphomycetes
  Order: Hyphomycetales
  Family: Dematiaceae
  Genus: *Alternaria*

The genus causes economically important diseases in various crops, mostly as necrotic lesions on leaves, stems and fruits.

Colonies: Effuse, usually grey, dark blackish brown or black.

Mycelium: Immersed or particularly superficial, hyphae colourless, olivaceous brown or brown.

Conidiophores: Macronematous, simple or irregularly and loosely branched, pale brown or brown, solitary or in fascicles.

Conidiogenous cells: Integrated, terminal becoming intercalary, polytretic, sympodial or sometimes monotretic, cicatrised.

Conidia: Catenate, solitary, dry, typically ovoid or obclavate, often rostrate, pale or mid-olivaceous brown to brown, smooth or verrucose, with transverse and frequently also oblique or longitudinal septa.

- *Helminthosporium* Link ex Fr.

  Kingdom: Protista
  Sub-kingdom: Mycota
  Division: Eumycota
  Sub-division: Deuteromycotina
  Class: Hyphomycetes
  Order: Hyphomycetales

Family: Dematiaceae

Genus: *Helminthosporium*

The economically important species occurring on Gramineae have been placed under Genus *Drechslera* which differs from *Helminthosporium* (Subramanian and Jain, 1966). Only few species cause diseases viz., silver scurf of Irish potato while rest are common on dead stems of herbaceous plants.

Colonies: effuse dark, hairy

Mycelium : Immersed

Stroma: Usually present, dark, often large

Conidiophores: Macronematous, mononematous, unbranched, straight or flexuous, cylindrical or subulate, mid to very dark brown, smooth or occasionally verruculose, with small pores at the apex and laterally beneath the septa.

Conidiogenous cells: Polytretic, integrated, terminal and intercalary, determinate, cylindrical.

Conidia: Solitary, acropleurogenous, developing laterally often in verticils through very small pores beneath the septa whilst the tip of the conidiophore is actively growing, growth of conidiophore ceasing with the formation of terminal conidia, simple, usually obclavate, sometimes, rostrate, subhyaline to brown, smooth, pseudoseptate, frequently with a prominent dark brown to black scar at the base.

- *Cercospora* Fresenius

  Kingdom Protista

  Sub-kingdom Mycota

  Division Eumycota

  Sub-division Deuteromycotina

  Class Hyphomycetes

  Order Hyphomycetales

  Family Dematiaceae

  Genus *Cercospora*

There are more than 2000 species but only few species causes the disease in economically important crops.

Colonies: Effuse, grayish brown, tufted

Mycelium: Mostly immersed

Stroma: Often present but not large

Conidiophores: Macronematous, mononematous, caespitose, straight or flexuous, sometimes geniculate, unbranched or rarely branched, olivaceous brown or brown, paler towards the apex, smooth.

Conidiogenous cells: Integrated, terminal, polyblastic, sympodial, cylindrical, cicatrized, scars usually conspicuous.

Conidia: Solitary, acropleurogenous, simple, obclavate or subulate, colourless or pale, pluriseptate, smooth.

- *Pyricularia* Sacc.

  Kingdom: Protista
  Sub-kingdom: Mycota
  Division: Eumycota
  Sub-division: Deuteromycotina
  Class: Hyphomycetes
  Order: Hyphomycetales
  Family: Dematiaceae
  Genus: *Pyricularia*

The *Pyricularia* spp. is mainly pathogenic to cereals and *P. oryzae* causes rice blast worldwide. Some species are host specific.

Colonies: Effuse, thin, hairy, grey to greyish, brown or olivaceous brown

Mycelium: Immersed

Chlamydospores: Sometimes found in culture

Stroma : None

Setae and Hyphopodia : Absent

Conidiophores: Macronematous, mononematous, slender, thin walled, usually emerging singly or in small groups through stromata, mostly unbranched, straight or flexuous, geniculate towards the apex, pale brown, smooth.

Conidiogenous cells: Polyblastic, integrated, terminal, sympodial, cylindrical, geniculate, denticulate, each dentical cylindrical, thin walled, cut off usually by septum to form a separating cell.

Conidia: Solitary, dry, acropleurogenous, simple, obpyriform, obturbinate or obclavate, hyaline to pale olivaceous brown, smooth, 1-3 (mostly 2) septate, hilum often protuberant.

- *Fusarium* Link ex Fr.

  Kingdom: Protista
  Sub-kingdom: Mycota
  Division: Eumycota
  Sub-division: Deuteromycotina
  Class: Hyphomycetes
  Order: Tuberculariales
  Family: Tuberculariaceae
  Genus: *Fusarium*

The genus is commonly associated with soil-borne diseases.

Macroconidia: Are fusoid, one or more septate, with a foot cell bearing heel.

Microconidia: Non-septate to one septate, ovoid to short cylindric, in short chains or more commonly in spore balls.

Chlamydospores: Globose with a thick walled, intercalary, solitary, in chains or clumps, or terminal as short lateral branches. They may also be formed from cell of the macroconidia.

- *Rhizoctonia* DC. ex Fr.

  Kingdom: Protista

  Sub-kingdom: Mycota

  Division: Eumycota

  Sub-division: Deuteromycotina

  Class: Hyphomycetes

  Order: Agonomycetales

  Family: Agonomycetaceae

  Genus: *Rhizoctonia*

Rhizoctonia species are mostly soil inhabitant. Disease symptoms caused are seed decay, damping off, stem lesion and canker, root rot, above ground rot, leaf blight and storage rot.

Colonies: Colourless, rapidly become brown, aerial mycelium variable, sparsely branched hyphae are frequently present.

Sclerotia: Irregular size and shape but of uniform structure, brown or black, more or less loosely packed.

Hyphae: Cells of hyphae at advancing edge usually 5-12 ìm wide and upto 250 ìm long.

Branches arise near distal end of cell, are constricted at point of origin and septate shortly above, cells multinucleate with conspicuous dolipore septa, the angle of branching approaches 90° and branches may arise at the various points along the cell length.

- *Sclerotium* Sacc.

  Kingdom: Protista

  Sub-kingdom: Mycota

  Division: Eumycota

  Sub-division: Deuteromycotina

  Class: Hyphomycetes

  Order: Agonomycetales

  Family: Agonomycetaceae

  Genus: *Sclerotium*

Sclerotium is an unspecialized parasite, a soil inhabitant and with very wide host range in warm and wet areas.

Colonies: Usually white with many mycelial strands in the aerial mycelium.

Sclerotia: Develop on colony surface, nearly spherical, mostly 1-2 mm across, with smooth or shallow pitted shiny surface,

Hyphae: Initially hyphal cell is usually 4.5-9 ìm wide and upto 350 ìm long with one or more clamp connections at septa.

- *Phyllosticta* Desm.

  Kingdom: Protista

  Sub-kingdom: Mycota

  Division: Eumycota

  Sub-division: Deuteromycotina

  Class: Coelomycetes

  Order: Sphaeropsidales

  Family: Sphaeropsidaceae

  Genus: *Phyllosticta*

The genus should not be confused with *Phoma* which has aseptate conidia without a slim layer or an apical appendage and a different development of the conidium. Only few spp. are economically important to crops causing leaf spots.

Stroma: Prosenchymatous or plectenchymatous, well developed in pure culture, often reduced or absent.

Pycnidia: Globose, pyriform or tympaniform, separate or in small groups, embaded in a subepidermal stroma, uni or multilocular, ostiolate, wall prosenchymatous or pseudoparenchymatous, variable in thickness, not sharply delimited from the stroma, outer cells mostly dark and thick walled, inner cells hyaline, isodiametric, giving rise to conidiogenous cells.

Conidiogenous cells: Short cylindrical or conical, forming blastoconidia in basipetal succession and abstracting them from a fixed locus with a broad base.

Conidia: Aseptate, hyaline, globose. obovoidal, ellipsoidal or clavate, broadly rounded apically, flattened, surrounded by slime layer and a apical appendage, usually containing characteristic greenish guttules, generally 8-20 x 5-10 ìm, often forming appresoria on germination.

- *Phoma* Sacc.

  Kingdom: Protista

  Sub-kingdom: Mycota

  Division: Eumycota

  Sub-division: Deuteromycotina

  Class: Coelomycetes

  Order: Sphaeropsidales

  Family: Sphaeropsidaceae

  Genus: *Phoma*

Phoma spp. occur as family specialized pathogen but some are generally weak and plurivorous parasites, saprophytes and soil fungi.

Pycnidia: Mostly glabrous but sometimes hairy, usually globose-subglobose or globose-ampulliform to obpyriform, separated or in small groups, usually subepidermal then erumpent with mostly one, but sometimes more with papillate openings (ostiole and porus), wall pseudoparenchymatous or prosenchymatous, the outer cells mostly dark and thick walled, the inner cells hyaline and more or less isodiametric.

Conidiogenous cells: Are usually indistinguishable from the inner cells of the pycnidial wall but for a single aperture.

Conidia: Hyaline or sometimes slightly coloured (yellow to pale brown), globose, obvoidal, ellipsoidal or clavate, mostly once or twice as long as wide, generally 2.5-10 x 1-3.5 ìm, aspetate but secondary separation may occur resulting in 2 celled conidia.

- *Colletotrichum* Corda.

  Kingdom: Protista

  Sub-kingdom: Mycota

  Division: Eumycota

  Sub-division: Deuteromycotina

  Class: Coelomycetes

  Order: Melanconiales

  Family: Melanconiaceae

  Genus: *Colletotricum*

The forms of *Colletotrichum* have a very wide range of behavioural patterns in nature, varying from saprophytes to specialized parasites with a narrow host range. The species generally survive for long periods on plant debris, in or on the soil.

Mycelium: Immersed, branched, septate, hyaline to pale brown, intracellular

Acervuli: Subcuticular, epidermal to subepidermal, septate or confluent, formed of hyaline or brown, thin or thick walled, pseudoparenchymatous.

Setae: Present or absent, originating irregularly, from the pseudoparenchyma, more or less straight, unbranched, tapered to an acute or obtuse apex, brown, smooth, thick walled, septate.

Conidiophores: Septate, branched at the base, hyaline to pale brown, smooth formed from upper cells of pseudoparenchyma.

Conidiogenous cells: Enteroblastic, phialidic, discrete or incorporated on conidiophores, determinate, cylindrical, hyaline to pale brown, smooth, narrow, periclinal wall thickened, collarette, sometimes distinct.

Conidia: Hyaline, aseptate, more or less guttulate, cylindrical, long clavate, falcate, fusiform or muticate, on germination become pale brown, septate and form appressoria.

Bacteria

Bacteria (sing. bacterium) are simplest prokaryotic unicellular microorganisms having the common chemical composition of DNA, RNA and protein. They are highly adaptable and can survive extremes of temperatures, pH, oxygen tension, osmotic and atmospheric pressures, and hence found in almost all natural conditions.

Morphological characters: Being unicellular organism, bacteria may form groups of cells as filaments. They are either motile or non-motile and lack the definitely organized nucleus. Bacterial cell possess five shapes

- Spherical (*Micrococcus*),
- Rod-like/bacilliform (*E. coli*) and
- Spiral-shaped (*Spirillum*) and
- Curved-rod (*Vibrio*) and
- Club-shaped (*Clavibacter*). Motile cells are having long, whip-like flagella which may arise from one or both ends of the cell (polar) or from all over the cell (peritrichous). Based on flagellar arrangement bacteria is classified into 6 groups:
- Monotrichous - single polar flagellum at one end (*Xanthomonas*)
- Amphitrichous - single polar flagellum at both ends (*Pseudomonas*)
- Cephalotrichous - several flagella at one end (*P. fluorescens*)
- Lophotrichous - several polar flagella at both ends (*Spirillum*)
- Sub-polar - single sub-polar flagellum (*Agrobacterium*)
- Peritrichous - all over the cells (*Erwinia*)

Many bacteria have filamentous appendages called fimbriae or pilli. The size of bacteria ranges from 1-5 ìm and normal range of volume of a structural unit lies within 5-50 ìm3. Structurally, a bacterial cell can be divided into following 5 regions as follows:

- *Surface appendages:* Flagella and pilli
- *Surface adherents:* Capsules and slime layers. The capsule in the outer most layer and composed of polysaccharide or disaccharide and in some cases polypeptides. When polysaccharide is more fluid in consistency, it forms a gelatinous slime layer around the cell wall.
- *Cell wall:* It provides shape to the cell and protects underlying protoplasm having cytoplasm, chromatin, vacuoles, globules etc. The bacterial cell wall is made up of mucopeptide (murein). On the basis of two types of chemical composition of cell wall, bacteria are grouped into two as Gram +ve (85% or more mucopeptide and rest is simple polysaccharide) and Gram -ve (only 3-12% mucopeptide and rest are lipo-protein and lipo-polysaccharides)
- *Cytoplasm and organelles:* They contain soluble cytoplasmic constituents, nucleoid, mesosomes, ribosomes, lamellae (thylakoid) or vesicles (=chromatophores, found in photosynthetic bacteria) and

some reserve materials like granules. Gas vacuoles or gas vesicles, chlorosomes, carboxysomes and magnetosomes are also special type of organelles found in some bacteria.

- *Special structures:* Some bacteria form sporulation structures. Most characteristic spore structures are endospores, exospores, conidia, spores(akinetes), myxospores, cysts, bdellocyst are also formed by some genera of bacteria .

Morphological Properties

Genera	Shape	Size (μm)	Motility
Xanthomonas	Rod	0.4-1.0 × 1.2-3	Single polar flagellum
Pseudomonas	Rod	0.5-1.0 × 1.5-4	One or many polar flagella
Erwinia	Rod	0.5-1.0 × 1.0-3	Peritrichous
Agrobacterium	Rod	0.8 × 1.5-3	Sub-polar or peritrichous(1-4)
Clavibacter	Club-shaped/Rod	0.5-0.9 × 1.5-4	Non-motile/motile with 1-2 polar flagella
Ralstonia	Rod		Single polar flagella
Streptomyces	Fila-mentous	0.5-2 (dia)	Non-motile
Xylella	Rod	0.3 × 1-4	Non-motile

Cultural Characteristics

Agrobacterium: Colonies are non-pigmented, smooth, gram -ve, oxidative metabolism

Clavibacter: Usually non pigmented, gram +ve, oxidative metabolism

Erwinia: Usually non-pigmented, gram -ve, fermentative metabolism

Pseudomonas: Green diffusible fluorescent, brown diffusible pigment or no pigment, gram - ve, oxidative metabolism.

Xanthomonas: Yellow, non-diffusible, gram -ve, oxidative metabolism

Ralstonia: Non-pigmented, creamy white colonies, oxidative metabolism, gram-ve

Streptomyces: Colonies are at first white coloured, small (1-10 mm dia), smooth, later become powdery velvetty due to weft of aerial mycelium, gram -ve, produce variety of pigments depending on the substrates, oxidative metabolism

Xyllela: Produce long filamentous strand when cultured, gram-ve, colonies are small, smooth or undulated margins, non-pigmented, arerobic/oxidative metabolism.

Taxonomy, classification and nomenclature of bacteria: Taxonomy is the art of biological classification which includes identification as well as description of the basic taxonomic units (species) as completely as possible; it also determines the correct way of arrangement (cataloguing) of these units.

- Major divisions of bacteria on the basis of cell wall structure
  - Kingdom : Prokaryotae
- *Division I :* Gracilicutes (thin skin/cell wall, gram negative bacteria)
  - Class : Scotobacteria
  - Anoxyphotobacteria
  - Oxyphotobacteria
- *Division II :* Firmicutes (strong/durable cell wall, gram positive bacteria)
  - *Class :* Firmibacteria
  - Thallobacteria

Division III : Tenericutes(soft/tender cell wall, mycoplasma)
Class : Mollicutes
Division IV : Mendosicutes (faulty cell wall)
Class : Archaebacteria

DIVISION AND GROUPS OF SYSTEMATICAL BACTERIOLOGY

- Kingdom : Prokaryotae
- Division I : Cyanobacteria(blue green algae, myxophyceae)
- Division II: Bacteria
- Phototrophic bacteria : 1Order, 3 Family, 18 Genera
- Gliding bacteria : 2 Orders, 8 Family,21 Genera
- Sheathed bacteria : 17 Genera
- Budding and Appendaged bacteria : 17 Genera
- Spirochetes : 1 Order, 1 Family, 5 Genera
- Spiral and curved bacteria : 1 Family, 2 Genera
- Gram-negative Aerobic rods and cocci : 5 Family, 14 Genera
- Gram-negative Facultative Anaerobic rods : 2 Family, 17 Genera
- Gram-negative Anaerobic bacteria : 1 Family, 3 Genera
- Gram-negative Cocci and Coccobacilli : 1 Family, 2 Genera
- Gram-negative Anaerobic Cocci : 1 Family, 3 Genera
- Gram-negative Chemolithotrophic bacteria : 2 Family, 17 Genera
- Methane producing bacteria : 1 Family, 3 Genera
- Gram-positive Cocci : 3 Family, 12 Genera
- Endospore forming Rods and Cocci : 1 Family, 5 Genera
- Gram- positive Asporogenous rod-shaped bacteria : 1 Family, 1 Genus
- Actinomycetes and related organisms : 4 Genera not assigned to any family; 1 Family with 2 Genera; 1 Order with 8 Family and 31 Genera
- Rickettsias : 2 Order, 4 Family, 18 Genera
- Mycoplasmas : 1 Class, 1 Order, 2 Family, 2 Genera.

Nutrition and Effect of Physiochemical Factors on Growth

Nutrition: Many organic substrates are the sources, two nutrients viz. carbon (C) and energy which are important for bacterial growth. Certain bacteria e.g. *Pseudomonas* can use more than 90% organic compounds as a sole source of C and energy. Some bacteria can use two substrates (methane and methanol by methane producing bacteria) or only one substrate (cellulose decomposing bacteria). Bacteria need CO_2 (5-10%) for satisfactory growth on organic media. Thiamin (Vitamin B1) is also required for the growth of bacteria. However, the bacterial species which can synthesise thaiamin, do not require any special compound.

While bacteria are grown on/in artificial medium, the medium should have balanced mixture of necessary nutrients. Synthetic (ingredients are chemically known) and complex (ingredients are chemically unknown) media are used for artifical culture of bacteria. Commonly used elements in synthetic medium are K, Mg, Fe, Ca, Mn, Mo, Co, Zn, NH_4 and glucose (for C). For nitrogen fixing bacteria, N is not needed in media. Complex medium viz. nutrient medium contains peptone and beef extract and is used to grow wide range of micro-organisms including those microbes whose precise requirements (growth factors) are not known. Based on nutritional requirement also bacterial classification is made using some specific terms like autotrophic, heterotrophic, phototrophic, chemotrophic, lithotrophic and organotrophic.

Growth and Reproduction: In all cellular organisms, growth is achieved through cell multiplication. Hence, multiplication of a multicellular organisms result in an increase in size, while the multiplication of unicellular organisms results in increase in number. Growth in bacteria takes place through multiplication where one bacterium doubles at regular intervals (doubling time or generation time is 20-30 minutes) by binary fission (asexual reproduction).

Thus number of bacterial cells increases exponentially. Formation of endospores, cysts, fragmentation, sporangiospores and conidia are some other means of asexual reproduction in bacteria. The sexual reproduction in bacteria is represented by transformation, conjugation, transduction and lysogenic conversions. The growth curve of bacteria can be plotted with four phases viz. lag phase (slow growth), log phase (exponential growth), stationary phase (no growth) and death phase (decline of living cells).

Effect of Physical Factors/Forces on Growth and Reproduction

- *Temperature:* Bacteria can survive temperatures of 0° to 85°C or even more depending upon the species. On the basis of temperature requirement, bacteria are divided into 3 catagories viz. psychrophilic (0-30°C, optimum 15°C), mesophilic (min. 5-25°C, opt. 18-45°C, max.30-50°C) and thermophilic (min. 25-45°C, opt. 55°C, max. 60-93°C) .

- *Moisture:* Bacteria are more aquatic than terrestrial and can survive in presence of high percentage of water.
- *Light:* Ordinary visible light does not affect bacterial activity. But different spectrum of light viz UV light, infra-red light have different effect on the activity of bacterial species.
- *Pressure:* Ordianary mechanical pressure can not affect bacterial cells. Principle of osmosis is the best used pressure for destruction of bacteria.
- *Hidrogen-ion concentration:* Suitable pH range for bacterial growth and reproduction is 5.0 to 9.0 .

Bacterial Genetics and Variability: Knowledge on the genetic system of bacteria was dull till 1940. Prior to this period, no definite nucleus had been demonstrated in bacteria although variability in bacterial cells was recognized well before. Only the development of science in Molecular Biology helped to recognize transfer of genetic material i.e. DNA to the daughter cells at the time of binary fission.

Variability among bacteria is resulted from the following processes:

- *Conjugation:* Two compatible bacterial cells come into contact. Then the recipient female cell (F-) receives the DNA from the donor male cell (Hfr). Thus genetic make up of both the cells is changed.
- *Transformation:* The bacterial cell absorbs DNA exuded by compatible cells or freed by dissolution of the cell-wall into the external medium.
- *Transduction:* This process is a "phage-mediated genetic transfer". The bacterial viruses (bacteriophages or phage) can acquire DNA from one cell and transfer it to the other cells attacked by them. If attacked cell is not destroyed due to infection by the phase, it reproduces to form new races with different genetic character.
- *Lysogeny:* It involves association of genetic material of a virus with that of bacterium. Although it is different from above three processes, it also provides a permanent genetic modification of the bacterial genome.

VIRUSES

Matthew (1981) defined a virus as "a set of one or more nucleic acid template molecules, normally encased in a protective coat, or coats of protein or lipoprotein, which is able to organize its own replication within suitable host cells. Within such cells, virus production is (a) dependent on the host's protein synthesizing machinery, (b) organized from pools of the required materials rather than by binary fission, and (c) located at sites which are non separated from the host cell contents by a lipoprotein, bilayer membrane".

Many plant diseases which are now known to be caused by viruses had been encountered long ago. The causes of those diseases were not known. The first breakthrough was made by Adolf Mayer in 1886, in the Netherlands, while studying the highly contagious, mysterious disease of tobacco which he called "Mosaikkrankheit" i.e. mosaic like disease. He found that healthy plants could be infected by injecting the sap of diseased plants.

He also observed that the unknown agent could be inactivated by boiling the sap. He concluded that the disease was the manifestation of a bacterium. In 1892, Ivanovsky confirmed Meyer's report and further showed the sap to remain infections even after passage through bacteria-proof filter.

However, he claimed the incitant to be a microbe. But Martinus Beijerinck realized the causal agent to be something novel. His results further confirmed the findings of Meyer and Ivanovsky and also showed that the incitant could diffuse into an agar gel. Based on all these findings, Beijerinck, in 1898, concluded that the mysterious pathogen was not a bacterium, but a *contagium vivum*

fluidum i.e. contagious infective material or infectious living fluid. He thought the contagium to be able to reproduce itself in living plants and referred it as a virus.

Architecture of Viruses and Viriods

Morphologically, virus particles are (i) isometric (spherical, polyhedral) and (ii) anisometric (rigid or flexuous rods, bacilliform or bullet-shaped). Many isometric viruses have symmetric polyhedra which are either of three cubic symmetry i.e. tetrahedral, octahedral or icosahedral. Isometric particles measure the diameter 17nm (satellite virus of tobacco necrosis virus) to 70nm (reoviruses).

The bacilliform viruses measure up to 300 nm length x 95 nm width (rhabdovirus group). The rod-shaped viruses having short rigid rod measure 114-215 nm length x 23 nm width (the tobraviruses) and those with long flexuous particles measure up to 2,000 nm length x 10 nm in width (the closteroviruses).

The rod shaped particles of tobacco mosaic virus (TMV) consist of protein sub-units (capsomeres) built up in a regular, helical array, with the RNA chain compactly coiled in a corresponding helix on the inside of the protein sub-units. The protein coat (capsid) and RNA genome surround an axial hole or canal. In membrane-bound viruses, the inner nucleoprotein core is called as nucleocapsid.

Viroids are smallest (1.1-1.3 x 105 mol. wt.), simplest and non-encapsidated RNA. They consist of a single molecular species of circular or linear form.

Chemical composition: Plant virus particles consist of infectious nucleic acid (the genome), which is encapsidated within a protective protein coat or shell. The genome, essential for virus replication, is composed of ribonucleic acid (RNA in most groups of viruses) and deoxyribonucleic acid (DNA in the caulimovirus and geminivirus groups). The RNA and DNA may be single stranded (ss) or double stranded (ds) Besides these two basic components, an envelop of lipid or lipoprotein membrane is present in some plant viruses. Other components are metallic ions and polyamines present in varying amounts. Some enzymes are found in reoviruses and rhabdoviruses. Water constitutes 10-50 per cent of the mass of virus particle.

The nucleic acid may be present as a single continuous strand (single molecular species) in a particle. It is called mono-partite genome. Some nucleic acid genomes have two or more pieces (molecular species) in different particles; usually they are not always encapsidated within separate protein shells. Such genomes are termed as bi-, tri-, or multi-partite or the viruses with divided genome.

In some RNA viruses, the genetic information is divided into two or more parts. They are called multi-component viruses and the individual components are not infectious alone. Hence two or more genomic elements are needed to cause infection and replication.

The genomic organization of viruses depicts structure and function of genes or cistrons. Some triplet bases called codons are responsible for expression of genes. There are two types of codons, (i) initiation codons (AUG, GUG) and (ii) termination codons (UAG, UAA, UGA) and they control functions of genes and translation products.

Nomenclature of viruses and classification: A number of addition and deletion was made in naming viral pathogens. Linnaean style of binomial nomenclature is not followed.

Instead, plant pathogenic viruses are named based on the common or vernacular name of the affected host plant such as tobacco mosaic virus, rice dwarf virus, caulimoviruses, potato virus X and Y etc. Later, these vernacular names were further simplified by abbreviating them such as TMV for tobacco mosaic virus, CMV for cucumber mosaic virus, potex virus for potato virus X, CaMV for cauliflower mosaic virus etc. In addition to such names, a system of cryptogams was introduced to give concise information on the properties (immediate summary) of one virus for example:

TMV = R/1 : 2/5 : E/E : S/O

(1st) (2nd) (3rd) (4th)

1st term : Type of nucleic acid (RNA, DNA)/number of strands of nuleic acid (1 = ss, 2 = ds)

2nd term : Molecular weight of nuleic acid in millions/% of nuleic acid in infective particle

3rd term : Outline of particle shape (E = elongate, S = spherical, B=bacilliform)/outline of nuclear capsid (E, S, B)

4th term : Type of host infected (B = bacteria, F=fungus, I = invertebrate, S = seed plant)/type of vector (Ap = aphid, Au = leafhopper, Cl = beetle, Fu = fungus, Ne = nematode, Th = thrips, W = whitefly, O = spread without vector, Se = seed transmitted)

A system of plant virus classification was introduced by International Committee on Taxonomy of Viruses (ICTV) based on the characteristics such as morphology of virus particle, type and quantity of nucleic acid, genomic structure and type of vector. For example

Classification based on particle morphology and type of nucleic acid:

I. Elongated, Helical, ss RNA
  A. Rigid
  (a) Monopartite
  (b) Multipartite
  B. Flexuous, all monopartite
II. Isometric
A. Single stranded RNA
  (a) Monopartite
  (b) Multipartite
  (b1) With envelope
B. Double stranded RNA
C. Single stranded DNA
D. Double stranded DNA
III Bacilliform
A. Without envelop
B. With envelop

Grouping of Plant Viruses

Like families and genera cited in the classification of fungi and bacteria, plant virologists have been using "groups" for viruses. A total of 26 groups have been accepted by ICTV and some of those are,

- Alfalfa mosaic virus group
- Bromoviruses
- Carlaviruses
- Caulimoviruses
- Comoviruses
- Dianthoviruses
- Geminiviruses etc.

There are about 11 unclassified virus groups are known, e.g.

- Barley yellow mosaic virus group
- Carnation mottle virus group
- Rice stripe virus group
- Satellite virus group etc.

CRITERIA FOR IDENTIFICATION OF VIRUSES

Viruses causing plant diseases can be identified correctly by number of ways:

- *Behaviour in host*: Host range, symptoms and their types, tissue restriction, type and location of intracellular inclusion bodies, seed transmissibility.

- *Vector relation*: Taxa, acquisition and inoculation thresholds, persistence in vector, multiplication in vector, modes of transmission (e.g. transovarial transmission).
- *Particle properties*: Shape and symmetry, size, presence or absence of envelope, capsomeric structure, sedimentation properties (number of components and sedimentation co-efficient); coat protein properties (no. of polypeptides and their molecular weight); properties of nucleic acid (RNA, DNA, strandedness, no. molecules and molecular weight, presence or absence of 5'-terminal M7 Gppp, 5'-terminal VPg, 3'-terminal Poly (A)); electrophoretic mobility; isoelectric point, serological relationship.
- *In vitro properties in crude sap*: This is determined by the tests viz. thermal inactivation point (TIP), dilution end point (DEP), longevity of virus *in vitro* (LIV).
- *Cross-Protection tests*: Virus identity and strain relationship can be done. However, this sort of test is less applied now-a-days.

Important Techniques for Virus Detection and Identification

- *Electron microscopy*: used to know shape, symmetry, size, presence or absence of envelope and capsomeric structure.
- *Immunosorbent electron microscopy* (ISEM).
- *Serology*: The technique includes
  - Precipitin tests (precipitin-tube test, precipitin-ring test, microprecipitin test).
  - Immunodiffusion tests (single, radial diffusion test, gel double-diffusion or Ouchterlony test)
  - Agglutination test (slide-agglutination or chloroplast-agglutination test).
  - Enzyme-linked immunosorbent assay (ELISA): It includes indirect ELISA and direct double antibody sandwich (DAS) ELISA.
  - Dot immunobinding assay (DIBA)
  - Molecular hybridization analysis (spot hybridization or dot-blot technique)
  - Monoclonal antibodies (MAb)

Multiplication and infection nature of plant viruses: The events in virus infection involve three steps: adsorption, penetration or entry and uncoating or disassembly. The initial contact between virus particle and host cell is referred to as adsorption or entry.

The process during which the virion or its nucleic acid passes into the cytoplasm of the cell is known as penetration or entry. Uncoating is the removal of various components of the mature virion and subsequent release of viral genome and other constituents that plays a major role in establishing infection.

- *Multiplication of virus in plants:* The replication of RNA and DNA plant viruses differ from diffent groups of viruses or for individial viruses. The mechanism known for different groups of plant viruses are – ssRNA virus- monopartite genomes, ssRNA virus-bipartite genome, ssRNA virus- tripartite genome, dsRNA virus- monopartite genome, satellite viruses, helper viruses and satellite RNAs, dsDNA viruses. The progeny RNA moves out to cytoplasm where the assembly sythensized proteins and encapsidation of virion take place. The process continues till the host is alive and a large number of new virus particles are formed.
- *Accumulation and movement of viruses in plant:* The nascent virus appears in the cell about 10 hours after infection. The concentration of virus varies based on the type of virus, temperature, nutrition and duration of light. The virions are found aggregated in amorphous or crystalline forms or they are dispersed in cytoplasm and nucleus.

In plant system, two stages of virus movement have been recorded. These are:

- *Cell to cell or short distance movement:* This type of virus movement takes place through protoplasmic bridges, the plasmodesmata. The plasmodesmata selectively allows passage to the macromolecules and thus virus can move through it with the help of virus-coded 'movement protein' mechanism.
- *Movement from one part to another part of plant:* This is a long distance movement of the viruses taking place through vascular system. The movement is faster in elongated young cells than in round and older cells. Moreover, virus moves fast at high temperature because the protoplasmic streaming and cellular activities are higher at high temperature. The nature of cells (parenchyma, xylem, phloem, sieve tubes) also determined rate movement of viruses in plant.

As a result of the multiplication and distribution of the viruses in the plant system, the host plant(s) are infected and exhibited varying degrees of disease symptoms. Transmission of plant viruses: Viruses are distributed or transmitted from the infected plants to the healthy ones in various ways in nature. As the plant viruses can not penetrate cuticle of their hosts and hence they can enter into the host tissue through wounds only. The means of transmission are:

- *Mechanical transmission*: The sap of the infected plant is manually transferred to the healthy plants. It is the easiest method of experimental inoculation.
- *Graft transmission*: In this practice, if either the scion (shoot portion)

or stock (root stock) is infected, the virus usually moves to the healthy partner which may later express visible symptoms of disease.

- *Transmission through vectors*
  - *Insects:* Some insect species are the vector of plant viruses which can carry/transmit viruses from infected plants to the healthy plants e.g. aphid (potato virus Y, PLVR), white flies (tobacco leaf curl), beetles (cowpea mosaic virus), mealy bugs (cacao mottle leaf), thrips (tomato spotted wilt), lace bugs (sugar beet viruses), mites (sterility mosaic of arhar), leaf hoppers (beet curly top, rice tungro etc.), plant-hoppers (maize mosaic, maize rough dwarf), tree hopper (tomato pseudo curly top).
  - *Nematodes:* Five genera of nematodues viz., *Xiphinema, Longidorus, Paralongidorus, Trichodorus* and *Paratrichodorus* can transmit plant viruses.
  - *Fungi:* Some species of fungi can also transmit viruses e.g. *Olpiduim brassicae* (tobacco necrosis), *O. cucurbitacearum* (cucumber necrosis), *Polymyxa graminis* (oat mosaic, wheat mosaic), *P. betae* (beet necrotic yellow vein) and *Spongospora subterranea* (potato mop top) etc.
- *Dodder transmission*: Many viruses can be transmitted through dodder (*Cuscuta* spp.). Dodder transmission is used in the laboratory to transfer viruses from the hosts.
- *Transmission through seeds and pollens:* Seed coat (testa), embryo, and also pollens of some plants can transmit viruses. e.g. alfalfa mosaic, barley stripe mosaic, bean common mosaic, lettuce mosaic are transmitted by both seeds and pollens of *Medicago sativa, Hordeum vulgare, Phaseolus vulgaris* and *Lactuca sativa*, respectively.

Basic characteristics of insect transmission: Virus transmission, specially in case of the aphid transmission, takes place by three ways viz. non-persistent (stylet borne), semi-persistant and persistent (circulative). These modes of transmission have been distinguished on the basis of acquisition feeding time, inoculation feeding period, latent period and multiplication or no-multiplication of the viruses within their vector etc.

Apart from aphids, semi-persistent mode is recorded in mealy bug (*Planococcoides njalensis* - cacao swollen shoot virus) and leaf hopper (*Graminella nigrifrons* - maize chlorotic dwarf virus; *Nephotettix impicticeps* - rice tungro virus) and persistent mode is recorded in leaf hopper (*N. cincticeps* - rice dwarf virus; *Agallia constricta* - wound tumer virus), plant hopper (*Peregrinus maidis* - maize mosaic virus), tree hopper (*Micrutalis malleifera* - tomato pseudo curly top virus), beetles (*Ceratoma trifurcata* – cowpea mosaic virus; *Phyllotreta* sp. – turnip yellow mosaic virus) and thrips (*Frankliniella fusca* – tomato spotted wilt virus).

Some viruses pathogenic to plant can be transmitted only in presence of a second virus (helper virus) in the host and this type is called dependent transmission. For example, aphid (*Myzus persicae*) transmits potato aucuba mosaic virus only if the source plant is also infected with potato virus A or Y.

Mycoplasma, Spiroplasma and Fastidious bacteria

Importance: Mycoplasma were known in animal pathology including human diseases. In 1988, they were first isolated and proved their Koch's postulates in bovine pleuropneumonia by Nocard and Roux. In plants, 'yellow' diseases were thought to be caused by some viral pathogens. Nullifying the role of viruses, Doi *et al.*, and Ishiie *et al.* (1967) from Japan reported the involvement of mycoplasma like organisms (MLO) in causing 'yellow' diseases in plants.

Little leaf of brinjal, grassy shoot of sugarcane, sandal spike are some of the important plant disease caused by MLOs (=Phytoplasma i.e. plant mycoplasma). In 1972, Davis *et al.*, observed a motile and helical microorganisms associated with corn stunt disease, they termed them as spiroplasma. Later, citrus stubborn disease was also reported fastidious bacteria from the phloem of clover and periwinkle plants infected with club leaf disease. The causal agents were called as rickettsia like organisms (RLO) or rickettsia like bacteria (RLB). Pierce's disease of grape, phony peach disease in peach, sugarcane ratoon stunting, alfalfa dwarf diseases are caused by fastidious vascular bacteria.

Morphological characters of Mycoplasma: Mycoplamas have a prokaryotic cellular organization except the cell wall which is bounded by triple layer unit membrane and lack the ability to synthesize cell wall material. They are highly pleomorphic with small coccoid bodies, ring forms, pear shaped and rarely filamentous (branched or unbreanched). Usually, they are non-motile and filterable through bacterial filters. They are very small and diameter ranges from 0.3-0.9 ìm. They produce typical 'fried egg' colonies in cell free solid media and require sterol for growth. They are insensitive to penicillin, but inhibited by tetracycline and some specific antibody.

Morphological characters of Spiroplasma: Spiroplasmas are helical (spiral) mycoplasmas. Branched non-helical filamentous are also found. They measure about 120 nm in diameter and 15 ìm long. They are motile, move by a slow undulation of the filament and probably by a rapid rotary a screw motion of the helix. Rest other characters are similar to those mycoplasmas.

Morhpological characters of Fastidious bacteria: Fastidious bacteria develop within the vascular system of plants and hence they are categorized into two types viz., fastidious phloem-limited and fastidious xylem-limited bacteria. They are generally rod-shaped cells of 0.2-0.5 μm in diameter by 1-4 ìm in length. They are bounded by a cell wall and a cell membrane, although in the phloem-inhabiting bacteria, cell wall appears more as a second membrane than as cell wall. The outer layer of the cell wall is usually undulating or rippled. They don't have any flagella. Nearly all fastidious vascular bacteria are Gram-negative and only few are Gram-positive.

Classification of Mycoplasma, Spiroplasma and Fastidious bacteria

A. Mycoplasma and Spiroplasma

Kingdom : Prokaryotae

Division : Tenericutes

Class : Mollicutes

Order : Mycoplasmatales

Family:

- Mycoplasmataceae
  - Genus: *Mycoplasma*
- Spiroplasmataceae
  - Genus: *Spiroplasma*
- Acholeplasmataceae
  - Genus: *Acholeplasma*

B. *Fastidious vascular bacteria:* There is no well accepted classification (taxonomy) made so far for these organisms. Hence classification for Rickettsia (RLO) and Fastidious bacteria (e.g. Xellella) are mentioned below:

B1: Rickettsia (RLO)

Kingdom: Prokaryotae

Division: Gracilicutes (Gram-ve bacteria)

Class: Proteobacteria

Sub-class: Alpha proteobacteria

Order: Rickettsiales

Family: Rickettsiaceae

Tribe: Rickettsiae

B2. Fastidious vascular bacteria

Kingdom: Prokaryotiae

Division: Gracilicutes (Gram-ve bacteria)

Class: Proteobacteria

Sub-class: Gamma proteobacteria

Order: Not classified

Family: Not classified

Tribe: Not classified

Genus: *Xelella*

Reproduction of mycoplasmas: MLOs can reproduce by binary fission, by formation of spores (elementary bodies), by filamentous growth and budding. The chromosome replication starts at a fixed site followed by bidirectional progression. Further; the outline of chromosome replication of mycoplasmas is somewhat similar to that of *E. coli*. However, the process of mycoplasmsa cell reproduction has not yet been well classified.

Reproduction of spiroplama: Cell division by budding, constriction followed by fission into unequal daughter cells are assumed to be the principal mode to reproduction in spiroplasma.

Reproduction of Fastidious bacteria: Binary fission gives rise to development of many hantle-like structures and forms in this organism. The resultants are pleomorphic bodies sometimes bounded by only one membrane.

Plant Pathogenic Nematodes

Nematodes belong to the Animal kingdom, and phylum Nematoda. Most of the important parasitic genera belong to the order Tylenchida and few under Dorylaimida. Important plant parasite species are *Meloidogyne incognita* (root knot nematodes), *Heterodera* and *Globodera* (cyst nematodes), *Tylenchulus semipenetrans* (citrus nematode), *Pratylenchus* (lesion neamtode), *Ditylenchus* (stem and bulb nematode), *Anguina* (seed-gall nematode), *Aphelenchoides* (foliar nematode) etc.

General morphology of nematodes: The body of the adult male is cylindrical, filiform, eel-shaped, round in cross section and tapering at each end. The anterior end is smooth, provided with papillae, leading to a buccal cavity and to oesophagus. Body is smooth, unsegmented without leg or other appendages. Plant parasitic nematodes mostly measure 300-1000 μm with some up to 4 mm long x 15-35 μm wide.

The females of some species become swollen at adult stage and have pear-shapes (pyriform) or spheroid bodies. Nematodes can be easily observed under microscope. A valve is located at the junction of oesophagus and the intestine, the latter opening into the rectum and anus at the posterior end of the body. The entire body is covered with a colourless, impermeable (permeable only to water) smooth or transversely striated cuticle with a sub-cuticular and muscular layer.

Life cycle and reproduction of nematodes: Nematodes produce eggs. Eggs hatch into larvae. Appearance and structure of larvae are usually similar to the adults. Larvae start to grow and each larval state is terminated by molt. Nematodes have four larval stages. Usually, first molt occurs in the egg and the final molt differentiates into adult male and female. Fertile eggs are produced by females after mating with a male, or parthenogenetically in absence of males or can produce sperm herself. A life cycle from egg to egg stage is completed with 3 to 4 weeks or requires slightly longer period in cooler temperature.

Behaviour of nematodes in soil: Except free living nematodes, all the plant parasitic nematodes complete a part of the life cycle in soil. Being soil-borne microfauna, the activities of these nematodes are affected by soil temperature, moisture, aeration, soil texture and pH, organic matter, rhizosphere and various cultural operations. Population of nematodes is high in soil layer of 0-15 cm depth and sometimes they can live upto the depth of 150 cm or more. Generally, concentration of nematodes is extremely high in the rhizosphere of susceptible host plants.

Movement of nematode in soil is very slow and a nematode can travel a maximum of one meter per season. However, they can move faster at certain soil moisture level when pores are lined with thin film of water under water logging conditions. In gall forming nematodes, temperature-moisture interaction determines the emergence of larvae from galls. Most nematodes eggs hatch freely in water in absence of any special stimulus. Nematodes are spread in a local areas by farm equipments, irrigation, flood or drainage water, animal feet and dust storms while spread to a longer distance through farm produce and nursery plants.

Nematodes as vector of plant pathogens: Some species of nematodes viz., dagger nematode (*Xiphinema* sp.), needle nematode (*Longidorus* spp. and *Paralongidorus* spp.) and stubby-root nematodes (*Trichodorus* spp. and *Paratrichodorus* spp.) can carry plant viruses. Members of *Longidorus* and *Xiphinema* (family Longidoridae) transmit the polyhedral nepoviruses (type member : tobacco ringspot virus) while *Trichodours* and *Paratrichodorus* (family Trichodoridae) transmit the straight tubular tobraviruses (type member: tobacoo rattle virus). Few examples of the nematodes as virus vectors are mentioned below:

S.No. Plant virus transmitted Nematode species involved as vector

- Arabis mosaic *Longidorus caespiticola, Paralongidorus maximus, Xiphinema index, Xiphenema* spp.
- Brome mosaic *L. macrosoma, X. coxi, X. diversicaudatum*
- Carnation ringspot *L. elongatus, X. diversicaudatum*
- Cowpea mosaic *X. basiri*
- Grapevine fan leaf *X. index* and *X. italiae*
- Grapevine vein banding *X. index*
- Mulberry ringspot *L. martini*
- Tobacco ringspot *X. americanum, X. coxi*
- Tomato ringspot *X. americanum, X. brevicolle*
- Tobacco rattle *Paratrichodorus minor, Trichodorus cylindricus, T. primitivus*

Phanerogamic Plant Parasites

Some flower and seed bearing higher plants (phanerogams) live parasitically on other living plants and can cause important diseases on agricultural crops and also in forest trees. They are classified in the following botanical families and genera,

- Cuscutaceae (stem parasite)
  Genus: *Cuscuta*, the dodders
- Viscaceae (stem parasites)

Genus: *Arceuthobium*, the dwarf mistletoes of conifers
Phoradendron, the American true mistletoes of broad leaved trees
Viscum, the European tree mistletoes
Dendrophthoe, the giant mistletoes

- Orobanchaceae (root parasite)
  Genus: *Orobanche*, the broomrapes
- Scrophulariaceae (root parasite)
  Genus: *Striga*, the witchweeds

Identifying Characters, Reproduction and Life Cycle

Dodder (Love-vine, Amarbel): Dodder is slender, twining plant. The stem is tough, succulent, threadlike, curling, leafless and bearing minute scales in place of leaves. Stem is usually yellowish or orange in colour. They grow and climb in abundance on the wild and cultivated plants. Their haustoria penetrate the stem or leave of the host plant and absorb foodstuffs and water.

Growth of the infected plants is suppressed and finally dies. Tiny whitish flowers arise in clusters from the stem and they produce numerous grey to reddish-brown seeds within few weeks of bloom. The seeds fall on the ground where they either germinate immediately or remain dormant until next season. The seeds may be spread to new areas by animals, water, equipments and by mixing with crop seeds.

Giant mistletoe (Loranthus): Mistletoes are semi-parasites of thee-trunks and branches. They have green leaves and many branches and hence grown like a small bush on the host.

They do not have true roots and hence develop haustoria to draw nutrients from the vascular system of the host plant. Flowers are long, tubular, greenish white to red in colour and borne in clusters. Seed is fleshy, contains one seed, sweet in taste and usually eaten by birds and animals. The parasite is spread mostly through birds. Droppings of bird containing seeds get deposited on the other trees, mainly at the junction of a branch and main trunk. These seeds germinate, develop haustoria and established on the host plant.

Broomrape: Broomrape is a complete root parasite. It affects tobacco mostly and many other Solanaceous and Cruciferous plants. The parasite has a stout, fleshy stem of 10-15 cm long. The stem is pale yellow or brownish red in colour and is covered by small, thin, browny, scaly leaves. Flowers are white and tubular and appear in the axil of leaves. Seeds are very small, black in colour and can remain viable in soil for several years. Roots are haustoria-like, can penetrate the roots of the host for nutrition. The affected plants become stunted and may die.

Witchweed: It is a semi root parasite but obligate in nature. Commonly affected plants by this parasitic plant are sugarcane, cereals, maize and millets.

Striga is a small plant with bright green, slightly hairy stem and leaves. The weed grows 15-60 cm high in clusters. Leaves are long and narrow in opposite pairs. Flowers are small, brick red or yellowish or almost white with yellow centers, appear throughout the season. Tiny brown seeds are produced in each capsule and thus a single plant can produce 50,000-500,000 seeds. The seeds can remain viable for 12-40 years. Roots are watery white in colour without root hairs and hence obtain nutrients by haustoria from the host plant. The life cycle from seed germination to first seed setting on the developed plant needs 90-120 days.

4

Investigation of Antioxidant Enzymes and Biochemical Techniques in Seed Production

In order to understand the true nature of a seed it is necessary to examine its origin and construction, and watch its development as far as possible from the earliest stages to the time when it gives rise to a completely formed young plant.

SEED

A seed is a small embryonic plant enclosed in a covering called the seed coat, usually with some stored food. It is the product of the ripened ovule of gymnosperm and angiosperm plants which occurs after fertilization and some growth within the mother plant. The formation of the seed completes the process of reproduction in seed plants (started with the development of flowers and pollination), with the embryo developed from the zygote and the seed coat from the integuments of the ovule. Seeds have been an important development in the reproduction and spread of flowering plants, relative to more primitive plants like mosses, ferns and liverworts, which do not have seeds and use other means to propagate themselves. This can be seen by the success of seed plants (both gymnosperms and angiosperms) in dominating biological niches on land, from forests to grasslands both in hot and cold climates. The term seed also has a general meaning that predates the above—anything that can be sown, *e.g.* "seed" potatoes, "seeds" of corn or sunflower "seeds". In the case of sunflower and corn "seeds", what is sown is the seed enclosed in a shell or hull, and the potato is a tuber.

SEED STRUCTURE

The embryo is an immature plant from which a new plant will grow under proper conditions. The embryo has one cotyledon or seed leaf in monocotyledons, two cotyledons in almost all dicotyledons and two or more in gymnosperms. The radicle is the embryonic root. The plumule is the embryonic shoot. The embryonic stem above the point of attachment of the cotyledon(s) is the epicotyl. The embryonic stem below the point of attachment is the hypocotyl.

A typical seed includes three basic parts:

1. An embryo,
2. A supply of nutrients for the embryo, and
3. A seed coat.

Within the seed, there usually is a store of nutrients for the seedling that will grow from the embryo. The form of the stored nutrition varies depending on the kind of plant. In angiosperms, the stored food begins as a tissue called the endosperm, which is derived from the parent plant via double fertilization. The usually triploid endosperm is rich in oil or starch and protein. In gymnosperms, such as conifers, the food storage tissue is part of the female gametophyte, a haploid tissue. In some species, the embryo is embedded in the endosperm or female gametophyte, which the seedling will use upon production.

In others, the endosperm is absorbed by the embryo as the latter grows within the developing seed, and the cotyledons of the embryo become filled with this stored food. At maturity, seeds of these species have no endosperm and are termed exalbuminous seeds. Some exalbuminous seeds are bean, pea, oak, walnut, squash, sunflower, and radish. Seeds with an endosperm at maturity are termed albuminous seeds. Most monocots (*e.g.* grasses and palms) and many dicots (*e.g.* brazil nut and castor bean) have albuminous seeds. All gymnosperm seeds are albuminous.

The seed coat (or testa) develops from the tissue, the integument, originally surrounding the ovule. The seed coat in the mature seed can be a paper–thin layer (*e.g.* peanut) or something more substantial (*e.g.* thick and hard in honey locust and coconut).

The seed coat helps protect the embryo from mechanical injury and from drying out. In addition to the three basic seed parts, some seeds have an appendage on the seed coat such an aril (as in yew and nutmeg) or an elaiosome (as in Corydalis) or hairs (as in cotton). There may also be a scar on the seed coat, called the hilum; it is where the seed was attached to the ovary wall by the funiculus.

Kinds of Seeds

Many structures commonly referred to as "seeds" are actually dry fruits. Sunflower seeds are sold commercially while still enclosed within the hard wall of the fruit, which must be split open to reach the seed. Different groups of plants have other modifications, the so–called stone fruits (such as the peach) have a hardened fruit layer (the endocarp) fused to and surrounding the actual seed. Nuts are the one–seeded, hard shelled fruit, of some plants, with an indehiscent seed, such as an acorn or hazelnut.

Seed Production

Seeds are produced in several related groups of plants, and their manner of production distinguishes the angiosperms ("enclosed seeds") from the gymnosperms ("naked seeds"). Angiosperm seeds are produced in a hard or fleshy structure called a fruit that encloses the seeds, hence the name. (Some fruits have layers of both hard and fleshy material). In gymnosperms, no special structure develops to enclose the seeds, which begin their development "naked" on the bracts of cones. However, the seeds do become covered by the cone scales as they develop in some species of conifer.

Seed production in natural plant populations vary widely from year–to–year in response to weather variables, insects and diseases, and internal cycles within the plants themselves. Over a 20–year period, for example, forests composed of loblolly pine and shortleaf pine produced from 0 to nearly 5 million sound pine seeds per hectare. Over this period, there were six bumper seeds, five poor seeds crops, and nine good seed crops, when evaluated in regard to producing adequate seedlings for natural forest reproduction.

Seed Development

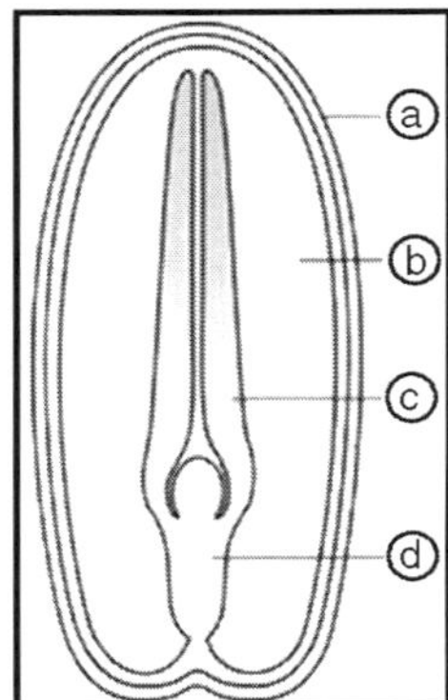

Fig. Diagram of the Internal Structure of a Dicot Seed and Embryo.

(a) Seed Coat,
(b) Endosperm,
(c) Cotyledon,
(d) Hypocotyl.

The seed, which is an embryo with two points of growth (one of which forms the stems the other the roots) is enclosed in a seed coat with some food reserves.

Angiosperm seeds consist of three genetically distinct constituents:

1. The embryo formed from the zygote,
2. The endosperm, which is normally triploid,
3. The seed coat from tissue derived from the maternal tissue of the ovule.

In angiosperms, the process of seed development begins with double fertilization and involves the fusion of the egg and sperm nuclei into a zygote. The second part of this process is the fusion of the polar nuclei with a second sperm cell nucleus, thus forming a primary endosperm. Right after fertilization, the zygote is mostly inactive but the primary endosperm divides rapidly to form the endosperm tissue.

This tissue becomes the food that the young plant will consume until the roots have developed after production or it develops into a hard seed coat. The seed coat forms from the two integuments or outer layers of cells of the ovule, which derive from tissue from the mother plant, the inner integument forms the tegmen and the outer forms the testa.

When the seed coat forms from only one layer it is also called the testa, though not all such testa are homologous from one species to the next. In gymnosperms, the two sperm cells transferred from the pollen do not develop seed by double fertilization but one sperm nucleus unites with the egg nucleus and the other sperm is not used. Sometimes each sperm fertilizes an egg cell and one zygote is then aborted or absorbed during early development. The seed is composed of the embryo (the result of fertilization) and tissue from the mother plant, which also form a cone around the seed in coniferous plants like Pine and Spruce.

The ovules after fertilization develop into the seeds; the main parts of the ovule are the funicle; which attaches the ovule to the placenta, the nucellus; the main region of the ovule were the embryo sac develops, the micropyle; A small pore or opening in the ovule where the pollen tube usually enters during the process of fertilization, and the chalaza; the base of the ovule opposite the micropyle, where integument and nucellus are joined together.

The shape of the ovules as they develop often affects the finale shape of the seeds. Plants generally produce ovules of four shapes: the most common shape is called anatropous, with a curved shape. Orthotropous ovules are straight with all the parts of the ovule lined up in a long row producing an uncurved seed. Campylotropous ovules have a curved embryo sac often giving the seed a tight "c" shape. The last ovule shape is called amphitropous, where the ovule is partly inverted and turned back 90 degrees on its stalk or funicle.

In the majority of flowering plants, the zygote's first division is transversely oriented in regards to the long axis, and this establishes the polarity of the embryo. The upper or chalazal pole becomes the main area of growth of the embryo, while the lower or micropylar pole produces the stalk–like suspensor that attaches to the micropyle. The suspensor absorbs and manufacturers nutrients from the endosperm that are utilized during the embryos growth.

The embryo is composed of different parts; the epicotyle will grow into the shoot, the radicle grows into the primary root, the hypocotyl connects the epicotyle and the radicle, the cotyledons form the seed leaves, the testa or seed coat forms the outer covering of the seed. Monocotyledonous plants like corn, have other structures; instead of the hypocotyle–epicotyle, it has a coleoptile that forms the first leaf and connects to the coleorhiza that connects to the primary root and adventitious roots form from the sides. The seeds of corn are constructed with these structures; pericarp, scutellum (single large cotyledon) that absorbs nutrients from the endosperm, endosperm, plumule, radicle, coleoptile and coleorhiza—these last two structures are sheath–like and enclose the plumule and radicle, acting as a protective covering. The testa or seed coats of both monocots and dicots are often marked with patterns and textured markings, or have wings or tufts of hair.

Seed Size and Seed Set

Seeds are very diverse in size. The dust–like orchid seeds are the smallest with about one million seeds per gram; they are often embryonic seeds with immature embryos and no significant energy reserves. Orchids and a few other groups of plants are myco–heterotrophs which depend on mycorrhizal fungi for nutrition during production and the early growth of the seedling. Some terrestrial Orchid seedlings, in fact, spend the first few years of their life deriving energy from the fungus and do not produce green leaves. At over 20 kg, the largest seed is the coco de mer.

Plants that produce smaller seeds can generate many more seeds per flower, while plants with larger seeds invest more resources into those seeds and normally produce fewer seeds. Small seeds are quicker to ripen and can be dispersed sooner, so fall blooming plants often have small seeds. Many annual plants produce great quantities of smaller seeds; this helps to ensure that at least a few will end in a favourable place for growth. Herbaceous perennials and woody plants often have larger seeds, they can produce seeds over many years, and larger seeds have more energy reserves for production and seedling growth and produce larger, more established seedlings after production.

Seed Functions

Seeds serve several functions for the plants that produce them. Key among these functions are nourishment of the embryo, dispersal to a new location, and dormancy during unfavourable conditions. Seeds fundamentally are a means of reproduction and most seeds are the product of sexual reproduction which produces a remixing of genetic material and phenotype variability that natural selection acts on.

Embryo Nourishment

Seeds protect and nourish the embryo or young plant. Seeds usually give a seedling a faster start than a sporeling from a spore, because of the larger food reserves in the seed and the multicellularity of the enclosed embryo.

SEED DISPERSAL

Unlike animals, plants are limited in their ability to seek out favourable conditions for life and growth. As a result, plants have evolved many ways to disperse their offspring by dispersing their seeds. A seed must somehow "arrive" at a location and be there at a time favourable for production and growth. When the fruits open and release their seeds in a regular way, it is called dehiscent, which is often distinctive for related groups of plants, these fruits include; Capsules, follicles, legumes, silicles and siliques. When fruits do not open and release their seeds in a regular fashion they are called indehiscent, which include the fruits achenes, caryopsis, nuts, samaras, and utricles. Seed dispersal is seen most obviously in fruits; however many seeds aid in their own dispersal. Some kinds of seeds are dispersed while still inside a fruit or cone, which later opens or disintegrates to release the seeds. Other seeds are expelled or released from the fruit prior to dispersal. For example, milkweeds produce a fruit type, known as a follicle, that splits open along one side to release the seeds. Iris capsules split into three "valves" to release their seeds.

By wind (anemochory):

- Many seeds (*e.g.* maple, pine) have a wing that aids in wind dispersal.
- The dustlike seeds of orchids are carried efficiently by the wind.
- Some seeds, (*e.g.* dandelion, milkweed, poplar) have hairs that aid in wind dispersal.

Some winged seeds have two, and some have only one wing.

By water (hydrochory):

- Some plants, such as Mucuna and Dioclea, produce buoyant seeds termed sea–beans or drift seeds because they float in rivers to the oceans and wash up on beaches.

By animals (zoochory):

- Seeds (burrs) with barbs or hooks (*e.g.* acaena, burdock, dock) which attach to animal fur or feathers, and then drop off later.
- Seeds with a fleshy covering (*e.g.* apple, cherry, juniper) are eaten by animals (birds, mammals, reptiles, fish) which then disperse these seeds in their droppings.
- Seeds (nuts) which are an attractive long–term storable food resource for animals (*e.g.* acorns, hazelnut, walnut); the seeds are stored some distance from the parent plant, and some escape being eaten if the animal forgets them.

MYRMECOCHORY

Myrmecochory is the dispersal of seeds by ants. Foraging ants disperse seeds which have appendages called elaiosomes (*e.g.* bloodroot, trilliums, Acacias, and many species of Proteaceae). Elaiosomes are soft, fleshy structures that contain nutrients for animals that eat them. The ants carry such seeds back to their nest, where the elaiosomes are eaten. The remainder of the seed, which is hard and inedible to the ants, then germinates either within the nest or at a removal site where the seed has been discarded by the ants.

This dispersal relationship is an example of mutualism, since the plants depend upon the ants to disperse seeds, while the ants depend upon the plants seeds for food. As a result, a drop in numbers of one partner can reduce success of the other. In South Africa, the Argentine ant (Linepithema humile) has invaded and displaced native species of ants. Unlike the native ant species, Argentine ants do not collect the seeds of Mimetes cucullatus or eat the elaiosomes. In areas where these ants have invaded, the numbers of Mimetes seedlings have dropped.

Seed Dormancy

Seed dormancy has two main functions: the first is synchronizing production with the optimal conditions for survival of the resulting seedling; the second is spreading production of a batch of seeds over time so that a catastrophe after production (*e.g.* late frosts, drought, herbivory) does not result in the death of all offspring of a plant (bet–hedging). Seed dormancy is defined as a seed failing to germinate under environmental conditions optimal for production, normally when the environment is at a suitable temperature with proper soil moisture. This true dormancy or innate dormancy is therefore caused by conditions within the seed that prevent production. Thus dormancy is a state of the seed, not of the environment. Induced dormancy, enforced dormancy or seed quiescence occurs when a seed fails to germinate because the external environmental conditions are inappropriate for production, mostly in response to conditions being too dark or light, too cold or hot, or too dry.

Seed dormancy is not the same as seed persistence in the soil or on the plant, though even in scientific publications dormancy and persistence are often confused or used as synonyms. Often seed dormancy is divided into four major categories: exogenous; endogenous; combinational; and secondary. A more recent system distinguishes five classes ofdormancy: morphological, physiological,morphophysiological, physical and combinational dormancy.

Exogenous dormancy is caused by conditions outside the embryo including:

- Physical dormancy or hard seed coats occurs when seeds are impermeable to water. At dormancy break a specialized structure, the 'water gap', is disrupted in response to environmental cues, especially temperature, so that water can enter the seed and production can occur. Plant families where physical dormancy occurs include Anacardiaceae, Cannaceae, Convulvulaceae, Fabaceae and Malvaceae.
- Chemical dormancy considers species that lack physiological dormancy, but where a chemical prevents production. This chemical

can be leached out of the seed by rainwater or snow melt or be deactivated somehow. Leaching of chemical inhibitors from the seed by rain water is often cited as an important cause of dormancy release in seeds of desert plants, however little evidence exists to support this claim.

Endogenous dormancy is caused by conditions within the embryo itself, including:

- Morphological dormancy where production is prevented due to morphological characteristics of the embryo. In some species the embryo is just a mass of cells when seeds are dispersed, it is not differentiated. Before production can take place both differentiation and growth of the embryo have to occur. In other species the embryo is differentiated but not fully grown (underdeveloped) at dispersal and embryo growth up to a species specific length is required before production can occur. Examples of plant families where morphological dormancy occurs are Apiaceae, Cycadaceae, Liliaceae, Magnoliaceae and Ranunculaceae.
- Morphophysiological dormancy seeds with underdeveloped embryos, and which in addition have physiological components to dormancy. These seeds therefore require a dormancy–breaking treatments as well as a period of time to develop fully grownem bryos. Plant families where morphophysiological dormancy occurs include Apiaceae, Aquifoliaceae, Liliaceae, Magnoliaceae, Papaveraceae and Ranunculaceae. Some plants with morphophysi-ological dormancy like Asarum or Trillium species have multiple types of dormancy, one affects radicle (root) growth while the other affects plumule (shoot) growth. The terms "double dormancy" and "2–year seeds" are used for species whose seeds need two years to complete production or at least two winters and one summer. Dormancy of the radicle (seedling root) is broken during the first winter after dispersal while dormancy of the shoot bud is broken during the second winter.
- Physiological dormancy means that the embryo can, due to physiological causes, not generate enough power to break through the seed coat, endosperm or other covering structures. Dormancy is typically broken at cool wet, warm wet or warm dry conditions. Abscisic acid is usually the growth inhibitor in seeds and its production can be affected by light.
  - Drying; some plants including a number of grasses and those from seasonally arid regions need a period of drying before they will germinate, the seeds are released but need to have a lower moisture content before production can begin. If the seeds remain moist after dispersal, production can be delayed for many

months or even years. Many herbaceous plants from temperate climate zones have physiological dormancy that disappears with drying of the seeds. Other species will germinate after dispersal only under very narrow temperature ranges, but as the seeds dry they are able to germinate over a wider temperature range.

- Combinational dormancy in seeds with combinational dormancy the seed or fruit coat is impermeable to water and the embryo has physiological dormancy. Depending on the species physical dormancy can be broken before or after physiological dormancy is broken.
- Secondary dormancy is caused by conditions after the seed has been dispersed and occurs in some seeds when non–dormant seed is exposed to conditions that are not favourable to production, very often high temperatures. The mechanisms of secondary dormancy are not yet fully understood but might involve the loss of sensitivity in receptors in the plasma membrane.

The following types of seed dormancy do not involve seed dormancy strictly spoken as lack of production is prevented by the environment not by characteristics of the seed itself:

- Photodormancy or light sensitivity affects production of some seeds. These photoblastic seeds need a period of darkness or light to germinate. In species with thin seed coats, light may be able to penetrate into the dormant embryo. The presence of light or the absence of light may trigger the production process, inhibiting production in some seeds buried too deeply or in others not buried in the soil.
- Thermodormancy is seed sensitivity to heat or cold. Some seeds including cocklebur and amaranth germinate only at high temperatures (30 °C or 86 °F) many plants that have seed that germinate in early to mid summer have thermodormancy and germinate only when the soil temperature is warm. Other seeds need cool soils to germinate, while others like celery are inhibited when soil temperatures are too warm. Often thermodormancy requirements disappear as the seed ages or dries.

Not all seeds undergo a period of dormancy. Seeds of some mangroves are viviparous, they begin to germinate while still attached to the parent. The large, heavy root allows the seed to penetrate into the ground when it falls. Many garden plants have seeds that will germinate readily as soon as they have water and are warm enough, though their wild ancestors may have had dormancy, these cultivated plants lack seed dormancy.

After many generations of selective pressure by plant breeders and gardeners dormancy has been selected out. For annuals, seeds are a way for the species to survive dry or cold seasons. Ephemeral plants are usually annuals that can go from seed to seed in as few as six weeks.

Seed Production

Seed production is a process by which a seed embryo develops into a seedling. It involves the reactivation of the metabolic pathways that lead to growth and the emergence of the radicle or seed root and plumule or shoot. The emergence of the seedling above the soil surface is the next phase of the plants growth and is called seedling establi-shment.

Three fundamental conditions must exist before production can occur:

1. The embryo must be alive, called seed viability.
2. Any dormancy requirements that prevent production must be overcome.
3. The proper environmental conditions must exist for production.

Seed viability is the ability of the embryo to germinate and is affected by a number of different conditions. Some plants do not produce seeds that have functional complete embryos or the seed may have no embryo at all, often called empty seeds. Predators and pathogens can damage or kill the seed while it is still in the fruit or after it is dispersed. Environmental conditions like flooding or heat can kill the seed before or during production. The age of the seed affects its health and production ability: since the seed has a living embryo, over time cells die and cannot be replaced. Some seeds can live for a long time before production, while others can only survive for a short period after dispersal before they die.

Seed vigour is a measure of the quality of seed, and involves the viability of the seed, the production percentage, production rate and the strength of the seedlings produced. The production percentage is simply the proportion of seeds that germinate from all seeds subject to the right conditions for growth. The production rate is the length of time it takes for the seeds to germinate. Production percentages and rates are affected by seed viability, dormancy and environmental effects that impact on the seed and seedling. In agriculture and horticulture quality seeds have high viability, measured by production percentage plus the rate of production.

This is given as a per cent of production over a certain amount of time, 90% production in 20 days, for example. 'Dormancy' is covered above; many plants produce seeds with varying degrees of dormancy, and different seeds from the same fruit can have different degrees of dormancy. It's possible to have seeds with no dormancy if they are dispersed right away and do not dry (if the seeds dry they go into physiological dormancy). There is great variation amongst plants and a dormant seed is still a viable seed even though the production rate might be very low. Environmental conditions effecting seed production include; water, oxygen, temperature and light. Three distinct phases of seed production occur: water imbibition; lag phase; and radicle emergence. In order for the seed coat to split, the embryo must imbibe (soak up water), which causes it to swell, splitting the seed coat. However, the nature of the seed coat determines how rapidly water can penetrate and subsequently initiate production. The rate of imbibition is dependent on the permeability of the seed coat, amount of water in the environment and the area of contact the seed has to the source of water. For some seeds, imbibing too much water too quickly can kill the seed. For some seeds, once water is imbibed the production process cannot be stopped, and drying then becomes fatal. Other seeds can imbibe and lose water a few times without causing ill effects, but drying can cause secondary dormancy.

Inducing Production

A number of different strategies are used by gardeners and horticulturists to break seed dormancy. Scarification which allows water and gases to penetrate into the seed, include methods that physically break the hard seed coats or soften them by chemicals.

Means of scarification include soaking in hot water or poking holes in the seed with a pin or rubbing them on sandpaper or cracking with a press or hammer. Soaking the seeds in solvents or acids is also effective for many seeds. Sometimes fruits are harvested while the seeds are still immature and the seed coat is not fully developed and sown right away before the seed coat become impermeable. Under natural conditions seed coats are worn down by rodents chewing on the seed, the seeds rubbing against rocks (seeds are moved by the wind or water currents), by undergoing freezing and thawing of surface water, or passing through an animal's digestive tract.

In the latter case, the seed coat protects the seed from digestion, while often weakening the seed coat such that the embryo is ready to sprout when it gets deposited (along with a bit of fertilizer) far from the parent plant. Microorganisms are often effective in breaking down hard seed coats and are sometimes used by people as a treatment, the seeds are stored in a moist warm sandy medium for several months under non–sterile conditions. Stratification also called moist–chilling is a method to break down physiological dormancy and involves the addition of moisture to the seeds so they imbibe water and then the seeds are subject to a period of moist chilling to after–ripen the embryo. Sowing outside in late summer and fall and allowing to overwinter outside under cool conditions is an effective way to stratify seeds, some seeds respond more favourably to periods of oscillating temperatures which are part of the natural environment. Leaching or the soaking in water removes chemical inhibitors in some seeds that prevent production. Rain and melting snow naturally accomplish this task. For seeds planted in gardens, running water is best—if soaked in a container, 12 to 24 hours of soaking is sufficient. Soaking longer, especially in stagnant water that is not changed, can result in oxygen starvation and seed death. Seeds with hard seed coats can be soaked in hot water to break open the impermeable cell layers that prevent water intake.

Other methods used to assist in the production of seeds that have dormancy include prechilling, predrying, daily alternation of temperature, light exposure, potassium nitrate, the use of plant growth regulators like gibberellins, cytokinins, ethylene, thiourea, sodium hypochlorite plus others. Some seeds germinate best after a fire, for some seeds fire cracks hard seed coats while in other seeds chemical dormancy is broken in reaction to the presence of smoke, liquid smoke is often used by gardeners to assist in the production of these species.

Origin and Evolution

The origin of seed plants is a problem that still remains unsolved. However, more and more data tends to place this origin in the middle Devonian. The description in 2004 of the proto–seed Runcaria heinzelinii in the Givetian of Belgium is an indication of that ancient origin of seed–plants. As with modern ferns, most land plants before this time reproduced by sending spores into the air, that would land and become whole new plants. The first "true" seeds are described from the upper Devonian, which is probably the theater of their true first evolutionary radiation. The seed plants progressively became one of the major elements of nearly all ecosystems.

PRODUCTION

When the pod of the bean is developing, the embryo in the seed is being fed by the parent and visibly grows until ripeness is attained. The young plant then assumes a dormant or resting state within the seed without showing any signs of life.

Under certain conditions, however, the plantlet begins to wake up, and soon escapes from its protective coat to lead a separate and independent life. This awakening from a resting condition to a state of active growth is called production and is dependent upon an adequate supply of

- Water
- Heat
- Air or oxygen.

It is also essential, of course, that the plantlet in the seed must be alive. The exact nature of the dormant state of seeds is not understood, but in old seeds and those which are gathered in an immature condition or badly stored the embryo is often weakened or actually dead; in the latter case no production is possible.

The exact length of time which seeds may be kept before death of the embryo takes place has never been satisfactorily determined; it varies with the species of the seed, its ripeness and composition, and also with the method of storage. In the case of most farm and garden seeds kept in the ordinary way, few of them are found capable of growth after ten years, and a large number die in two or three years.

For present purposes it will suffice merely to mention that age is a determining factor in the production of seeds. Water is necessary is well known, as beans may be kept indefinitely in a sack or drawer at various temperatures and with access to air without production taking place.

When placed in moist ground, or between damp blotting-paper, they absorb water very readily. This is most easily observed when beans are soaked for twelve hours in a dish containing water. The water is transmitted through all parts of the coat, but much more quickly and easily through the micropyle and the line of softer material which runs the whole length of the centre of the hilum.

It is rapidly brought into contact with the part of the embryo which grows first, namely, the radicle. The soft spongy thicker part of the inside of the testa lying beneath the hilum stores up a considerable amount of water for the benefit of the developing plant, and the whole of the embryo and the seed-coat absorb water and become softer and larger in consequence; it is only after this swelling has happened that a bean begins to show any signs of production.

The need of an adequate temperature for production is a matter of common knowledge among those accustomed to sow seeds. If soaked beans are placed in the ground in midwinter they show little or no signs of waking from their dormant condition, yet when placed under a glass on damp blotting-paper indoors, the radicle makes its exit from the seed in a few days.

Seeds differ in the temperature which is necessary to induce them to germinate, the embryos in some commence to extend their radicles and push their way through the seed-coat even if just kept above freezing point; others require a temperature of 9° or 10°C to start growth.

If attempts are made to grow beans at 45°C it will be found to be too hot, and they make little or no progress. Between this high temperature at which growth appears impossible, and the freezing point where the development of the embryo of the bean is also suspended, there is a temperature at which the embryo makes the most rapid progress, and emerges from the seed-coat in the shortest possible time; this most favourable temperature is about 28°C, both above it and below it the production of the bean is retarded. The supply of fresh air is also an essential condition for growth of the young plant from the bean seed, but the evidence for its need is not so manifest or so generally recognised as the necessity for moisture and warmth.

It will be found, however, when beans are placed in a flask or bottle containing carbon dioxide or hydrogen gas they refuse to germinate, even when they are supplied with a proper amount of water. The peculiar extension or growth of the parts of the interior of the bean seed, and the fact that a suitable supply of water, air and heat is necessary for the manifestation of these changes, suggests to us that we are dealing with a living structure.

This becomes all the more«apparent when we observe that the oxygen of the air is absorbed, and in its place carbon dioxide is given off into the surrounding air, for this is what happens in the breathing of a living animal. Carbon dioxide is produced when beans germinate.

Place twenty soaked beans in a wide-mouthed bottle, and cork them up after showing that a match burns freely in the bottle. Leave them in a warm place for twenty-four hours, and try if a match will now burn in the bottle. The carbon dioxide gas can be poured out into a beaker containing lime water; on shaking, its presence is proved by the lime water becoming 1 milky ' owing to the precipitation of carbonate of lime. The particular use of the water, heat and air to the plant we cannot at present discuss.

Without water the embryo would have little chance of becoming free from the tough and hard seed-coat surrounding it; water softens the latter, and makes it more easily torn by the extending radicle and plumule. In the early stages of the life of the bean plant, from the commencement of production up to the time when the first green leaves are unfolded, the development and building up of the elongating rootlet and shoot depend upon the thick cotyledons.

At first the latter are thick and fleshy, but as the radicle and plumule grow the cotyledons become softer and thinner, ultimately shrivelling considerably. The cotyledons are leaves, the interior of which is packed with food for the rest of the growing embryo, and a large amount of the water absorbed by the seed is used for the purpose of dissolving the nutrient material in them, and carrying it from them to the various parts of the root and shoot of the young plant where growth is going on the cotyledons are esential to the development of the root and shoot of the embryo by cutting them off as soon as the two latter parts have emerged from the seed-coat. Try separating one cotyledon and then two at various stages of development, and see if the axis (root and shoot) can be made to develop without them.

The growth should be allowed to continue some time in order to obtain well-marked effects. Not only do the changes observed innhe embryo of a germinating bean point to the conclusion that it is a living structure and like an animal dependent on a proper supply of water, heat and air for the manifestation of its life, but the parts of a young bean plant after emerging from the seed soon give evidence of the possession of peculiarities which are associated with life. When put in the ground, the radicle, in coming out of the seed, turns straight downwards and continues to grow in this direction. This is the case no matter in what position the seed is placed. If, after production has commenced, it is taken and replanted with the primary root pointing to the surface of the soil, the tip of the root soon begins to curve downwards again, and will maintain this course until again disturbed.

The plumule behaves in exactly the opposite manner; after emerging from the seed-coat its bent tip grows upwards and away from the root; if the seed is reversed and replanted the plumule begins to curve in such a manner that its tip is driven upwards towards the surface of the ground.

That these peculiarities are somehow connected with life is clear, as dead embryos show no such behaviour. Sow soaked beans in a flower pot or box filled with ordinary garden soil placing them in various positions in it, some laid on the flat side, some with the hilum directed upwards, and others with the hilum downwards. Allow them to grow in a warm place: take them up as soon as signs of production are noticed, and observe the direction the root and shoot have taken. The peculiar tendency for the root always to go downwards and stem upwards can be investigated by sowing beans in ordinary garden soil and afterwards reversing them.

To avoid error all should be taken up, and then placed again in the soil in various positions—some as they were, a few with their roots and stems reversed, and others laid in a horizontal position. They may be re-examined at the end of a week. When the roots have extended about half an inch take two seeds and suspend them by means of thread side by side in a bottle with their roots downwards and stem upwards. The bottle should contain a little water to keep the air damp.

When the roots have grown about two inches reverse one of the seeds so that its root points upwards tad stem downwards. Notice that the tip of the root of the reversed seed in about twelve hours begins to turn downwards, while the plumule more slowly bends in such a way as to assume the position it had before it was reversed.

The bottle should be placed in a dark box or cupboard to avoid the influence of light on the plant, and fresh air should be blown into the bottle twice a day. Although seeds vary almost indefinitely in regard to size and shape they are similar to the bean in so far as they all contain a young plant packed away within the seed-coats.

In this essential feature all seeds agree with few exceptions, and it is on account of the existence of a young plant within them that they are of use in the raising of crops or plants. The manner, however, in which the embryo is arranged, and the relative size and appearance of its various parts, differ considerably in seeds; moreover, the growth during and after production is not the same in all cases. A few of the more important and common variations in these respects must be noticed.

THE COMMON BEAN

A broad bean is one of the largest seeds met with in ordinary farm or garden practice, and as its parts are all sufficiently large to be observed without the special aid of anything more than an ordinary pocket lens, it is especially fitted for study. When a nearly ripe pod of a broad bean plant is opened, each seed within it is found attached to. The inside by means and it is through this stalk that all the nourishment passes from the parent to enable the young seed to develop. A1 first the pod exists in a rudimentary form in the centre of a flower and its parts and contents are very small; they are nevertheless readily seen with a pocket lens. After the fading of the flower, the pod and seeds within it grow larger and larger at the expense of food supplied by the rest of the plant, and ultimately when ripe the funicles wither and dry up, and the seeds become detached from the parent which has produced them. When dry and ripe each bean seed is hard, with an uneven surface, but its internal construction cannot be clearly examined in this condition.

On soaking in water for twelve hours, however, it becomes softer, and the parts can then be easily investigated. The outside, which is a pale buff colour, is smooth, and has at one end a narrow elongated black scar called the hilum of the seed. It is known popularly as the ' eye' of the bean, and marks the place where the broad end of the funicle separated from the seed when it ripened in the pod. Quite close to one end of the hilum is a very minute hole known as the micropyh, easily seen with a lens, and through which water oozes out usually accompanied by bubbles of air when soaked beans are squeezed between the finger and thumb.

This opening communicates with the interior of the seed, and is the only one it possesses. The rest of the seed after the testa is removed, is of oval flattened shape similar to the complete bean, and is divisible into two large fleshy halves called cotyledons which, however, are not completely separate from each other, but connected at the side with a conical projecting body, one end of which is found to fit into a hollow cavity in the seed-coat exactly opposite the micropyle; the other end is bent and turned inwards between the fleshy cotyledons.

The extent and shape of this small curved structure is most easily observed when one of the cotyledons is removed completely; it remains attached to the other. Soak some broad beans in water and keep them in a warm place all night. Examine them next day and make drawings of the various parts seen both before and after stripping off the testa. Observe the relative position of the parts of the embryo in reference to each other and to the seed-coat.

Examine and compare the structure of the following seeds after soaking in the same way:—Pea, scarlet runner beans, vetches, and red clover. The bean seed contains nothing more than what has already been described; the nature and relationship of its component parts only become intelligible when the seed is placed in the ground or maintained under certain conditions, and allowed to grow.

When growth commences the lower end of the small curved structure elongates and breaks its way through the coat of the seed at a point very close to the micropyle, but not, as often erroneously stated, through the micropyle itself. It soon assumes the form and is recognised as a root of a young bean plant The upper bent half, which lies between the cotyledons, also pushes its way out of the same opening in the seed-coat and develops into a stem, from the tip of which leaves are gradually unfolded.

It is thus seen that the seed of a broad bean is a packet containing a bean plant in a rudimentary condition. This plantlet is callai an embryo, and the portion of it which becomes root and stem is its primary axis.

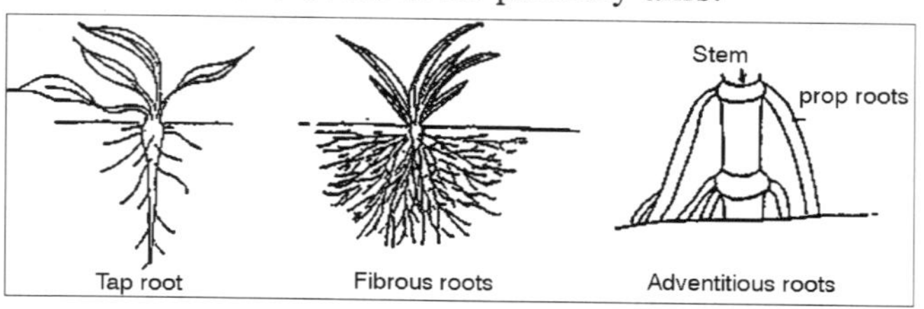

Fig. Root System

The part of the primary axis which is below the point where the cotyledons are attached consists of a very small piece of stem, the hypocotyl at the end of which is the radicle or primary root. Where the stem ends and the root begins cannot be determined in the bean seedling without the aid of the microscope and examination of the internal structure of the axis of the plant. The curved end of the primary axis above the cotyledons is the plumule of the embryo, and consists of a very short piece of stem, the tpicotyl on the top of which is a bud.

From the latter is derived the ordinary stem which comes above ground with its green leaves and flowers. In the early stages of the growth of the embryo from the seed the hypocotyl grows very little. The upper part of the stem bearing the plumule comes out of the seed bent and it maintain this curved shape for some time after emerging. By this behaviour the delicate leaves of the plumule are protected from injury during their progress upwards when a seed is sown in earth or sand. Fold up some soaked beans in two thicknesses of white flannel made damp, and place them on a plate. Cover them with another plate placed upside down, and leave them in a warm room.

Examine them twice a day, leaving them, exposed to the fresh air for a few minutes each time, and keeping the flannel damp, not wet. When they sprout notice the place where the radicle has come out of the seed-coat. Let some grow till the radicle and plumule are well out of the seed, and compare the various parts of the sprouted seeds with unsprouted ones.

SEED OF WHITE MUSTARD

The seed of white mustard (Brassica alba Boiss.) contains an embryo which like that of the bean consists of a radicle, plumule and two cotyledons; the latter, which are folded together, are relatively thinner than those of the bean and deeply notched. The radicle is bent round and lies in the fold of the cotyledons, between which is the very small, almost invisible plumule. On production the cotyledons, instead of remaining within the seed-coat and below the ground as in the case of the broad bean, escape from the enclosing coats altogether and grow up out of the ground, enlarging at the same time, and becoming green like ordinary leaves.

They are the first ' smooth' leaves of the seedling mustard plant. After a short time the plumule grows up from between the cotyledons and forms a stem upon which are gradually unfolded the ordinary divided ' rough' leaves. Soak white mustard seeds, and examine their structure, noting especially the way in which the embryos are packet in them.

Allow some to germinate and grow for a week or more on damp flannel, and examine them in various stages of development, noting the notched cotyledons with small. The term hypogean is applied to cotyledons which remain below ground, those coming above being epigean, the relative amount of growth in the hypocotyl and epicotyl determining their position. If the hypocotyl grows vigorously during or after production the cotyledons are forced above ground; when only the epicotyl grows the plumule is lifted up above the soil, but the cotyledons remain below where the seed is placed. In the broad bean the hypocotyl is very short, and the point where it ends and the root begins is not clearly denned. In a mustard seedling, however, the point of separation between the root and stem is somewhat swollen and readily distinguished.

All plants whose embryos, like those of the bean and mustard, possess two cotyledons, are known as Dicotyledons they form a very large, well-marked class of the flowering or seed-bearing plants. The seeds contain within their coats nothing but an embryo plant, which depends for the development of its root and shoot upon the substances stored up in some part of its body, its cotyledons chiefly.

This is true even in the case of seeds like those of white mustard, in which the cotyledons of the embryo are comparatively thin. There are, however, a number of plants, such as the ash, mangel and potato, which, although belonging to the Dicotyledons, have seeds in which there are stores of food inside the seed-coat, but free from the embryo and its cotyledons.

Such separate reserve-food is stored in that part of the seed known as the endosperm or 'albumen,1 and seeds in which it is present are called endospermous or albuminous seeds. Those like the bean, pea, and vetch, mustard, and turnip, which have no separate reserve-food, are known as exendospermous or exalbuminous seeds.

Take out a seed from the fruit of the ash tree in autumn; carefully cut thin shavings from the flat side of the seed, stafting about the middle of the seed and cutting towards the narrow end. Note the white embryo with its well-marked radicle, hypocotyl and two flat cotyledons lying within the semi-transparent endosperm. Some of the most commonly occurring endospermous seeds will be found to have embryos within them which are not dicotyledonous, and whose structure is in many respects. A good example is met with in the onion.

WHEAT GRAIN

A wheat grain, which may be taken as an example, is not a seed, but a kind of nut with a single seed within it. The seed grows in such a way as to completely fill up the interior of the nut, and become practically united with its inside wall. The embryo occupies only a small part of the grain, the rest being taken up by the floury endosperm of the seed. The embryo is easily seen at the base of a soaked grain on the side opposite the furrow.

When removed it has the appearance. The part of it which lies close up to the endosperm is a flattened somewhat fleshy shield-shaped structure called the scutcllum attached to the front of the scutellum is the plumule, consisting of a bud formed of an extremely short stem, upon which are sheath-like leaves enclosing each other.

The embryo generally possesses five roots, one primary and two pairs of secondary. They are all completely enclosed by a sheath which continuous wjtn the scutellum, and are consequently not visible from outside; their position, however, is marked by projecting bosses. The sheath round the roots is termed the coleorhiza, and when production takes place it expands and bursts the coats of the grain, the roots about the same time breaking through the enclosing coleo-rhiza.

When a wheat grain is sown in the ground it remains there, but the plumule grows upwards, its first leaf, the coleoptile, appearing above the soil as a single pale tube-like structure; from a slit in the tip of the latter the first flat green blade soon appears, and is followed by a succession of single green leaves, the younger ones growing from within the older ones in regular order. With a sharp knife or razor cut through from back to front, so as to divide the grain into two longitudinal halves, and note the floury endosperm and the shape and parts of the divided embryo. Place a folded sheet of damp blotting paper on a plate, sow some soaked wheat grains on it, and cover with a tumbler. The grains will germinate. Watch their development up to the time the first green leaf appears, taking out the embryo and examining it at different stages of its growth. There is difference of opinion as to which part of the embryo is to be considered the cotyledon.

Soine authorities regard the scutellum as the cotyledon, while others give this name to the coleoptile or first sheathing leaf which comes above ground, and which has no green blade. Others, again, consider that the first sheathing leaf is an extension of the scutellum, and the two combined is therefore the cotyledon.

In any case, there is only one cotyledon present, and wheat therefore belongs to the class of monocotyledonous plants. During the growth of the embryo of a wheat grain, it will be noticed that the endosperm becomes soft and decreases in quantity as the roots and plumule expand and develop; the endosperm is the food upon which the young plant depends during the early staggs of its life, the scutellum acting as a structure for changing, absorbing, and transferring this reserve-food to the growing parts which need it. Remove the embryos from well-soaked grains, and grow them without the endosperm on damp blotting-paper.

Allow ordinary uninjured grains to grow with them. Both the embryos in the grains and those removed from the grains develop, but there is a great difference in the results after a few days.

The store of reserve-food on which the young plant depends for its early development is sufficient to enable it to form a root, stem, and several leaves, as is evident when seeds are allowed to germinate upon damp flannel or blotting-paper, from which nothing but water is absorbed. No food-materials or manures are needed for this primary development, and seeds germinate and the seedlings grow for a considerable time as well in poor soil or sand as in good rich ground.

As soon as the reserve store is exhausted hunger becomes apparent, and unless the plants are then supplied with suitable nutriment from the soil and air, and are also placed under conditions favourable for growth, weakness and death are likely to occur.

Among the larger seeds, such as beans and peas, where there is an abundant store of reserve-food, the young seedlings begin to manufacture food for themselves from materials absorbed from the soil and air, long before their reserve is exhausted.

In small seeds the reserve is sometimes almost consumed before the roots and leaves are sufficiently developed to carry on their work properly, in which cases a more or less temporary starvation and check to growth ensues. Especially does this happen when seeds are sown too deeply, for a large amount of food is then used in the production of a stem long enough to lift the leaves up into the air.

ONION

The seed is black, somewhat oval in outline, with one side convex, the other almost flat. Each contains within it endosperm and an embryo which lies curled up inside in the form seen. When production commences, the curved part imbedded in the middle of the endosperm grows and forces the end of the embryo out of the seed.

From this exposed end, which is the radicle, a straight, slender, primary root develops. The part of the young seedling which extends from the root into the interior of the seed, grows very rapidly at first, at the same time assuming a sharply bent outline.

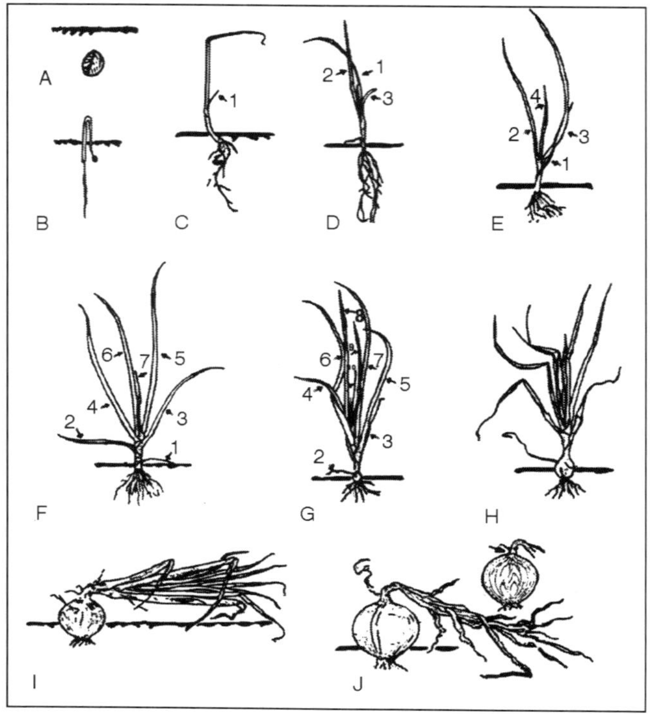

Fig. Production of Onion

It comes above ground in the form of a close loop but on further growth the end within the seed is pulled out of the soil, and grows up in the air. The tip within the seed changes and absorbs the endosperm, and usually remains there until all the nutrient material has been transferred from it to the various centres of growth in the young plant. After the food-reserve is exhausted the tip withers and becomes free from the seed coat.

In loose soils the latter is pulled above ground before the endosperm is exhausted, and remains on the end of the tip for some time. In other cases where the soil is damper and of a stiffer nature the seed-coat remains below ground altogether. The curved part of the embryo which comes above ground is a leaf. It is the cotyledon of the embryo, and is in reality a thin hollow leaf like those of the full grown onion plant: within it is the plumule, which consists of a series of hollow conical leaves arranged one inside the other.

After one leaf emerges others soon follow, the younger ones coming out in regular order through slits in the sides of those immediately older than themselves. Soak fresh onion seeds in water for a few hours. With a razor cut through some parallel to their flat sides in order to show the embryo within.

Sow others in damp blotting-paper; allow them to germinate and the seedlings to develop; make observations of them at different stages of growth.

Watch the production of seeds sown in boxes or pots containing ordinary garden soil. Plants whose embryos possess only one cotyledon are known as Monocotyledons, and form the second large class of seed-bearing plants. Few of the representatives of this class with which we are ordinarily familiar have true seeds large enough for examination.

The onion is probably one of the best commonly occurring examples which may be considered typical of the monocotyledons and easily obtainable. To this class, however, belong all the grasses, but their seeds and embryos are so different in many respects from those of the onion that it is necessary to examine one of them in detail.

PRIMARY AND SECONDARY BOOTS

It was noticed, when dealing with the bean seedling, that its two ends always grow in opposite directions; the plantlet can be considered as an elongated axis, one end of which bears the leaves and invariably comes above ground, while the opposite end never bears leaves, and persistently follows the plumb-line downwards. The descending part is known as the root. As will be pointed out later, all roots do not behave in this manner, and it is to be specially noted that many of the underground parts of plants are not roots; the exceptions, however, may be left for future consideration. The first or primary root which the bean plant possesses is merely an extension of the radicle of the embryo which exists nrithin the seed.

Soon after making its exit from the seed, it takes a downward course, and elongates by growth taking place near its tip. Another point to be noted is that the lateral roots do not arise as up-growths on the surface of the primary root, but come from within it, and are described as endogenous. The slits which they make in the substance of the primary root, and through which they emerge, can be readily seen in a bean seedling.

The secondary lateral roots are connected with its central more solid core; the three lowest ones, although they have just begun to grow, have not yet penetrated the outer layer of the root, and would not be visible on the outside of the latter. This mode of origin is characteristic of lateral roots generally wherever they are met with. Many dicotyledonous plants have roots similar to those of the bean plant.

When, as in this case, the primary one continues to grow, keeping distinctly larger than the lateral ones, it is called a tap root. Very good examples are met with among cultivated plants in the carrot, mangel, red clover, and mustard in shep-herd's-purse, poppy, and many other weeds, as well as in most broad-leaved trees. A number of plants have swollen fleshy roots in which food materials are stored for future use; they are described as tuberous roots, and must be distinguished from tubers, which are fleshy underground stems. To designate the different forms of thickened roots various special terms are in use. The typical carrot root is conical; that of the turnip napiform. The root of the radish is spoken of as fusiform.

In some instances the primary root is soon rivalled in size by its branches; it may even cease growth altogether. Such plants, on being pulled out of the ground, exhibit a bunch of slender roots, the chief of which are all much the same diameter and length; roots of this character are described as fibrous and are well exemplified in common groundsel and grasses.

ADVENTITIOUS ROOTS OF MONOCOTYLEDONOUS PLANTS

The roots of monocotyledonous plants differ in their development from those of dicotyledons. The single primary root of the onion, lasts but a short time, and is succeeded by others which do not arise as branches upon the primary one, but spring from the very short stem of the plant. Roots which arise on stems and leaves, or on various parts of the roots of plants, but not in acropetal succession are described as adventitious. They are very common upon all monoco-tyledonous plants of the farm and garden, and may be considered the chief ones which such plants possess.

In wheat, the embryo within the grain possesses three roots; in barley, five or six. Young barley plant show- use during the earlier stages of growth. Although common upon monocotyledons adventitious roots are not confined to this class. Examples are abundant upon many kinds of dicotyledonous plants.

Good instances are met with on the underground stems of mint, potato and hop, and on the runners of the strawberry, stems of creeping crowfoot and white clover, as well as many others. They are generally produced at the joints where leaves grow upon the stem, and may arise in some plants from internal causes apart from any external influences; in other instances their development depends upon contact of the stem with water or moist soil.

Almost all parts of certain plants may be made to produce them, and the propagation of plants Buch as gooseberries, currants, roses and hops by means of slips and cuttings depends upon their development. Pieces of stem cut off just below a leaf, and placed in moist earth soon develop adventitious roots near the cut end. Advantage is taken of their formation in the propagation of plants by means of layers.

Examine the roots upon any cuttings or slips which can be obtained, and observe whether they arise on the cut surface or at a point some distance away from it. Usually adventitious roots are thin and fibrous, but those of the dahlia are tuberous. The complete root-systems of plants vary enormously in extent, but in all cases the total length is much greater than is usually anticipated.

That of an ordinary oat plant, although not spreading through more than a cubic yard or two of soil, measured in one instance over one hundred and fifty yards in length. A tree uprooted by the wind exposes to view a small number of stout roots similar to the thicker branches of the crown, and from these are given off a larger number of a finer texture. By far the greater bulk, however, which the tree possessed remain in the ground in the form of extremely fine rootlets or fibrils extending outwards generally as far, or a little farther, than the branches and leaves of the tree, but in some instances to much greater distances.

Not only do the roots grow out horizontally and near the surface of the soil, but they extend downwards as well. In isolated instances, where an adequate supply of air has been maintained along open cracks and fissures, roots have been found to descend many yards into the ground, but in general the roots of the tallest trees rarely go down more than seven or eight feet. The want of air and presence of noxious substances in the lower regions of the soil checks further growth in this direction.

With many plants almost every cubic inch of soil immediately beneath their shade contains fine delicate rootlets, and the extent of their root-branching is very rarely realised on account of the ease with which these hair-like fibrils are broken off when the plant is pulled up or disturbed. Many forest trees have a natural habit of sending their roots several feet into the soil Examples of fruit-trees belonging to this class, and requiring a deep soil for proper growth, are the cherry and wild pear. Some trees, such as larch, keep their root-system nearer the surface, spreading out more horizontally in the ground.

The quince, used as a stock on which to graft pears, has roots which remain in the upper regions of the soil. Similar surface-rooting habit is very marked in the ' Paradise' apple on which apples are grafted, and in Prunus Mahaleb upon which cherries are often grafted. The root-system of wheat penetrates more deeply than that of barley; the mangel sends its fine rootlets more extensively into the deeper layers of the soil than cabbage or turnip red clover roots more deeply than white clover, and almost all plants have distinct and peculiar habits in this respect.

The character and extent of root-development is not, however, altogether dependent upon the species of the plant concerned, but is materially influenced by external circumstances, such as the texture of the soil, and the amount of water in it. In deep open soils and loose sands the root-system of a plant is much larger than that of a similar plant grown in compact heavy ground. In soils which are not water-logged, increase of moisture up to a certain extent increases the branching of the root, and excellent examples of the influence of water, coupled with good air supply, are seen in well managed plants in pots, and also among plants growing near wells and in drain pipes; the latter in some instances become completely blocked by the large number of fine rootlets of trees growing in their neighbourhood. The root-system is also considerably modified by the totyl amount and kind of the manures or food-materials present in the soil. Up to a certain point an increase of nutrient substances increases root-development; an excess hinders it. Mutilation influences the development of the root-system. If the growing-point of the tap root of a cabbage or tree is cut off its further elongation is prevented, but the secondary roots make up for the loss by more vigourous growth and often many adventitious roots arise near the cut end. In order to properly cultivate plants of all kinds it is very important to study the habit or manner of branching of their roots and the relative proportions of the thick tap and secondary roots to the fine ramifications to which they give rise and which spread through the soil in all directions.

The proportion of root-system below ground to the branches and leaves above is also worthy of attention. The adaptability of plants to the various kinds of soils, their need of water, the cultivation which the ground should have and the rational application of manures to the plant are best understood and appreciated after a careful study of these points. Tap-rooted crops, such as sugar-beet, mangel, carrot, and parsnip, require the soil to be well-worked to some considerable depth. Surface-rooting plants, such as barley, can be grown upon comparatively thin soils.

The same is true of pears grafted on the quince and apples on the Paradise stock, and such plants respond more quickly to, and are more benefited by, top-dressings of soluble manures than plants with a deeply-penetrating root-system.

FACTORS AFFECTING SEED PRODUCTION

Seed production depends on both internal and external conditions. The most important external factors include temperature, water, oxygen and sometimes light or darkness. Various plants require different variables for successful seed production, often this depends on the individual seed variety and is closely linked to the ecological conditions of a plant's natural habitat.

For some seeds, their future production response is affected by environmental conditions during seed formation; most often these responses are types of seed dormancy.

WATER

Water is required for production. Mature seeds are often extremely dry and need to take in significant amounts of water, relative to the dry weight of the seed, before cellular metabolism and growth can resume.

Most seeds need enough water to moisten the seeds but not enough to soak them. The uptake of water by seeds is called imbibition, which leads to the swelling and the breaking of the seed coat. When seeds are formed, most plants store a food reserve with the seed, such as starch, proteins, or oils. This food reserve provides nourishment to the growing embryo. When the seed imbibes water, hydrolytic enzymes are activated which break down these stored food resources into metabolically useful chemicals.

After the seedling emerges from the seed coat and starts growing roots and leaves, the seedling's food reserves are typically exhausted; at this point photosynthesis provides the energy needed for continued growth and the seedling now requires a continuous supply of water, nutrients, and light.

OXYGEN

Oxygen is required by the germinating seed for metabolism. Oxygen is used in aerobic respiration, the main source of the seedling's energy until it grows leaves.

Oxygen is an atmospheric gas that is found in soil pore spaces; if a seed is buried too deeply within the soil or the soil is waterlogged, the seed can be oxygen starved. Some seeds have impermeable seed coats that prevent oxygen from entering the seed, causing a type of physical dormancy which is broken when the seed coat is worn away enough to allow gas exchange and water uptake from the environment.

TEMPERATURE

Temperature affects cellular metabolic and growth rates. Seeds from different species and even seeds from the same plant germinate over a wide range of temperatures. Seeds often have a temperature range within which they will germinate, and they will not do so above or below this range.

Many seeds germinate at temperatures slightly above room–temperature 60–75 °F (16–24 °C), while others germinate just above freezing and others germinate only in response to alternations in temperature between warm and cool. Some seeds germinate when the soil is cool 28–40 °F (–2—4 °C), and some when the soil is warm 76–90 °F (24–32 °C). Some seeds require exposure to cold temperatures (vernalization) to break dormancy. Seeds in a dormant state will not germinate even if conditions are favourable.

Seeds that are dependent on temperature to end dormancy have a type of physiological dormancy. For example, seeds requiring the cold of winter are inhibited from germinating until they take in water in the fall and experience cooler temperatures.

Four degrees Celsius is cool enough to end dormancy for most cool dormant seeds, but some groups, especially within the family Ranunculaceae and others, need conditions cooler than –5 C.

Some seeds will only germinate after hot temperatures during a forest fire which cracks their seed coats; this is a type of physical dormancy.

Most common annual vegetables have optimal production temperatures between 75–90 °F (24–32 °C), though many species (*e.g.* radishes or spinach) can germinate at significantly lower temperatures, as low as 40 °F (4 °C), thus allowing them to be grown from seed in cooler climates. Suboptimal temperatures lead to lower success rates and longer production periods.

LIGHT OR DARKNESS

Light or darkness can be an environmental trigger for production and is a type of physiological dormancy. Most seeds are not affected by light or darkness, but many seeds, including species found in forest settings, will not germinate until an opening in the canopy allows sufficient light for growth of the seedling.

Scarification mimics natural processes that weaken the seed coat before production. In nature, some seeds require particular conditions to germinate, such as the heat of a fire (*e.g.*, many Australian native plants), or soaking in a body of water for a long period of time. Others need to be passed through an animal's digestive tract to weaken the seed coat enough to allow the seedling to emerge.

DORMANCY

Some live seeds are dormant and need more time, and/or need to be subjected to specific environmental conditions before they will germinate. Seed dormancy can originate in different parts of the seed, for example, within the embryo; in other cases the seed coat is involved. Dormancy breaking often involves changes in membranes, initiated by dormancy–breaking signals.

This generally occurs only within hydrated seeds. Factors affecting seed dormancy include the presence of certain plant hormones, notably abscisic acid, which inhibits production, and gibberellin, which ends seed dormancy. In brewing, barley seeds are treated with gibberellin to ensure uniform seed production for the production of barley malt.

SEEDLING ESTABLISHMENT

In some definitions, the appearance of the radicle marks the end of production and the beginning of "establishment", a period that ends when the seedling has exhausted the food reserves stored in the seed.

Production and establishment as an independent organism are critical phases in the life of a plant when they are the most vulnerable to injury, disease, and water stress. The production index can be used as an indicator of phytotoxicity in soils. The mortality between dispersal of seeds and completion of establishment can be so high that many species have adapted to produce huge numbers of seeds.

Production Rate

In agriculture and gardening, the production rate describes how many seeds of a particular plant species, variety or seedlot are likely to germinate. It is usually expressed as a percentage, *e.g.*, an 85% production rate indicates that about 85 out of 100 seeds will probably germinate under proper conditions. The production rate is useful for calculating the seed requirements for a given area or desired number of plants.

Dicot Production

The part of the plant that first emerges from the seed is the embryonic root, termed the radicle or primary root. It allows the seedling to become anchored in the ground and start absorbing water.

After the root absorbs water, an embryonic shoot emerges from the seed. This shoot comprises three main parts: the cotyledons (seed leaves), the section of shoot below the cotyledons (hypocotyl), and the section of shoot above the cotyledons (epicotyl). The way the shoot emerges differs among plant groups.

Epigeous

In epigeous (or epigeal) production, the hypocotyl elongates and forms a hook, pulling rather than pushing the cotyledons and apical meristem through the soil. Once it reaches the surface, it straightens and pulls the cotyledons and shoot tip of the growing seedlings into the air. Beans, tamarind, and papaya are examples of plants that germinate this way.

Hypogeous

Another way of production is hypogeous (or hypogeal), where the epicotyl elongates and forms the hook. In this type of production, the cotyledons stay underground where they eventually decompose. Peas, for example, germinate this way.

Monocot Production

In monocot seeds, the embryo's radicle and cotyledon are covered by a coleorhiza and coleoptile, respectively. The coleorhiza is the first part to grow out of the seed, followed by the radicle. The coleoptile is then pushed up through the ground until it reaches the surface. There, it stops elongating and the first leaves emerge.

Precocious Production

While not a class of production, precocious production refers to seed production before the fruit has released seed. The seeds of the green apple commonly germinate in this manner.

Pollen Production

Another production event during the life cycle of gymnosperms and flowering plants is the production of a pollen grain after pollination. Like seeds, pollen grains are severely dehydrated before being released to facilitate their dispersal from one plant to another.

They consist of a protective coat containing several cells (up to 8 in gymnosperms, 2–3 in flowering plants). One of these cells is a tube cell.

Once the pollen grain lands on the stigma of a receptive flower (or a female cone in gymnosperms), it takes up water and germinates. Pollen production is facilitated by hydration on the stigma, as well as by the structure and physiology of the stigma and style.

Pollen can also be induced to germinate in vitro (in a petri dish or test tube). During production, the tube cell elongates into a pollen tube. In the flower, the pollen tube then grows towards the ovule where it discharges the sperm produced in the pollen grain for fertilization. The germinated pollen grain with its two sperm cells is the mature male microgametophyte of these plants.

SELF–INCOMPATIBILITY

Since most plants carry both male and female reproductive organs in their flowers, there is a high risk of self–pollination and thus inbreeding. Some plants use the control of pollen production as a way to prevent this self–pollination. Production and growth of the pollen tube involve molecular signaling between stigma and pollen. In self–incompatibility in plants, the stigma of certain plants can molecularly recognize pollen from the same plant and prevent it from germinating.

Ferns and Mosses

In plants such as bryophytes, ferns, and a few others, spores germinate into independent gametophytes. In the bryophytes (*e.g.*, mosses and liverworts), spores germinate into protonemata, similar to fungal hyphae, from which the gametophyte grows. In ferns, the gametophytes are small, heart–shaped prothalli that can often be found underneath a spore–shedding adult plant.

Spore Production

Production can also refer to the emergence of cells from resting spores and the growth of sporeling hyphae or thalli from spores in fungi, algae and some plants. Conidia are asexual reproductive spores of fungi which germinate under specific conditions. A variety of cells can be formed from the germinating conidia. The most common are germ tubes which grow and develop into hyphae. Another type of cell is a conidial anastomosis tube (CAT); these differ from germ tubes in that they are thinner, shorter, lack branches, exhibit determinate growth and home towards each other. Each cell is of a tubular shape, but the conidial anastomosis tube forms a bridge that allows fusion between conidia.

Resting Spores

In resting spores, production that involves cracking the thick cell wall of the dormant spore. For example, in zygomycetes the thick–walled zygosporangium cracks open and the zygospore inside gives rise to the emerging sporangiophore. In slime molds, production refers to the emergence of amoeboid cells from the hardened spore. After cracking the spore coat, further development involves cell division, but not necessarily the development of a multicellular organism (for example in the free–living amoebas of slime molds).

5

Protein Purification

WESTERN BLOT

The western blot (sometimes called the protein immunoblot) is a widely used analytical technique used to detect specific proteins in a sample of tissue homogenate or extract. It uses gel electrophoresis to separate native proteins by 3-D structure or denatured proteins by the length of the polypeptide. The proteins are then transferred to a membrane (typically nitrocellulose or PVDF), where they are stained with antibodies specific to the target protein. The gel electrophoresis step is included in western blot analysis to resolve the issue of the cross-reactivity of antibodies.

There are many reagent companies that specialize in providing antibodies (both monoclonal and polyclonal antibodies) against tens of thousands of different proteins. Commercial antibodies can be expensive, although the unbound antibody can be reused between experiments. This method is used in the fields of molecular biology, immunogenetics and other molecular biology disciplines. A number of search engines, such as CiteAb, Antibodypedia, and SeekProducts, are available that can help researchers find suitable antibodies for use in western blotting. Other related techniques include dot blot analysis, immunohistochemistry and immunocytochemistry where antibodies are used to detect proteins in tissues and cells by immunostaining, and enzyme-linked immunosorbent assay (ELISA).

The method originated in the laboratory of Harry Towbin at the Friedrich Miescher Institute. The name *western blot* was given to the technique by W. Neal Burnette and is a play on the name Southern blot, a technique for DNA detection developed earlier by Edwin Southern. Detection of RNA is termed northern blot and was developed by George Stark at Stanford.

STEPS

Tissue preparation

Samples can be taken from whole tissue or from cell culture. Solid tissues are first broken down mechanically using a blender (for larger sample volumes), using a homogenizer (smaller volumes), or by sonication. Cells may also be broken open by one of the above mechanical methods. However, virus or environmental samples can be the source of protein and thus western blotting is not restricted to cellular studies only.

Assorted detergents, salts, and buffers may be employed to encourage lysis of cells and to solubilize proteins. Protease and phosphatase inhibitors are often added to prevent the digestion of the sample by its own enzymes. Tissue preparation is often done at cold temperatures to avoid protein denaturing and degradation.

A combination of biochemical and mechanical techniques – comprising various types of filtration and centrifugation – can be used to separate different cell compartments and organelles.

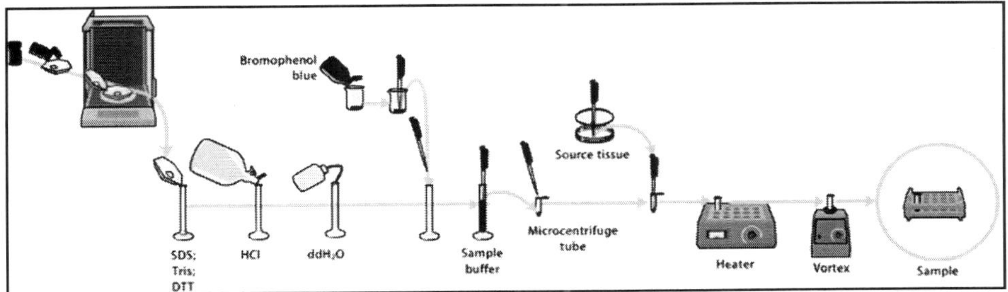

Gel electrophoresis

The proteins of the sample are separated using gel electrophoresis. Separation of proteins may be by isoelectric point (pI), molecular weight, electric charge, or a combination of these factors. The nature of the separation depends on the treatment of the sample and the nature of the gel. This is a very useful way to identify a protein.

By far the most common type of gel electrophoresis employs polyacrylamide gels and buffers loaded with sodium dodecyl sulfate (SDS). SDS-PAGE (SDS polyacrylamide gel electrophoresis) maintains polypeptides in a denatured state once they have been treated with strong reducing agents to remove secondary and tertiary structure (e.g. disulfide bonds [S-S] to sulfhydryl groups [SH and SH]) and thus allows separation of proteins by their molecular weight. Sampled proteins become covered in the negatively charged SDS and move to the positively charged electrode through the acrylamide mesh of the gel. Smaller proteins migrate faster through this mesh and the proteins are thus separated according to size (usually measured in kilodaltons, kDa). The concentration of acrylamide determines the resolution of the gel - the greater the acrylamide concentration, the better the resolution of lower molecular weight proteins. The lower the acrylamide concentration, the better the resolution of higher molecular weight proteins. Proteins travel only in one dimension along the gel for most blots.

Samples are loaded into *wells* in the gel. One lane is usually reserved for a *marker* or *ladder*, a commercially available mixture of proteins having defined molecular weights, typically stained so as to form visible, coloured bands. When voltage is applied along the gel, proteins migrate through it at different speeds dependent on their size.

These different rates of advancement (different *electrophoretic mobilities*) separate into *bands* within each *lane*.

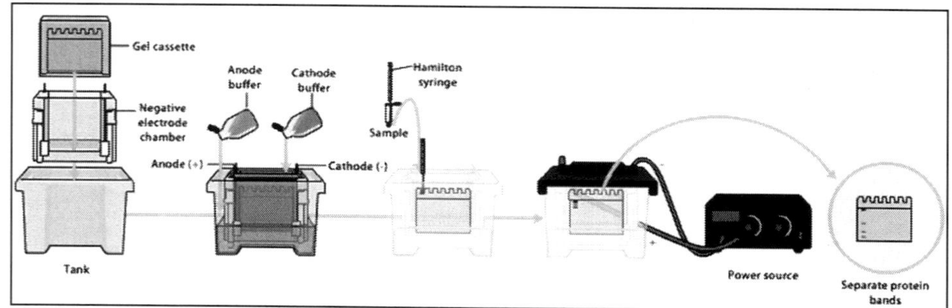

It is also possible to use a two-dimensional (2-D) gel which spreads the proteins from a single sample out in two dimensions. Proteins are separated according to isoelectric point (pH at which they have neutral net charge) in the first dimension, and according to their molecular weight in the second dimension.

Transfer

In order to make the proteins accessible to antibody detection, they are moved from within the gel onto a membrane made of *nitrocellulose or polyvinylidene difluoride (PVDF*). The primary method for transferring the proteins is called electroblotting and uses an electric current to pull proteins from the gel into the PVDF or nitrocellulose membrane. The proteins move from within the gel onto the membrane while maintaining the organization they had within the gel.

An older method of transfer involves placing a membrane on top of the gel, and a stack of filter papers on top of that. The entire stack is placed in a buffer solution which moves up the paper by capillary action, bringing the proteins with it.

In practice this method is not used as it takes too much time; electroblotting is preferred. As a result of either "blotting" process, the proteins are exposed on a thin surface layer for detection (see below). Both varieties of membrane are chosen for their non-specific protein binding properties (i.e. binds all proteins equally well). Protein binding is based upon hydrophobic interactions, as well as charged interactions between the membrane and protein. Nitrocellulose membranes are cheaper than PVDF, but are far more fragile and do not stand up well to repeated probings.

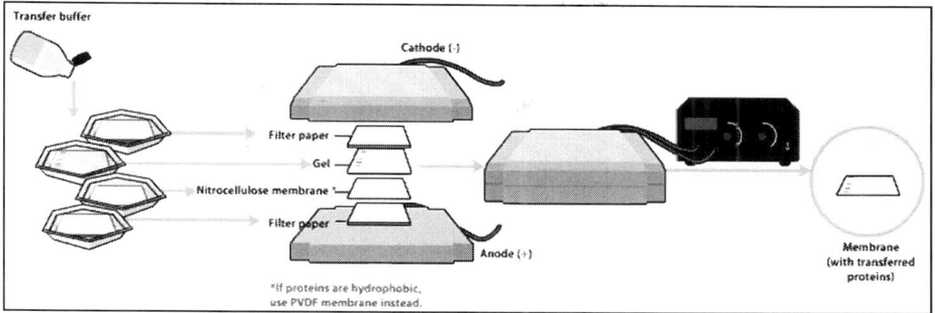

The uniformity and overall effectiveness of transfer of protein from the gel to the membrane can be checked by staining the membrane with Coomassie Brilliant Blue or Ponceau S dyes. Ponceau S is the more common of the two, due to its higher sensitivity and water solubility, the latter making it easier to subsequently destain and probe the membrane, as described below.

Blocking

Since the membrane has been chosen for its ability to bind protein and as both antibodies and the target are proteins, steps must be taken to prevent the interactions between the membrane and the antibody used for detection of the target protein. Blocking of non-specific binding is achieved by placing the membrane in a dilute solution of protein - typically 3-5% Bovine serum albumin (BSA) or non-fat dry milk (both are inexpensive) in Tris-Buffered Saline (TBS) or I-Block, with a minute percentage (0.1%) of detergent such as Tween 20 or Triton X-100. The protein in the dilute solution attaches to the membrane in all places where the target proteins have not attached. Thus, when the antibody is added, there is no room on the membrane for it to attach other than on the binding sites of the specific target protein. This reduces background in the final product of the western blot, leading to clearer results, and eliminates false positives.

Incubation

During the detection process the membrane is "probed" for the protein of interest with a modified antibody which is linked to a reporter enzyme; when exposed to an appropriate substrate, this enzyme drives a colourimetric reaction and produces a color. For a variety of reasons, this traditionally takes place in a two-step process, although there are now one-step detection methods available for certain applications.

Two steps

Primary antibody

The primary antibodies are generated when a host species or immune cell culture is exposed to protein of interest (or a part thereof). Normally, this is part of the immune response, whereas here they are harvested and used as sensitive and specific detection tools that bind the protein directly.

After blocking, a dilute solution of primary antibody (generally between 0.5 and 5 micrograms/mL) is incubated with the membrane under gentle agitation. Typically, the solution is composed of buffered saline solution with a small percentage of detergent, and sometimes with powdered milk or BSA. The antibody solution and the membrane can be sealed and incubated together for anywhere from 30 minutes to overnight. It can also be incubated at different temperatures, with higher temperatures being associated with more binding, both specific (to the target protein, the "signal") and non-specific ("noise").

Secondary antibody

After rinsing the membrane to remove unbound primary antibody, the membrane is exposed to another antibody, directed at a species-specific portion of the primary antibody. Antibodies come from animal sources (or animal sourced hybridoma cultures); an anti-mouse secondary will bind to almost any mouse-sourced primary antibody, which allows some cost savings by allowing an entire lab to share a single source of mass-produced antibody, and provides far more consistent results. This is known as a secondary antibody, and due to its targeting properties, tends to be referred to as "anti-mouse," "anti-goat," etc. The secondary antibody is usually linked to biotin or to a reporter enzyme such as alkaline phosphatase or horseradish peroxidase. This means that several secondary antibodies will bind to one primary antibody and enhance the signal.

Most commonly, a horseradish peroxidase-linked secondary is used to cleave a chemiluminescent agent, and the reaction product produces luminescence in proportion to the amount of protein. A sensitive sheet of photographic film is placed against the membrane, and exposure to the light from the reaction creates an image of the antibodies bound to the blot. A cheaper but less sensitive approach utilizes a 4-chloronaphthol stain with 1% hydrogen peroxide; reaction of peroxide radicals with 4-chloronaphthol produces a dark purple stain that can be photographed without using specialized photographic film.

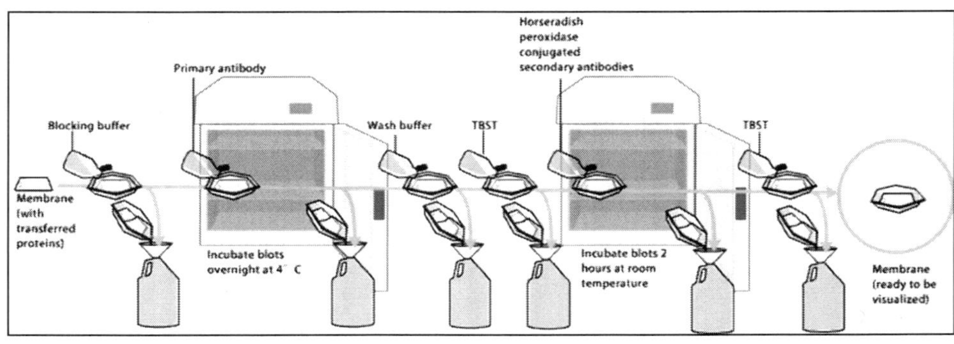

As with the ELISPOT and ELISA procedures, the enzyme can be provided with a substrate molecule that will be converted by the enzyme to a colored reaction product that will be visible on the membrane (see the figure below with blue bands).

Another method of secondary antibody detection utilizes a near-infrared (NIR) fluorophore-linked antibody. Light produced from the excitation of a fluorescent dye is static, making fluorescent detection a more precise and accurate measure of the difference in signal produced by labeled antibodies bound to proteins on a western blot.

Proteins can be accurately quantified because the signal generated by the different amounts of proteins on the membranes is measured in a static state, as compared to chemiluminescence, in which light is measured in a dynamic state.

A third alternative is to use a radioactive label rather than an enzyme coupled to the secondary antibody, such as labeling an antibody-binding protein like *Staphylococcus* Protein A or Streptavidin with a radioactive isotope of iodine. Since other methods are safer, quicker, and cheaper, this method is now rarely used; however, an advantage of this approach is the sensitivity of auto-radiography-based imaging, which enables highly accurate protein quantification when combined with optical software (e.g. Optiquant).

One step

Historically, the probing process was performed in two steps because of the relative ease of producing primary and secondary antibodies in separate processes. This gives researchers and corporations huge advantages in terms of flexibility, and adds an amplification step to the detection process. Given the advent of high-throughput protein analysis and lower limits of detection, however, there has been interest in developing one-step probing systems that would allow the process to occur faster and with fewer consumables. This requires a probe antibody which both recognizes the protein of interest and contains a detectable label, probes which are often available for known protein tags.

The primary probe is incubated with the membrane in a manner similar to that for the primary antibody in a two-step process, and then is ready for direct detection after a series of wash steps.

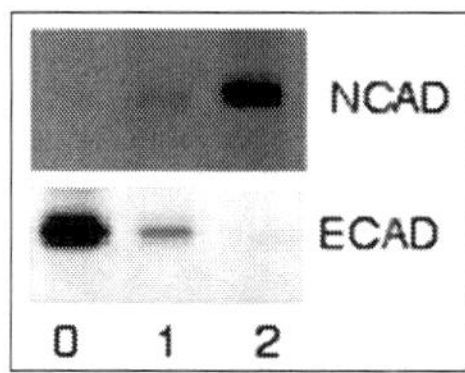

Fig. Western blot using radioactive detection system

DETECTION/ VISUALIZATION

After the unbound probes are washed away, the western blot is ready for detection of the probes that are labeled and bound to the protein of interest. In practical terms, not all westerns reveal protein only at one band in a membrane. Size approximations are taken by comparing the stained bands to that of the marker or ladder loaded during electrophoresis. The process is commonly repeated for a structural protein, such as actin or tubulin, that should not change between samples. The amount of target protein is normalized to the structural protein to control between groups. A superior strategy is the normalization to the total protein visualized with trichloroethanol or epicocconone. This practice ensures correction for the amount of total protein on the membrane in case of errors or incomplete transfers.

COLORIMETRIC DETECTION

The colorimetric detection method depends on incubation of the western blot with a substrate that reacts with the reporter enzyme (such as peroxidase) that is bound to the secondary antibody. This converts the soluble dye into an insoluble form of a different color that precipitates next to the enzyme and thereby stains the membrane. Development of the blot is then stopped by washing away the soluble dye. Protein levels are evaluated through densitometry (how intense the stain is) or spectrophotometry.

CHEMILUMINESCENT DETECTION

Chemiluminescent detection methods depend on incubation of the western blot with a substrate that will luminesce when exposed to the reporter on the secondary antibody. The light is then detected by CCD cameras which capture a digital image of the western blot or photographic film. The use of film for western blot detection is slowly disappearing because of non linearity of the image (non accurate quantification).

The image is analysed by densitometry, which evaluates the relative amount of protein staining and quantifies the results in terms of optical density. Newer software allows further data analysis such as molecular weight analysis if appropriate standards are used.

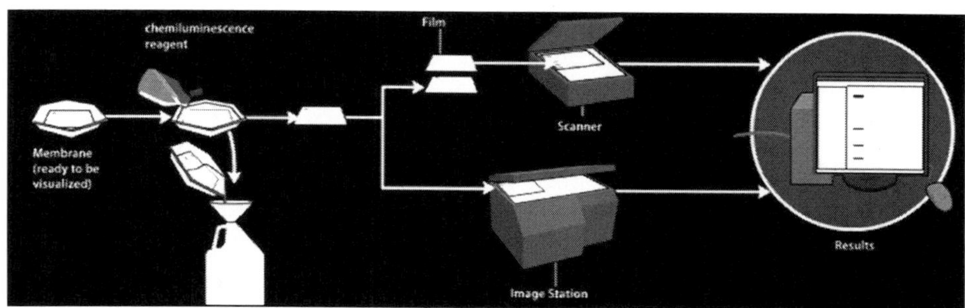

The main companies offering Chemiluminescence imaging systems are Analytik Jena, Syngene, UVP, Azure Biosystems, UVITEC, GE, Biorad and Vilber Lourmat.

RADIOACTIVE DETECTION

Radioactive labels do not require enzyme substrates, but rather, allow the placement of medical X-ray film directly against the western blot, which develops as it is exposed to the label and creates dark regions which correspond to the protein bands of interest (see image to the right). The importance of radioactive detections methods is declining due to its hazardous radiation, because it is very expensive, health and safety risks are high, and ECL (enhanced chemiluminescence) provides a useful alternative.

FLUORESCENT DETECTION

The fluorescently labeled probe is excited by light and the emission of the excitation is then detected by a photosensor such as a CCD camera equipped with appropriate emission filters which captures a digital image of the western blot and allows further data analysis such as molecular weight analysis and a quantitative western blot analysis. Fluorescence is considered to be one of the best methods for quantification but is less sensitive than chemiluminescence.

SECONDARY PROBING

One major difference between nitrocellulose and PVDF membranes relates to the ability of each to support "stripping" antibodies off and reusing the membrane for subsequent antibody probes. While there are well-established protocols available for stripping nitrocellulose membranes, the sturdier PVDF allows for easier stripping, and for more reuse before background noise limits experiments. Another difference is that, unlike nitrocellulose, PVDF must be soaked in 95% ethanol, isopropanol or methanol before use. PVDF membranes also tend to be thicker and more resistant to damage during use.

2-D GEL ELECTROPHORESIS

2-dimensional SDS-PAGE uses the principles and techniques outlined above. 2-D SDS-PAGE, as the name suggests, involves the migration of polypeptides in 2 dimensions. For example, in the first dimension, polypeptides are separated according to isoelectric point, while in the second dimension, polypeptides are separated according to their molecular weight. The isoelectric point of a given protein is determined by the relative number of positively (e.g. lysine, arginine) and negatively (e.g. glytamate, aspartate) charged amino acids, with negatively charged amino acids contributing to a low isoelectric point and positively charged amino acids contributing to a high isoelectric point. Samples could also be separated first under nonreducing conditions using SDS-PAGE, and under reducing conditions in the second dimension, which breaks apart disulfide bonds that hold subunits together. SDS-PAGE might also be coupled with urea-PAGE for a 2-dimensional gel.

In principle, this method allows for the separation of all cellular proteins on a single large gel. A major advantage of this method is that it often distinguishes between different isoforms of a particular protein - e.g. a protein that has been phosphorylated (by addition of a negatively charged group). Proteins that have been separated can be cut out of the gel and then analysed by mass spectrometry, which identifies the protein.

MEDICAL DIAGNOSTIC APPLICATIONS

- The confirmatory HIV test employs a western blot to detect anti-HIV antibody in a human serum sample. Proteins from known HIV-infected cells are separated and blotted on a membrane as above. Then, the serum to be tested is applied in the primary antibody incubation step; free antibody is washed away, and a secondary anti-human antibody linked to an enzyme signal is added. The stained bands then indicate the proteins to which the patient's serum contains antibody.
- A western blot is also used as the definitive test for bovine spongiform encephalopathy (BSE, commonly referred to as 'mad cow disease').
- Some forms of Lyme disease testing employ western blotting.
- A western blot can also be used as a confirmatory test for Hepatitis B infection.
- In veterinary medicine, a western blot is sometimes used to confirm FIV+ status in cats.

CHROMATOGRAPHY

Chromatography (*chroma* which means "color" and γραφειν *graphein* "to write") is the collective term for a set of laboratory techniques for the separation of mixtures. The mixture is dissolved in a fluid called the *mobile phase,* which carries it through a structure holding another material called the *stationary phase.* The various constituents of the mixture travel at different speeds, causing them to separate. The separation is based on differential partitioning between the mobile and stationary phases. Subtle differences in a compound's partition coefficient result in differential retention on the stationary phase and thus changing the separation. Chromatography may be preparative or analytical. The purpose of preparative chromatography is to separate the components of a mixture for more advanced use (and is thus a form of purification). Analytical chromatography is done normally with smaller amounts of material and is for measuring the relative proportions of analytes in a mixture. The two are not mutually exclusive.

Chromatography was first employed in Russia by the Italian-born scientist Mikhail Tsvet in 1900. He continued to work with chromatography in the first decade of the 20th century, primarily for the separation of plant pigments such as chlorophyll, carotenes, and xanthophylls. Since these components have different colors (green, orange, and yellow, respectively) they gave the technique its name. New types of chromatography developed during the 1930s and 1940s made the technique useful for many separation processes.

Chromatography technique developed substantially as a result of the work of Archer John Porter Martin and Richard Laurence Millington Synge during the 1940s and 1950s. They established the principles and basic techniques of partition chromatography, and their work encouraged the rapid development of several chromatographic methods: paper chromatography, gas chromatography, and what would become known as high performance liquid chromatography. Since then, the technology has advanced rapidly. Researchers found that the main principles of Tsvet's chromatography could be applied in many different ways, resulting in the different varieties of chromatography described below. Advances are continually improving the technical performance of chromatography, allowing the separation of increasingly similar molecules.

CHROMATOGRAPHY TERMS

- The analyte is the substance to be separated during chromatography. It is also normally what is needed from the mixture.
- Analytical chromatography is used to determine the existence and possibly also the concentration of analyte(s) in a sample.
- A bonded phase is a stationary phase that is covalently bonded to the support particles or to the inside wall of the column tubing.

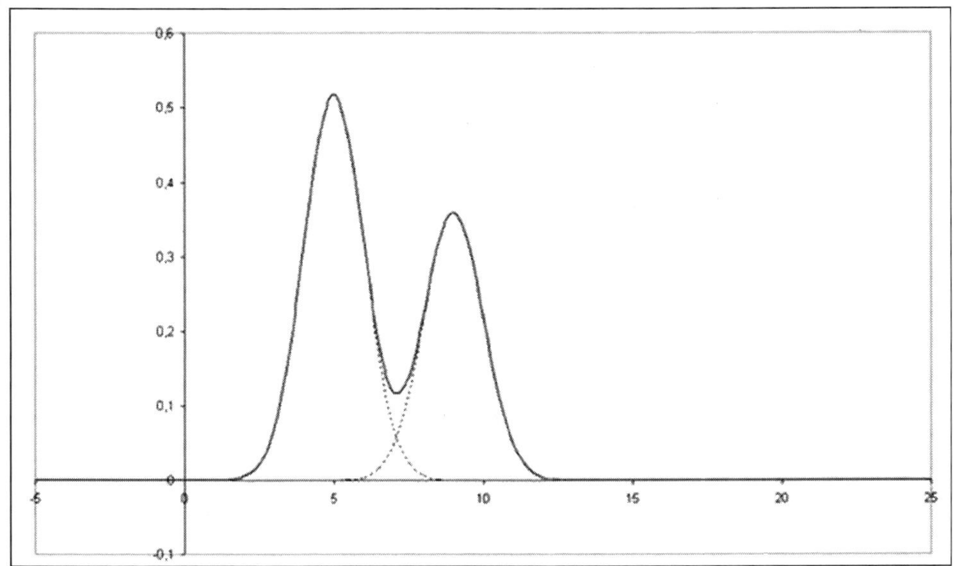

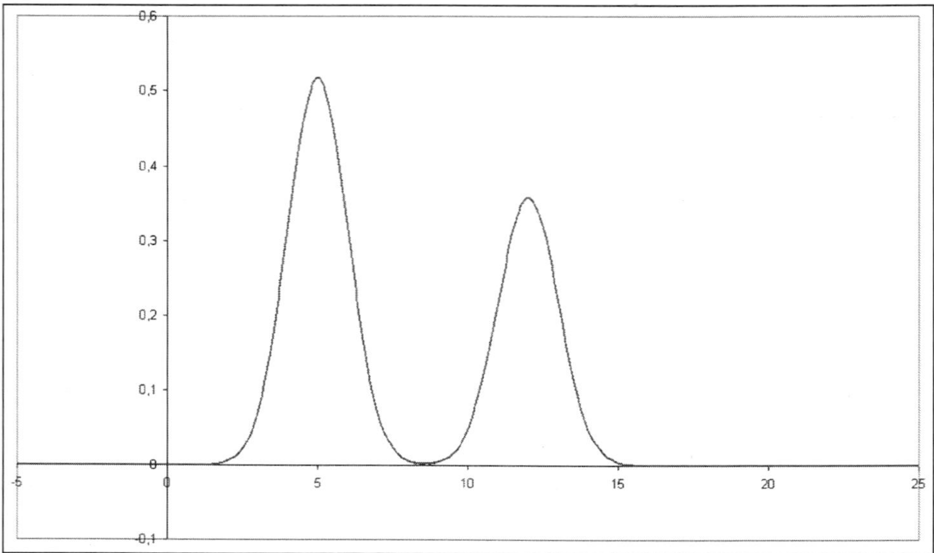

- A chromatogram is the visual output of the chromatograph. In the case of an optimal separation, different peaks or patterns on the chromatogram correspond to different components of the separated mixture.

Plotted on the x-axis is the retention time and plotted on the y-axis a signal (for example obtained by a spectrophotometer, mass spectrometer or a variety of other detectors) corresponding to the response created by the analytes exiting the system. In the case of an optimal system the signal is proportional to the concentration of the specific analyte separated.

- A chromatograph is equipment that enables a sophisticated separation, e.g. gas chromatographic or liquid chromatographic separation.
- Chromatography is a physical method of separation that distributes components to separate between two phases, one stationary (stationary phase), the other (the mobile phase) moving in a definite direction.
- The eluate is the mobile phase leaving the column.
- The eluent is the solvent that carries the analyte.
- An eluotropic series is a list of solvents ranked according to their eluting power.
- An immobilized phase is a stationary phase that is immobilized on the support particles, or on the inner wall of the column tubing.
- The mobile phase is the phase that moves in a definite direction. It may be a liquid (LC and Capillary Electrochromatography (CEC)), a gas (GC), or a supercritical fluid (supercritical-fluid chromatography, SFC). The mobile phase consists of the sample being separated/ analyzed and the solvent that moves the sample through the column.

In the case of HPLC the mobile phase consists of a non-polar solvent(s) such as hexane in normal phase or polar solvents in reverse phase chromatography and the sample being separated. The mobile phase moves through the chromatography column (the stationary phase) where the sample interacts with the stationary phase and is separated.

- Preparative chromatography is used to purify sufficient quantities of a substance for further use, rather than analysis.
- The retention time is the characteristic time it takes for a particular analyte to pass through the system (from the column inlet to the detector) under set conditions. See also: Kovats' retention index
- The sample is the matter analyzed in chromatography. It may consist of a single component or it may be a mixture of components. When the sample is treated in the course of an analysis, the phase or the phases containing the analytes of interest is/are referred to as the sample whereas everything out of interest separated from the sample before or in the course of the analysis is referred to as waste.
- The solute refers to the sample components in partition chromatography.
- The solvent refers to any substance capable of solubilizing another substance, and especially the liquid mobile phase in liquid chromatography.
- The stationary phase is the substance fixed in place for the chromatography procedure. Examples include the silica layer in thin layer chromatography
- The detector refers to the instrument used for qualitative and quantitative detection of analytes after separation.

Chromatography is based on the concept of partition coefficient. Any solute partitions between two immiscible solvents. When we make one solvent immobile (by adsorption on a solid support matrix) and another mobile it results in most common applications of chromatography. If matrix support is polar (e.g. paper, silica etc.) it is forward phase chromatography, and if it is non-polar (C-18) it is reverse phase.

TECHNIQUES BY CHROMATOGRAPHIC BED SHAPE

Column chromatography

Column chromatography is a separation technique in which the stationary bed is within a tube. The particles of the solid stationary phase or the support coated with a liquid stationary phase may fill the whole inside volume of the tube (packed column) or be concentrated on or along the inside tube wall leaving an open, unrestricted path for the mobile phase in the middle part of the tube (open tubular column). Differences in rates of movement through the medium are calculated to different retention times of the sample.

In 1978, W. Clark Still introduced a modified version of column chromatography called flash column chromatography (flash). The technique is very similar to the traditional column chromatography, except for that the solvent is driven through the column by applying positive pressure. This allowed most separations to be performed in less than 20 minutes, with improved separations compared to the old method. Modern flash chromatography systems are sold as pre-packed plastic cartridges, and the solvent is pumped through the cartridge. Systems may also be linked with detectors and fraction collectors providing automation. The introduction of gradient pumps resulted in quicker separations and less solvent usage.

In expanded bed adsorption, a fluidized bed is used, rather than a solid phase made by a packed bed. This allows omission of initial clearing steps such as centrifugation and filtration, for culture broths or slurries of broken cells.

Phosphocellulose chromatography utilizes the binding affinity of many DNA-binding proteins for phosphocellulose. The stronger a protein's interaction with DNA, the higher the salt concentration needed to elute that protein.

Planar chromatography

Planar chromatography is a separation technique in which the stationary phase is present as or on a plane. The plane can be a paper, serving as such or impregnated by a substance as the stationary bed (paper chromatography) or a layer of solid particles spread on a support such as a glass plate (thin layer chromatography). Different compounds in the sample mixture travel different distances according to how strongly they interact with the stationary phase as compared to the mobile phase. The specific Retention factor (R_f) of each chemical can be used to aid in the identification of an unknown substance.

Paper chromatography

Paper chromatography is a technique that involves placing a small dot or line of sample solution onto a strip of *chromatography paper*. The paper is placed in a container with a shallow layer of solvent and sealed. As the solvent rises through the paper, it meets the sample mixture, which starts to travel up the paper with the solvent. This paper is made of cellulose, a polar substance, and the compounds within the mixture travel farther if they are non-polar. More polar substances bond with the cellulose paper more quickly, and therefore do not travel as far.

Thin layer chromatography (TLC)

Thin layer chromatography (TLC) is a widely employed laboratory technique and is similar to paper chromatography. However, instead of using a stationary phase of paper, it involves a stationary phase of a thin layer of adsorbent like silica gel, alumina, or cellulose on a flat, inert substrate. Compared to paper, it has the advantage of faster runs, better separations, and the choice between different adsorbents. For even better resolution and to allow for quantification, high-performance TLC can be used. An older popular use had been to differentiate chromosomes by observing distance in gel (separation of was a separate step).

DISPLACEMENT CHROMATOGRAPHY

The basic principle of displacement chromatography is: A molecule with a high affinity for the chromatography matrix (the displacer) competes effectively for binding sites, and thus displace all molecules with lesser affinities. There are distinct differences between displacement and elution chromatography. In elution mode, substances typically emerge from a column in narrow, Gaussian peaks. Wide separation of peaks, preferably to baseline, is desired for maximum purification. The speed at which any component of a mixture travels down the column in elution mode depends on many factors. But for two substances to travel at different speeds, and thereby be resolved, there must be substantial differences in some interaction between the biomolecules and the chromatography matrix. Operating parameters are adjusted to maximize the effect of this difference. In many cases, baseline separation of the peaks can be achieved only with gradient elution and low column loadings. Thus, two drawbacks to elution mode chromatography, especially at the preparative scale, are operational complexity, due to gradient solvent pumping, and low throughput, due to low column loadings. Displacement chromatography has advantages over elution chromatography in that components are resolved into consecutive zones of pure substances rather than "peaks". Because the process takes advantage of the nonlinearity of the isotherms, a larger column feed can be separated on a given column with the purified components recovered at significantly higher concentrations.

TECHNIQUES BY PHYSICAL STATE OF MOBILE PHASE

Gas chromatography

Gas chromatography (GC), also sometimes known as gas-liquid chromatography, (GLC), is a separation technique in which the mobile phase is a gas. Gas chromatographic separation is always carried out in a column, which is typically "packed" or "capillary". Packed columns are the routine work horses of gas chromatography, being cheaper and easier to use and often giving adequate performance. Capillary columns generally give far superior resolution and although more expensive are becoming widely used, especially for complex mixtures. Both types of column are made from non-adsorbent and chemically inert materials. Stainless steel and glass are the usual materials for packed columns and quartz or fused silica for capillary columns.

Gas chromatography is based on a partition equilibrium of analyte between a solid or viscous liquid stationary phase (often a liquid silicone-based material) and a mobile gas (most often helium). The stationary phase is adhered to the inside of a small-diameter (commonly 0.53 – 0.18mm inside diameter) glass or fused-silica tube (a capillary column) or a solid matrix inside a larger metal tube (a packed column). It is widely used in analytical chemistry; though the high temperatures used in GC make it unsuitable for high molecular weight biopolymers or proteins (heat denatures them), frequently encountered in biochemistry, it is well suited for use in the petrochemical, environmental monitoring and remediation, and industrial chemical fields. It is also used extensively in chemistry research.

LIQUID CHROMATOGRAPHY

Liquid chromatography (LC) is a separation technique in which the mobile phase is a liquid. It can be carried out either in a column or a plane. Present day liquid chromatography that generally utilizes very small packing particles and a relatively high pressure is referred to as high performance liquid chromatography (HPLC).

In HPLC the sample is forced by a liquid at high pressure (the mobile phase) through a column that is packed with a stationary phase composed of irregularly or spherically shaped particles, a porous monolithic layer, or a porous membrane. HPLC is historically divided into two different sub-classes based on the polarity of the mobile and stationary phases. Methods in which the stationary phase is more polar than the mobile phase (e.g., toluene as the mobile phase, silica as the stationary phase) are termed normal phase liquid chromatography (NPLC) and the opposite (e.g., water-methanol mixture as the mobile phase and C18 = octadecylsilyl as the stationary phase) is termed reversed phase liquid chromatography (RPLC).

AFFINITY CHROMATOGRAPHY

Affinity chromatography is based on selective non-covalent interaction between an analyte and specific molecules. It is very specific, but not very robust. It is often used in biochemistry in the purification of proteins bound to tags. These fusion proteins are labeled with compounds such as His-tags, biotin or antigens, which bind to the stationary phase specifically. After purification, some of these tags are usually removed and the pure protein is obtained.

Affinity chromatography often utilizes a biomolecule's affinity for a metal (Zn, Cu, Fe, etc.). Columns are often manually prepared. Traditional affinity columns are used as a preparative step to flush out unwanted biomolecules. However, HPLC techniques exist that do utilize affinity chromatogaphy properties. Immobilized Metal Affinity Chromatography (IMAC) is useful to separate aforementioned molecules based on the relative affinity for the metal (I.e. Dionex IMAC). Often these columns can be loaded with different metals to create a column with a targeted affinity.

Supercritical fluid chromatography

Supercritical fluid chromatography is a separation technique in which the mobile phase is a fluid above and relatively close to its critical temperature and pressure.

TECHNIQUES BY SEPARATION MECHANISM

Ion exchange chromatography

Ion exchange chromatography (usually referred to as ion chromatography) uses an ion exchange mechanism to separate analytes based on their respective charges. It is usually performed in columns but can also be useful in planar mode. Ion exchange chromatography uses a charged stationary phase to separate charged compounds including anions, cations, amino acids, peptides, and proteins. In conventional methods the stationary phase is an ion exchange resin that carries charged functional groups that interact with oppositely charged groups of the compound to retain. Ion exchange chromatography is commonly used to purify proteins using FPLC.

Size-exclusion chromatography

Size-exclusion chromatography (SEC) is also known as gel permeation chromatography (GPC) or gel filtration chromatography and separates molecules according to their size (or more accurately according to their hydrodynamic diameter or hydrodynamic volume). Smaller molecules are able to enter the pores of the media and, therefore, molecules are trapped and removed from the flow of the mobile phase. The average residence time in the pores depends upon the effective size of the analyte molecules. However, molecules that are larger than the average pore size of the packing are excluded and thus suffer essentially no retention; such species are the first to be eluted. It is generally a low-resolution chromatography technique and thus it is often reserved for the final, "polishing" step of a purification. It is also useful for determining the tertiary structure and quaternary structure of purified proteins, especially since it can be carried out under native solution conditions.

Expanded Bed Adsorption (EBA) Chromatographic Separation

Expanded Bed Adsorption (EBA) is also known as gel permeation chromatography (GPC) or gel filtration chromatography and separates molecules according to their size (or more accurately according to their hydrodynamic diameter or hydrodynamic volume). Smaller molecules are able to enter the pores of the media and, therefore, molecules are trapped and removed from the flow of the mobile phase. The average residence time in the pores depends upon the effective size of the analyte molecules. However, molecules that are larger than the average pore size of the packing are excluded and thus suffer essentially no retention; such species are the first to be eluted. It is generally a low-resolution chromatography technique and thus it is often reserved for the final, "polishing" step of a purification. It is also useful for determining the tertiary structure and quaternary structure of purified proteins, especially since it can be carried out under native solution conditions.

SPECIAL TECHNIQUES

Reversed-phase chromatography

Reversed-phase chromatography (RPC) is any liquid chromatography procedure in which the mobile phase is significantly more polar than the stationary phase. It is so named because in normal-phase liquid chromatography, the mobile phase is significantly less polar than the stationary phase. Hydrophobic molecules in the mobile phase tend to adsorb to the relatively hydrophobic stationary phase. Hydrophilic molecules in the mobile phase will tend to elute first. Separating columns typically comprise a C8 or C18 carbon-chain bonded to a silica particle substrate.

Hydrophobic interactions between proteins and the chromatographic matrix can be exploited to purify the proteins. In hydrophobic interaction chromatography, the matrix material is lightly substituted with octyl or phenyl groups.

At high salt concentrations, nonpolar groups on the surface on proteins "interact" with the hydrophobic groups; that is, both types of groups are excluded by the polar solvent (hydrophobic effects are augmented by increased ionic strength). The eluant is typically an aqueous buffer with decreasing salt concentrations, increasing concentrations of detergent (which disrupts hydrophobic interactions), or changes in pH.

Two-dimensional chromatography

In some cases, the chemistry within a given column can be insufficient to separate some analytes. It is possible to direct a series of unresolved peaks onto a second column with different physico-chemical (Chemical classification) properties. Since the mechanism of retention on this new solid support is different from the first dimensional separation, it can be possible to separate compounds that are indistinguishable by one-dimensional chromatography. The sample is spotted at one corner of a square plate,developed, air-dried, then rotated by 90° and usually redeveloped in a second solvent system.

Simulated moving-bed chromatography

The simulated moving bed (SMB) technique is a variant of high performance liquid chromatography; it is used to separate particles and/or chemical compounds that would be difficult or impossible to resolve otherwise. This increased separation is brought about by a valve-and-column arrangement that is used to lengthen the stationary phase indefinitely. In the moving bed technique of preparative chromatography the feed entry and the analyte recovery are simultaneous and continuous, but because of practical difficulties with a continuously moving bed, simulated moving bed technique was proposed. In the simulated moving bed technique instead of moving the bed, the sample inlet and the analyte exit positions are moved continuously, giving the impression of a moving bed. True moving bed chromatography (TMBC) is only a theoretical concept. Its simulation, SMBC is achieved by the use of a multiplicity of columns in series and a complex valve arrangement, which provides for sample and solvent feed, and also analyte and waste takeoff at appropriate locations of any column, whereby it allows switching at regular intervals the sample entry in one direction, the solvent entry in the opposite direction, whilst changing the analyte and waste takeoff positions appropriately as well.

Pyrolysis gas chromatography

Pyrolysis gas chromatography mass spectrometry is a method of chemical analysis in which the sample is heated to decomposition to produce smaller molecules that are separated by gas chromatography and detected using mass spectrometry.

Pyrolysis is the thermal decomposition of materials in an inert atmosphere or a vacuum. The sample is put into direct contact with a platinum wire, or placed in a quartz sample tube, and rapidly heated to 600–1000 °C. Depending on the application even higher temperatures are used. Three different heating techniques are used in actual pyrolyzers: Isothermal furnace, inductive heating (Curie Point filament), and resistive heating using platinum filaments. Large molecules cleave at their weakest points and produce smaller, more volatile fragments. These fragments can be separated by gas chromatography. Pyrolysis GC chromatograms are typically complex because a wide range of different decomposition products is formed. The data can either be used as fingerprint to prove material identity or the GC/MS data is used to identify individual fragments to obtain structural information. To increase the volatility of polar fragments, various methylating reagents can be added to a sample before pyrolysis.

Besides the usage of dedicated pyrolyzers, pyrolysis GC of solid and liquid samples can be performed directly inside Programmable Temperature Vaporizer (PTV) injectors that provide quick heating (up to 30 °C/s) and high maximum temperatures of 600–650 °C. This is sufficient for some pyrolysis applications. The main advantage is that no dedicated instrument has to be purchased and pyrolysis can be performed as part of routine GC analysis. In this case quartz GC inlet liners have to be used. Quantitative data can be acquired, and good results of derivatization inside the PTV injector are published as well.

Fast protein liquid chromatography

Fast protein liquid chromatography (FPLC), is a form of liquid chromatography that is often used to analyze or purify mixtures of proteins. As in other forms of chromatography, separation is possible because the different components of a mixture have different affinities for two materials, a moving fluid (the "mobile phase") and a porous solid (the stationary phase). In FPLC the mobile phase is an aqueous solution, or "buffer". The buffer flow rate is controlled by a positive-displacement pump and is normally kept constant, while the composition of the buffer can be varied by drawing fluids in different proportions from two or more external reservoirs. The stationary phase is a resin composed of beads, usually of cross-linked agarose, packed into a cylindrical glass or plastic column. FPLC resins are available in a wide range of bead sizes and surface ligands depending on the application.

Countercurrent chromatography

Countercurrent chromatography (CCC) is a type of liquid-liquid chromatography, where both the stationary and mobile phases are liquids. The operating principle of CCC equipment requires a column consisting of an open tube coiled around a bobbin. The bobbin is rotated in a double-axis gyratory motion (a cardioid), which causes a variable gravity (G) field to act on the column during each rotation. This motion causes the column to see one partitioning step per revolution and components of the sample separate in the column due to their partitioning coefficient between the two immiscible liquid phases used. There are many types of CCC available today. These include HSCCC (High Speed CCC) and HPCCC (High Performance CCC). HPCCC is the latest and best performing version of the instrumentation available currently.

Chiral chromatography

Chiral chromatography involves the separation of stereoisomers. In the case of enantiomers, these have no chemical or physical differences apart from being three-dimensional mirror images. Conventional chromatography or other separation processes are incapable of separating them. To enable chiral separations to take place, either the mobile phase or the stationary phase must themselves be made chiral, giving differing affinities between the analytes. Chiral chromatography HPLC columns (with a chiral stationary phase) in both normal and reversed phase are commercially available.

ELISA

The enzyme-linked immunosorbent assay (ELISA) is a test that uses antibodies and color change to identify a substance.

ELISA is a popular format of "wet-lab" type analytic biochemistry assay that uses a solid-phase enzyme immunoassay (EIA) to detect the presence of a substance, usually an antigen, in a liquid sample or wet sample.

The ELISA has been used as a diagnostic tool in medicine and plant pathology, as well as a quality-control check in various industries.

Antigens from the sample are attached to a surface. Then, a further specific antibody is applied over the surface so it can bind to the antigen. This antibody is linked to an enzyme, and, in the final step, a substance containing the enzyme's substrate is added. The subsequent reaction produces a detectable signal, most commonly a color change in the substrate.

Performing an ELISA involves at least one antibody with specificity for a particular antigen. The sample with an unknown amount of antigen is immobilized on a solid support (usually a polystyrene microtiter plate) either non-specifically (via adsorption to the surface) or specifically (via capture by another antibody specific to the same antigen, in a "sandwich" ELISA). After the antigen is immobilized, the detection antibody is added, forming a complex with the antigen.

The detection antibody can be covalently linked to an enzyme, or can itself be detected by a secondary antibody that is linked to an enzyme through bioconjugation. Between each step, the plate is typically washed with a mild detergent solution to remove any proteins or antibodies that are non-specifically bound. After the final wash step, the plate is developed by adding an enzymatic substrate to produce a visible signal, which indicates the quantity of antigen in the sample.

Of note, ELISA can perform other forms of ligand binding assays instead of strictly "immuno" assays, though the name carried the original "immuno" because of the common use and history of development of this method. The technique essentially requires any ligating reagent that can be immobilized on the solid phase along with a detection reagent that will bind specifically and use an enzyme to generate a signal that can be properly quantified.

In between the washes, only the ligand and its specific binding counterparts remain specifically bound or "immunosorbed" by antigen-antibody interactions to the solid phase, while the nonspecific or unbound components are washed away. Unlike other spectrophotometric wet lab assay formats where the same reaction well (e.g. a cuvette) can be reused after washing, the ELISA plates have the reaction products immunosorbed on the solid phase which is part of the plate, and so are not easily reusable.

PRINCIPLE

As an analytic biochemistry assay, ELISA involves detection of an "analyte" (i.e. the specific substance whose presence is being quantitatively or qualitatively analyzed) in a liquid sample by a method that continues to use liquid reagents during the "analysis" (i.e. controlled sequence of biochemical reactions that will generate a signal which can be easily quantified and interpreted as a measure of the amount of analyte in the sample) that stays liquid and remains inside a reaction chamber or well needed to keep the reactants contained; It is opposed to "dry lab" that can use dry strips – and even if the sample is liquid (e.g. a measured small drop), the final detection step in "dry" analysis involves reading of a dried strip by methods such as reflectometry and does not need a reaction containment chamber to prevent spillover or mixing between samples.

As a heterogenous assay, ELISA separates some component of the analytical reaction mixture by adsorbing certain components onto a solid phase which is physically immobilized. In ELISA, a liquid sample is added onto a stationary solid phase with special binding properties and is followed by multiple liquid reagents that are sequentially added, incubated and washed followed by some optical change (e.g. color development by the product of an enzymatic reaction) in the final liquid in the well from which the quantity of the analyte is measured.

The qualitative "reading" usually based on detection of intensity of transmitted light by spectrophotometry, which involves quantitation of transmission of some specific wavelength of light through the liquid (as well as the transparent bottom of the well in the multiple-well plate format). The sensitivity of detection depends on amplification of the signal during the analytic reactions. Since enzyme reactions are very well known amplification processes, the signal is generated by enzymes which are linked to the detection reagents in fixed proportions to allow accurate quantification – thus the name "enzyme linked".

The analyte is also called the ligand because it will specifically bind or ligate to a detection reagent, thus ELISA falls under the bigger category of ligand binding assays. The ligand-specific binding reagent is "immobilized", i.e., usually coated and dried onto the transparent bottom and sometimes also side wall of a well (the stationary "solid phase'/"solid substrate" here as opposed to solid microparticle/beads that can be washed away), which is usually constructed as a multiple-well plate known as the "ELISA plate". Conventionally, like other forms of immunoassays, the specificity of antigen-antibody type reaction is used because it is easy to raise an antibody specifically against an antigen in bulk as a reagent. Alternatively, if the analyte itself is an antibody, its target antigen can be used as the binding reagent.

HISTORY

Before the development of the ELISA, the only option for conducting an immunoassay was radioimmunoassay, a technique using radioactively labeled antigens or antibodies.

In radioimmunoassay, the radioactivity provides the signal, which indicates whether a specific antigen or antibody is present in the sample. Radioimmunoassay was first described in a scientific paper by Rosalyn Sussman Yalow and Solomon Berson published in 1960.

Because radioactivity poses a potential health threat, a safer alternative was sought. A suitable alternative to radioimmunoassay would substitute a nonradioactive signal in place of the radioactive signal. When enzymes (such as horseradish peroxidase) react with appropriate substrates (such as ABTS or TMB), a change in color occurs, which is used as a signal. However, the signal has to be associated with the presence of antibody or antigen, which is why the enzyme has to be linked to an appropriate antibody.

This linking process was independently developed by Stratis Avrameas and G. B. Pierce. Since it is necessary to remove any unbound antibody or antigen by washing, the antibody or antigen has to be fixed to the surface of the container; i.e., the immunosorbent must be prepared. A technique to accomplish this was published by Wide and Jerker Porath in 1966.

In 1971, Peter Perlmann and Eva Engvall at Stockholm University in Sweden, and Anton Schuurs and Bauke van Weemen in the Netherlands independently published papers that synthesized this knowledge into methods to perform EIA/ELISA.

Traditional ELISA typically involves chromogenic reporters and substrates that produce some kind of observable color change to indicate the presence of antigen or analyte. Newer ELISA-like techniques use fluorogenic, electrochemiluminescent, and quantitative PCR reporters to create quantifiable signals.

These new reporters can have various advantages, including higher sensitivities and multiplexing. In technical terms, newer assays of this type are not strictly ELISAs, as they are not "enzyme-linked", but are instead linked to some nonenzymatic reporter. However, given that the general principles in these assays are largely similar, they are often grouped in the same category as ELISAs.

In 2012 an ultrasensitive, enzyme-based ELISA test using nanoparticles as a chromogenic reporter was able to give a naked-eye colour signal from the detection of mere attograms of analyte.

A blue color appears for positive results and red color for negative. Note that this detection only can confirm the presence or the absence of analyte not the actual concentration.

TYPES

Direct ELISA

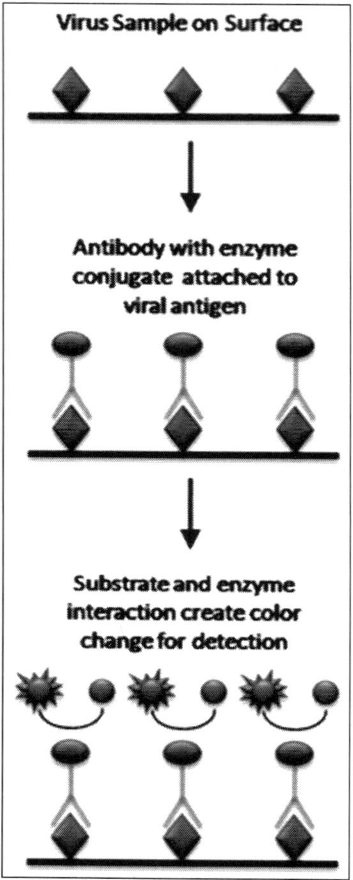

Fig. Direct ELISA diagram

The steps of direct ELISA follows the mechanism below:-

- A buffered solution of the antigen to be tested for is added to each well of a microtiter plate, where it is given time to adhere to the plastic through charge interactions.
- A solution of nonreacting protein, such as bovine serum albumin or casein, is added to well (usually 96-well plates) in order to cover any plastic surface in the well which remains uncoated by the antigen.
- The primary antibody with an attached (conjugated) enzyme is added, which binds specifically to the test antigen coating the well.
- A substrate for this enzyme is then added. Often, this substrate changes color upon reaction with the enzyme.

- The higher the concentration of the primary antibody present in the serum, the stronger the color change. Often, a spectrometer is used to give quantitative values for color strength.

The enzyme acts as an amplifier; even if only few enzyme-linked antibodies remain bound, the enzyme molecules will produce many signal molecules. Within common-sense limitations, the enzyme can go on producing color indefinitely, but the more antibody is bound, the faster the color will develop. A major disadvantage of the direct ELISA is the method of antigen immobilization is not specific; when serum is used as the source of test antigen, all proteins in the sample may stick to the microtiter plate well, so small concentrations of analyte in serum must compete with other serum proteins when binding to the well surface. The sandwich or indirect ELISA provides a solution to this problem, by using a "capture" antibody specific for the test antigen to pull it out of the serum's molecular mixture.

ELISA may be run in a qualitative or quantitative format. Qualitative results provide a simple positive or negative result (yes or no) for a sample. The cutoff between positive and negative is determined by the analyst and may be statistical. Two or three times the standard deviation (error inherent in a test) is often used to distinguish positive from negative samples. In quantitative ELISA, the optical density (OD) of the sample is compared to a standard curve, which is typically a serial dilution of a known-concentration solution of the target molecule. For example, if a test sample returns an OD of 1.0, the point on the standard curve that gave OD = 1.0 must be of the same analyte concentration as the sample.

The use and meaning of the names "direct ELISA" and "indirect ELISA" differs in the literature and on web sites depending on the context of the experiment. When the presence of an antigen is analyzed, the name "direct ELISA" refers to an ELISA in which only a labelled primary antibody is used, and the term "indirect ELISA" refers to an ELISA in which the antigen is bound by the primary antibody which then is detected by a labelled secondary antibody. In the latter case a sandwich ELISA is clearly distinct from an indirect ELISA. When the 'primary' antibody is of interest, e.g. in the case of immunization analyses, this antibody is directly detected by the secondary antibody and the term "direct ELISA" applies to a setting with two antibodies.

Sandwich ELISA

A "sandwich" ELISA, is used to detect sample antigen. The steps are:

1. A surface is prepared to which a known quantity of capture antibody is bound.
2. Any nonspecific binding sites on the surface are blocked.
3. The antigen-containing sample is applied to the plate, and captured by antibody.
4. The plate is washed to remove unbound antigen.

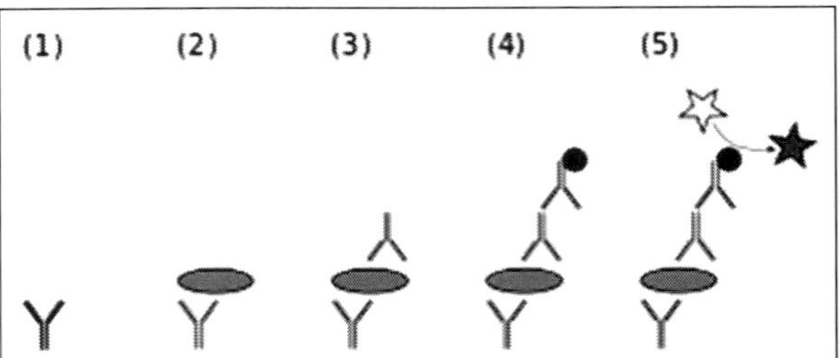

Fig. A sandwich ELISA. (1) Plate is coated with a capture antibody; (2) sample is added, and any antigen present binds to capture antibody; (3) detecting antibody is added, and binds to antigen; (4) enzyme-linked secondary antibody is added, and binds to detecting antibody; (5) substrate is added, and is converted by enzyme to detectable form.

5. A specific antibody is added, and binds to antigen (hence the 'sandwich': the Ag is stuck between two antibodies). This primary antibody could also be in the serum of a donor to be tested for reactivity towards the antigen.
6. Enzyme-linked secondary antibodies are applied as detection antibodies that also bind specifically to the antibody's Fc region (nonspecific).
7. The plate is washed to remove the unbound antibody-enzyme conjugates.
8. A chemical is added to be converted by the enzyme into a color or fluorescent or electrochemical signal.
9. The absorbance or fluorescence or electrochemical signal (e.g., current) of the plate wells is measured to determine the presence and quantity of antigen.

The image to the right includes the use of a secondary antibody conjugated to an enzyme, though, in the technical sense, this is not necessary if the primary antibody is conjugated to an enzyme (which would be direct ELISA). However, the use of a secondary-antibody conjugate avoids the expensive process of creating enzyme-linked antibodies for every antigen one might want to detect. By using an enzyme-linked antibody that binds the Fc region of other antibodies, this same enzyme-linked antibody can be used in a variety of situations. Without the first layer of "capture" antibody, any proteins in the sample (including serum proteins) may competitively adsorb to the plate surface, lowering the quantity of antigen immobilized.

Use of the purified specific antibody to attach the antigen to the plastic eliminates a need to purify the antigen from complicated mixtures before the measurement, simplifying the assay, and increasing the specificity and the sensitivity of the assay.

Competitive ELISA

A third use of ELISA is through competitive binding.

The steps for this ELISA are somewhat different from the first two examples:

1. Unlabeled antibody is incubated in the presence of its antigen (sample).
2. These bound antibody/antigen complexes are then added to an antigen-coated well.
3. The plate is washed, so unbound antibodies are removed. (The more antigen in the sample, the more Ag-Ab complexes are formed and so there are less unbound antibodies available to bind to the antigen in the well, hence "competition".)
4. The secondary antibody, specific to the primary antibody, is added. This second antibody is coupled to the enzyme.
5. A substrate is added, and remaining enzymes elicit a chromogenic or fluorescent signal.
6. The reaction is stopped to prevent eventual saturation of the signal.

Some competitive ELISA kits include enzyme-linked antigen rather than enzyme-linked antibody. The labeled antigen competes for primary antibody binding sites with the sample antigen (unlabeled). The less antigen in the sample, the more labeled antigen is retained in the well and the stronger the signal.

Commonly, the antigen is not first positioned in the well.

For the detection of HIV antibodies, the wells of microtiter plate are coated with the HIV antigen. Two specific antibodies are used, one conjugated with enzyme and the other present in serum (if serum is positive for the antibody). Cumulative competition occurs between the two antibodies for the same antigen, causing a stronger signal to be seen. Sera to be tested are added to these wells and incubated at 37 °C, and then washed. If antibodies are present, the antigen-antibody reaction occurs.

No antigen is left for the enzyme-labelled specific HIV antibodies. These antibodies remain free upon addition and are washed off during washing. Substrate is added, but there is no enzyme to act on it, so positive result shows no color change.

APPLICATIONS

Because the ELISA can be performed to evaluate either the presence of antigen or the presence of antibody in a sample, it is a useful tool for determining serum antibody concentrations (such as with the HIV test or West Nile virus). It has also found applications in the food industry in detecting potential food allergens, such as milk, peanuts, walnuts, almonds, and eggs and as serological blood test for coeliac disease.

ELISA can also be used in toxicology as a rapid presumptive screen for certain classes of drugs.

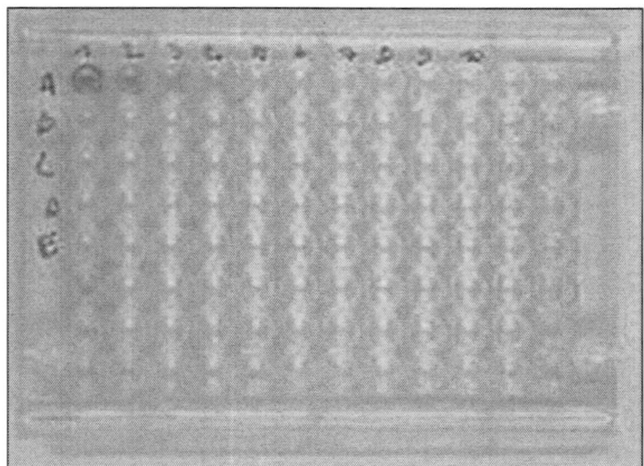

Fig. Enzyme-linked immunosorbent assay plate

The ELISA was the first screening test widely used for HIV because of its high sensitivity. In an ELISA, a person's serum is diluted 400 times and applied to a plate to which HIV antigens are attached. If antibodies to HIV are present in the serum, they may bind to these HIV antigens. The plate is then washed to remove all other components of the serum. A specially prepared "secondary antibody" — an antibody that binds to other antibodies — is then applied to the plate, followed by another wash. This secondary antibody is chemically linked in advance to an enzyme.

Thus, the plate will contain enzyme in proportion to the amount of secondary antibody bound to the plate. A substrate for the enzyme is applied, and catalysis by the enzyme leads to a change in color or fluorescence. ELISA results are reported as a number; the most controversial aspect of this test is determining the "cut-off" point between a positive and a negative result.

A cut-off point may be determined by comparing it with a known standard. If an ELISA test is used for drug screening at workplace, a cut-off concentration, 50 ng/ml, for example, is established, and a sample containing the standard concentration of analyte will be prepared. Unknowns that generate a stronger signal than the known sample are "positive." Those that generate weaker signal are "negative".

Dr Dennis E Bidwell and Alister Voller created the ELISA test to detect various kind of diseases, such as malaria, Chagas disease, and Johne's disease. ELISA tests also are used as in *in vitro* diagnostics in medical laboratories. The other uses of ELISA include:

- Detection of *Mycobacterium* antibodies in tuberculosis
- Detection of rotavirus in feces
- Detection of hepatitis B markers in serum
- Detection of enterotoxin of *E. coli* in feces
- Detection of HIV antibodies in blood samples

6

Structural Determination

X-RAY CRYSTALLOGRAPHY

X-ray crystallography is a tool used for identifying the atomic and molecular structure of a crystal, in which the crystalline atoms cause a beam of incident X-rays to diffract into many specific directions. By measuring the angles and intensities of these diffracted beams, a crystallographer can produce a three-dimensional picture of the density of electrons within the crystal. From this electron density, the mean positions of the atoms in the crystal can be determined, as well as their chemical bonds, their disorder and various other information.

Since many materials can form crystals—such as salts, metals, minerals, semiconductors, as well as various inorganic, organic and biological molecules—X-ray crystallography has been fundamental in the development of many scientific fields. In its first decades of use, this method determined the size of atoms, the lengths and types of chemical bonds, and the atomic-scale differences among various materials, especially minerals and alloys. The method also revealed the structure and function of many biological molecules, including vitamins, drugs, proteins and nucleic acids such as DNA.

X-ray crystallography is still the chief method for characterizing the atomic structure of new materials and in discerning materials that appear similar by other experiments. X-ray crystal structures can also account for unusual electronic or elastic properties of a material, shed light on chemical interactions and processes, or serve as the basis for designing pharmaceuticals against diseases.

In a single-crystal X-ray diffraction measurement, a crystal is mounted on a goniometer. The goniometer is used to position the crystal at selected orientations. The crystal is bombarded with a finely focused monochromatic beam of X-rays, producing a diffraction pattern of regularly spaced spots known as *reflections*. The two-dimensional images taken at different rotations are converted into a three-dimensional model of the density of electrons within the crystal using the mathematical method of Fourier transforms, combined with chemical data known for the sample. Poor resolution (fuzziness) or even errors may result if the crystals are too small, or not uniform enough in their internal makeup.

X-ray crystallography is related to several other methods for determining atomic structures. Similar diffraction patterns can be produced by scattering electrons or neutrons, which are likewise interpreted by Fourier transformation. If single crystals of sufficient size cannot be obtained, various other X-ray methods can be applied to obtain less detailed information; such methods include fiber diffraction, powder diffraction and (if the sample is not crystallized) small-angle X-ray scattering (SAXS). If the material under investigation is only available in the form of nanocrystalline powders or suffers from poor crystallinity, the methods of electron crystallography can be applied for determining the atomic structure.

For all above mentioned X-ray diffraction methods, the scattering is elastic; the scattered X-rays have the same wavelength as the incoming X-ray. By contrast, *inelastic* X-ray scattering methods are useful in studying excitations of the sample, rather than the distribution of its atoms.

HISTORY

Early scientific history of crystals and X-rays

Crystals have long been admired for their regularity and symmetry, but they were not investigated scientifically until the 17th century. Johannes Kepler hypothesized in his work *Strena seu de Nive Sexangula* (A New Year's Gift of Hexagonal Snow) (1611) that the hexagonal symmetry of snowflake crystals was due to a regular packing of spherical water particles.

Crystal symmetry was first investigated experimentally by Danish scientist Nicolas Steno (1669), who showed that the angles between the faces are the same in every exemplar of a particular type of crystal, and by René Just Haüy (1784), who discovered that every face of a crystal can be described by simple stacking patterns of blocks of the same shape and size. Hence, William Hallowes Miller in 1839 was able to give each face a unique label of three small integers, the Miller indices which are still used today for identifying crystal faces. Haüy's study led to the correct idea that crystals are a regular three-dimensional array (a Bravais lattice) of atoms and molecules; a single unit cell is repeated indefinitely along three principal directions that are not necessarily perpendicular. In the 19th century, a complete catalog of the possible symmetries of a crystal was worked out by Johan Hessel, Auguste Bravais, Evgraf Fedorov, Arthur Schönflies and (belatedly) William Barlow. From the available data and physical reasoning, Barlow proposed several crystal structures in the 1880s that were validated later by X-ray crystallography; however, the available data were too scarce in the 1880s to accept his models as conclusive.

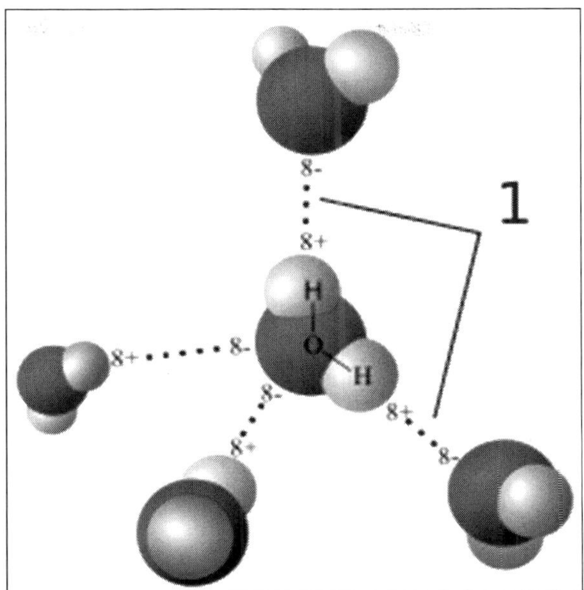

Fig. X-ray crystallography shows the arrangement of water molecules in ice, revealing the hydrogen bonds (1) that hold the solid together. Few other methods can determine the structure of matter with such precision (*resolution*).

X-rays were discovered by Wilhelm Conrad Röntgen in 1895, just as the studies of crystal symmetry were being concluded. Physicists were initially uncertain of the nature of X-rays, although it was soon suspected (correctly) that they were waves of electromagnetic radiation, in other words, another form of light.

At that time, the wave model of light—specifically, the Maxwell theory of electromagnetic radiation—was well accepted among scientists, and experiments by Charles Glover Barkla showed that X-rays exhibited phenomena associated with electromagnetic waves, including transverse polarization and spectral lines akin to those observed in the visible wavelengths. Single-slit experiments in the laboratory of Arnold Sommerfeld suggested the wavelength of X-rays was about 1 angstrom. However, X-rays are composed of photons, and thus are not only waves of electromagnetic radiation but also exhibit particle-like properties.

The photon concept was introduced by Albert Einstein in 1905, but it was not broadly accepted until 1922, when Arthur Compton confirmed it by the scattering of X-rays from electrons. Therefore, these particle-like properties of X-rays, such as their ionization of gases, caused William Henry Bragg to argue in 1907 that X-rays were *not* electromagnetic radiation. Nevertheless, Bragg's view was not broadly accepted and the observation of X-ray diffraction by Max von Laue in 1912 confirmed for most scientists that X-rays were a form of electromagnetic radiation.

X-ray analysis of crystals

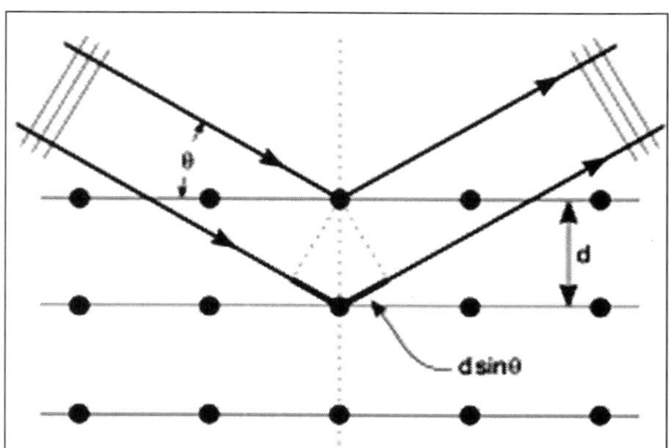

Fig. The incoming beam (coming from upper left) causes each scatterer to re-radiate a small portion of its intensity as a spherical wave. If scatterers are arranged symmetrically with a separation *d*, these spherical waves will be in sync (add constructively) only in directions where their path-length difference $2d \sin \theta$ equals an integer multiple of the wavelength λ. In that case, part of the incoming beam is deflected by an angle 2θ, producing a *reflection* spot in the diffraction pattern.

Crystals are regular arrays of atoms, and X-rays can be considered waves of electromagnetic radiation. Atoms scatter X-ray waves, primarily through the atoms' electrons. Just as an ocean wave striking a lighthouse produces secondary circular waves emanating from the lighthouse, so an X-ray striking an electron produces secondary spherical waves emanating from the electron. This phenomenon is known as elastic scattering, and the electron (or lighthouse) is known as the *scatterer*. A regular array of scatterers produces a regular array of spherical waves. Although these waves cancel one another out in most directions through destructive interference, they add constructively in a few specific directions, determined by Bragg's law:

$$2d \sin\theta = n\lambda$$

Here *d* is the spacing between diffracting planes, θ is the incident angle, *n* is any integer, and λ is the wavelength of the beam. These specific directions appear as spots on the diffraction pattern called *reflections*. Thus, X-ray diffraction results from an electromagnetic wave (the X-ray) impinging on a regular array of scatterers (the repeating arrangement of atoms within the crystal).

X-rays are used to produce the diffraction pattern because their wavelength λ is typically the same order of magnitude (1–100 angstroms) as the spacing *d* between planes in the crystal. In principle, any wave impinging on a regular array of scatterers produces diffraction, as predicted first by Francesco Maria Grimaldi in 1665.

To produce significant diffraction, the spacing between the scatterers and the wavelength of the impinging wave should be similar in size. For illustration, the diffraction of sunlight through a bird's feather was first reported by James

Gregory in the later 17th century. The first artificial diffraction gratings for visible light were constructed by David Rittenhouse in 1787, and Joseph von Fraunhofer in 1821. However, visible light has too long a wavelength (typically, 5500 angstroms) to observe diffraction from crystals. Prior to the first X-ray diffraction experiments, the spacings between lattice planes in a crystal were not known with certainty.

The idea that crystals could be used as a diffraction grating for X-rays arose in 1912 in a conversation between Paul Peter Ewald and Max von Laue in the English Garden in Munich.

Ewald had proposed a resonator model of crystals for his thesis, but this model could not be validated using visible light, since the wavelength was much larger than the spacing between the resonators. Von Laue realized that electromagnetic radiation of a shorter wavelength was needed to observe such small spacings, and suggested that X-rays might have a wavelength comparable to the unit-cell spacing in crystals. Von Laue worked with two technicians, Walter Friedrich and his assistant Paul Knipping, to shine a beam of X-rays through a copper sulfate crystal and record its diffraction on a photographic plate.

After being developed, the plate showed a large number of well-defined spots arranged in a pattern of intersecting circles around the spot produced by the central beam. Von Laue developed a law that connects the scattering angles and the size and orientation of the unit-cell spacings in the crystal, for which he was awarded the Nobel Prize in Physics in 1914.

Scattering

As described in the mathematical derivation below, the X-ray scattering is determined by the density of electrons within the crystal. Since the energy of an X-ray is much greater than that of a valence electron, the scattering may be modeled as Thomson scattering, the interaction of an electromagnetic ray with a free electron. This model is generally adopted to describe the polarization of the scattered radiation.

The intensity of Thomson scattering for one particle with mass m and charge q is:

$$I_o = I_e \left(\frac{q^4}{m^2c^4}\right)\frac{1+\cos^2 2\theta}{2} = I_e 7.94.10^{-26}\frac{1+\cos^2 2\theta}{2} = I_e f$$

Hence the atomic nuclei, which are much heavier than an electron, contribute negligibly to the scattered X-rays.

Development from 1912 to 1920

After Von Laue's pioneering research, the field developed rapidly, most notably by physicists William Lawrence Bragg and his father William Henry

Bragg. In 1912–1913, the younger Bragg developed Bragg's law, which connects the observed scattering with reflections from evenly spaced planes within the crystal.

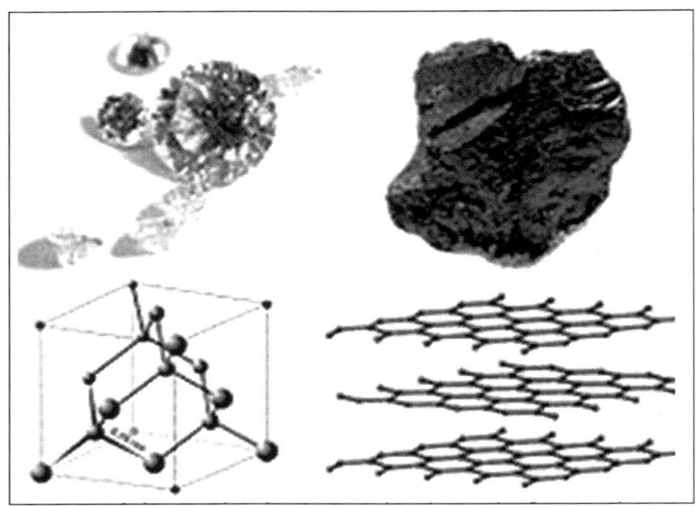

Fig. Although diamonds (top left) and graphite (top right) are identical in chemical composition—being both pure carbon—X-ray crystallography revealed the arrangement of their atoms (bottom) accounts for their different properties. In diamond, the carbon atoms are arranged tetrahedrally and held together by single covalent bonds, making it strong in all directions. By contrast, graphite is composed of stacked sheets. Within the sheet, the bonding is covalent and has hexagonal symmetry, but there are no covalent bonds between the sheets, making graphite easy to cleave into flakes.

The Braggs, father and son, shared the 1915 Nobel Prize in Physics for their work in crystallography. The earliest structures were generally simple and marked by one-dimensional symmetry. However, as computational and experimental methods improved over the next decades, it became feasible to deduce reliable atomic positions for more complicated two- and three-dimensional arrangements of atoms in the unit-cell.

The potential of X-ray crystallography for determining the structure of molecules and minerals—then only known vaguely from chemical and hydrodynamic experiments—was realized immediately. The earliest structures were simple inorganic crystals and minerals, but even these revealed fundamental laws of physics and chemistry. The first atomic-resolution structure to be "solved" (i.e., determined) in 1914 was that of table salt. The distribution of electrons in the table-salt structure showed that crystals are not necessarily composed of covalently bonded molecules, and proved the existence of ionic compounds.

The structure of diamond was solved in the same year, proving the tetrahedral arrangement of its chemical bonds and showing that the length of C–C single bond was 1.52 angstroms. Other early structures included copper, calcium fluoride (CaF_2, also known as *fluorite*), calcite ($CaCO_3$) and pyrite (FeS_2)

in 1914; spinel ($MgAl_2O_4$) in 1915; the rutile and anatase forms of titanium dioxide (TiO_2) in 1916; pyrochroite $Mn(OH)_2$ and, by extension, brucite $Mg(OH)_2$ in 1919;. Also in 1919 sodium nitrate ($NaNO_3$) and caesium dichloroiodide ($CsICl_2$) were determined by Ralph Walter Graystone Wyckoff, and the wurtzite (hexagonal ZnS) structure became known in 1920.

The structure of graphite was solved in 1916 by the related method of powder diffraction, which was developed by Peter Debye and Paul Scherrer and, independently, by Albert Hull in 1917. The structure of graphite was determined from single-crystal diffraction in 1924 by two groups independently. Hull also used the powder method to determine the structures of various metals, such as iron and magnesium.

Cultural and aesthetic importance of X-ray crystallography

In what has been called his scientific autobiography, *The Development of X-ray Analysis*, Sir William Lawrence Bragg mentioned that he believed the field of crystallography was particularly welcoming to women because the techno-aesthetics of the molecular structures resembled textiles and household objects. Bragg was known to compare crystal formation to "curtains, wallpapers, mosaics, and roses."

In 1951, the Festival Pattern Group at the Festival of Britain hosted a collaborative group of textile manufacturers and experienced crystallographers to design lace and prints based on the X-ray crystallography of insulin, china clay, and hemoglobin. One of the leading scientists of the project was Dr. Helen Megaw (1907–2002), the Assistant Director of Research at the Cavendish Laboratory in Cambridge at the time. Megaw is credited as one of the central figures who took inspiration from crystal diagrams and saw their potential in design. In 2008, the Wellcome Collection in London curated an exhibition on the Festival Pattern Group called "From Atom to Patterns."

CONTRIBUTIONS TO CHEMISTRY AND MATERIAL SCIENCE

X-ray crystallography has led to a better understanding of chemical bonds and non-covalent interactions. The initial studies revealed the typical radii of atoms, and confirmed many theoretical models of chemical bonding, such as the tetrahedral bonding of carbon in the diamond structure, the octahedral bonding of metals observed in ammonium hexachloroplatinate (IV), and the resonance observed in the planar carbonate group and in aromatic molecules.

Kathleen Lonsdale's 1928 structure of hexamethylbenzene established the hexagonal symmetry of benzene and showed a clear difference in bond length between the aliphatic C–C bonds and aromatic C–C bonds; this finding led to the idea of resonance between chemical bonds, which had profound consequences for the development of chemistry. Her conclusions were anticipated by William Henry Bragg, who published models of naphthalene and anthracene in 1921 based on other molecules, an early form of molecular replacement. Also in the 1920s, Victor Moritz Goldschmidt and later Linus Pauling developed rules for eliminating chemically unlikely structures and for determining the relative sizes of atoms. These rules led to the structure of brookite (1928) and an understanding of the relative stability of the rutile, brookite and anatase forms of titanium dioxide.

The distance between two bonded atoms is a sensitive measure of the bond strength and its bond order; thus, X-ray crystallographic studies have led to the discovery of even more exotic types of bonding in inorganic chemistry, such as metal-metal double bonds, metal-metal quadruple bonds, and three-center, two-electron bonds. X-ray crystallography—or, strictly speaking, an inelastic Compton scattering experiment—has also provided evidence for the partly covalent character of hydrogen bonds. In the field of organometallic chemistry, the X-ray structure of ferrocene initiated scientific studies of sandwich compounds, while that of Zeise's salt stimulated research into "back bonding" and metal-pi complexes. Finally, X-ray crystallography had a pioneering role in the development of supramolecular chemistry, particularly in clarifying the structures of the crown ethers and the principles of host-guest chemistry.

In material sciences, many complicated inorganic and organometallic systems have been analyzed using single-crystal methods, such as fullerenes, metalloporphyrins, and other complicated compounds. Single-crystal diffraction is also used in the pharmaceutical industry, due to recent problems with polymorphs.

The major factors affecting the quality of single-crystal structures are the crystal's size and regularity; recrystallization is a commonly used technique to improve these factors in small-molecule crystals. The Cambridge Structural Database contains over 500,000 structures; over 99% of these structures were determined by X-ray diffraction.

Mineralogy and metallurgy

Since the 1920s, X-ray diffraction has been the principal method for determining the arrangement of atoms in minerals and metals. The application of X-ray crystallography to mineralogy began with the structure of garnet, which was determined in 1924 by Menzer. A systematic X-ray crystallographic study of the silicates was undertaken in the 1920s. This study showed that, as the Si/O ratio is altered, the silicate crystals exhibit significant changes in their atomic arrangements. Machatschki extended these insights to minerals in which aluminium substitutes for the silicon atoms of the silicates. The first application of X-ray crystallography to metallurgy likewise occurred in the mid-1920s. Most notably, Linus Pauling's structure of the alloy Mg_2Sn led to his theory of the stability and structure of complex ionic crystals.

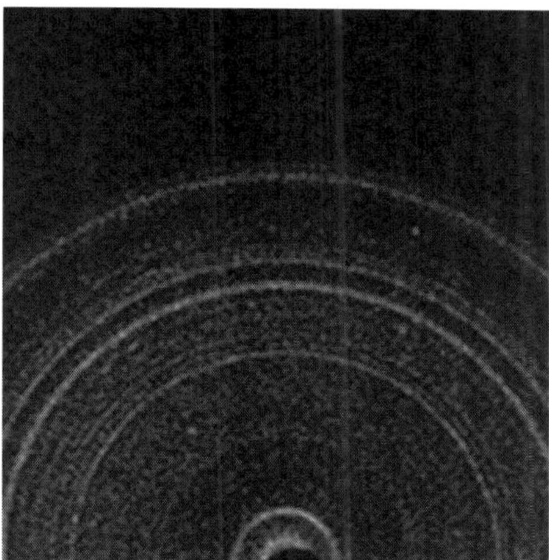

Fig. First X-ray diffraction view of Martian soil - CheMin analysis reveals feldspar, pyroxenes, olivine and more (Curiosity rover at "Rocknest", October 17, 2012).

On October 17, 2012, the Curiosity rover on the planet Mars at "Rocknest" performed the first X-ray diffraction analysis of Martian soil. The results from the rover's CheMin analyzer revealed the presence of several minerals, including feldspar, pyroxenes and olivine, and suggested that the Martian soil in the sample was similar to the "weathered basaltic soils" of Hawaiian volcanoes.

Early organic and small biological molecules

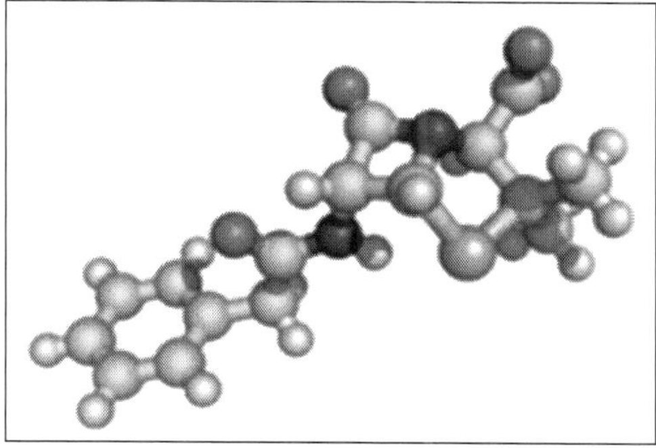

Fig. The three-dimensional structure of penicillin, solved by Dorothy Crowfoot Hodgkin in 1945. The green, white, red, yellow and blue spheres represent atoms of carbon, hydrogen, oxygen, sulfur and nitrogen, respectively.

The first structure of an organic compound, hexamethylenetetramine, was solved in 1923. This was followed by several studies of long-chain fatty acids, which are an important component of biological membranes. In the 1930s, the structures of much larger molecules with two-dimensional complexity began to be solved. A significant advance was the structure of phthalocyanine, a large planar molecule that is closely related to porphyrin molecules important in biology, such as heme, corrin and chlorophyll. X-ray crystallography of biological molecules took off with Dorothy Crowfoot Hodgkin, who solved the structures of cholesterol (1937), penicillin (1946) and vitamin B12 (1956), for which she was awarded the Nobel Prize in Chemistry in 1964.

In 1969, she succeeded in solving the structure of insulin, on which she worked for over thirty years.

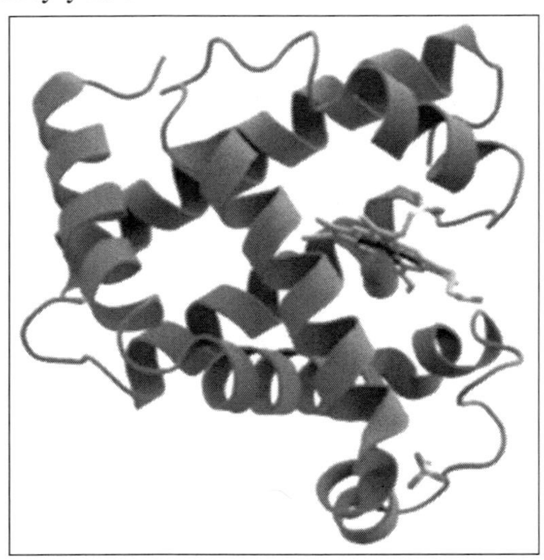

Fig. Ribbon diagram of the structure of myoglobin, showing colored alpha helices. Such proteins are long, linear molecules with thousands of atoms; yet the relative position of each atom has been determined with sub-atomic resolution by X-ray crystallography. Since it is difficult to visualize all the atoms at once, the ribbon shows the rough path of the protein polymer from its N-terminus (blue) to its C-terminus (red).

Biological macromolecular crystallography

Crystal structures of proteins (which are irregular and hundreds of times larger than cholesterol) began to be solved in the late 1950s, beginning with the structure of sperm whale myoglobin by Sir John Cowdery Kendrew, for which he shared the Nobel Prize in Chemistry with Max Perutz in 1962. Since that success, over 86817 X-ray crystal structures of proteins, nucleic acids and other biological molecules have been determined. For comparison, the nearest competing method in terms of structures analyzed is nuclear magnetic resonance (NMR) spectroscopy, which has resolved 9561 chemical structures.

Moreover, crystallography can solve structures of arbitrarily large molecules, whereas solution-state NMR is restricted to relatively small ones (less than 70 kDa). X-ray crystallography is now used routinely by scientists to determine how a pharmaceutical drug interacts with its protein target and what changes might improve it.

However, intrinsic membrane proteins remain challenging to crystallize because they require detergents or other means to solubilize them in isolation, and such detergents often interfere with crystallization. Such membrane proteins are a large component of the genome and include many proteins of great physiological importance, such as ion channels and receptors. Helium cryogenics are used to prevent radiation damage in protein crystals.

RELATIONSHIP TO OTHER SCATTERING TECHNIQUES

Elastic vs. inelastic scattering

X-ray crystallography is a form of elastic scattering; the outgoing X-rays have the same energy, and thus same wavelength, as the incoming X-rays, only with altered direction. By contrast, *inelastic scattering* occurs when energy is transferred from the incoming X-ray to the crystal, e.g., by exciting an inner-shell electron to a higher energy level. Such inelastic scattering reduces the energy (or increases the wavelength) of the outgoing beam. Inelastic scattering is useful for probing such excitations of matter, but not in determining the distribution of scatterers within the matter, which is the goal of X-ray crystallography.

X-rays range in wavelength from 10 to 0.01 nanometers; a typical wavelength used for crystallography is 1 Å (0.1 nm), which is on the scale of covalent chemical bonds and the radius of a single atom. Longer-wavelength photons (such as ultraviolet radiation) would not have sufficient resolution to determine the atomic positions. At the other extreme, shorter-wavelength photons such as gamma rays are difficult to produce in large numbers, difficult to focus, and interact too strongly with matter, producing particle-antiparticle pairs. Therefore, X-rays are the "sweetspot" for wavelength when determining atomic-resolution structures from the scattering of electromagnetic radiation.

Other X-ray techniques

Other forms of elastic X-ray scattering include powder diffraction, SAXS and several types of X-ray fiber diffraction, which was used by Rosalind Franklin in determining the double-helix structure of DNA. In general, single-crystal X-ray diffraction offers more structural information than these other techniques; however, it requires a sufficiently large and regular crystal, which is not always available.

These scattering methods generally use *monochromatic* X-rays, which are restricted to a single wavelength with minor deviations. A broad spectrum of X-rays (that is, a blend of X-rays with different wavelengths) can also be used to carry out X-ray diffraction, a technique known as the Laue method. This is the method used in the original discovery of X-ray diffraction. Laue scattering provides much structural information with only a short exposure to the X-ray beam, and is therefore used in structural studies of very rapid events (Time resolved crystallography). However, it is not as well-suited as monochromatic scattering for determining the full atomic structure of a crystal and therefore works better with crystals with relatively simple atomic arrangements.

The Laue back reflection mode records X-rays scattered backwards from a broad spectrum source. This is useful if the sample is too thick for X-rays to transmit through it. The diffracting planes in the crystal are determined by knowing that the normal to the diffracting plane bisects the angle between the incident beam and the diffracted beam. A Greninger chart can be used to interpret the back reflection Laue photograph.

Electron and neutron diffraction

Other particles, such as electrons and neutrons, may be used to produce a diffraction pattern. Although electron, neutron, and X-ray scattering are based on different physical processes, the resulting diffraction patterns are analyzed using the same coherent diffraction imaging techniques.

As derived below, the electron density within the crystal and the diffraction patterns are related by a simple mathematical method, the Fourier transform, which allows the density to be calculated relatively easily from the patterns. However, this works only if the scattering is *weak*, i.e., if the scattered beams are much less intense than the incoming beam. Weakly scattered beams pass through the remainder of the crystal without undergoing a second scattering event. Such re-scattered waves are called "secondary scattering" and hinder the analysis.

Any sufficiently thick crystal will produce secondary scattering, but since X-rays interact relatively weakly with the electrons, this is generally not a significant concern. By contrast, electron beams may produce strong secondary scattering even for relatively thin crystals (>100 nm). Since this thickness corresponds to the diameter of many viruses, a promising direction is the electron diffraction of isolated macromolecular assemblies, such as viral capsids and molecular machines, which may be carried out with a cryo-electron microscope. Moreover, the strong interaction of electrons with matter (about 1000 times stronger than for X-rays) allows determination of the atomic structure of extremely small volumes. The field of applications for electron crystallography ranges from bio molecules like membrane proteins over organic thin films to the complex structures of (nanocrystalline) intermetallic compounds and zeolites.

Neutron diffraction is an excellent method for structure determination, although it has been difficult to obtain intense, monochromatic beams of neutrons in sufficient quantities. Traditionally, nuclear reactors have been used, although the new Spallation Neutron Source holds much promise in the near future. Being uncharged, neutrons scatter much more readily from the atomic nuclei rather than from the electrons. Therefore, neutron scattering is very useful for observing the positions of light atoms with few electrons, especially hydrogen, which is essentially invisible in the X-ray diffraction. Neutron scattering also has the remarkable property that the solvent can be made invisible by adjusting the ratio of normal water, H_2O, and heavy water, D_2O.

METHODS

Overview of single-crystal X-ray diffraction

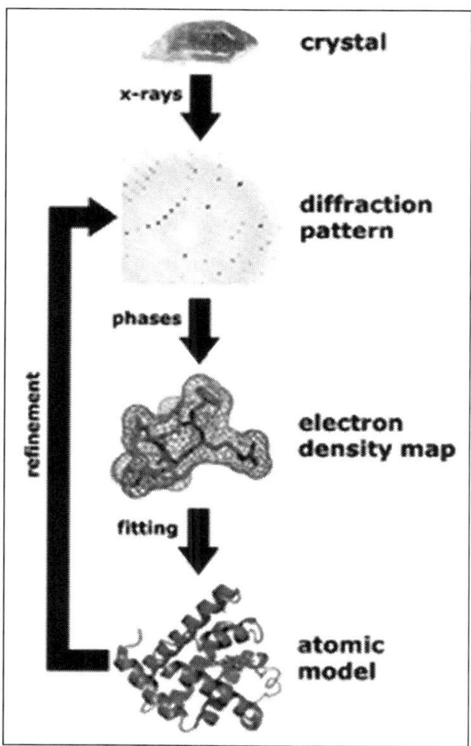

Fig. Workflow for solving the structure of a molecule by X-ray crystallography.

The oldest and most precise method of X-ray crystallography is *single-crystal X-ray diffraction*, in which a beam of X-rays strikes a single crystal, producing scattered beams. When they land on a piece of film or other detector, these beams make a *diffraction pattern* of spots; the strengths and angles of these beams are recorded as the crystal is gradually rotated. Each spot is called a *reflection*, since it corresponds to the reflection of the X-rays from one set of evenly spaced planes within the crystal. For single crystals of sufficient purity and regularity, X-ray diffraction data can determine the mean chemical bond lengths and angles to within a few thousandths of an angstrom and to within a few tenths of a degree, respectively. The atoms in a crystal are not static, but oscillate about their mean positions, usually by less than a few tenths of an angstrom. X-ray crystallography allows measuring the size of these oscillations.

Procedure

The technique of single-crystal X-ray crystallography has three basic steps. The first—and often most difficult—step is to obtain an adequate crystal of the material under study. The crystal should be sufficiently large (typically larger than 0.1 mm in all dimensions), pure in composition and regular in structure, with no significant internal imperfections such as cracks or twinning.

In the second step, the crystal is placed in an intense beam of X-rays, usually of a single wavelength (*monochromatic X-rays*), producing the regular pattern of reflections. As the crystal is gradually rotated, previous reflections disappear and new ones appear; the intensity of every spot is recorded at every orientation of the crystal. Multiple data sets may have to be collected, with each set covering slightly more than half a full rotation of the crystal and typically containing tens of thousands of reflections.

In the third step, these data are combined computationally with complementary chemical information to produce and refine a model of the arrangement of atoms within the crystal. The final, refined model of the atomic arrangement—now called a *crystal structure*—is usually stored in a public database.

Limitations

As the crystal's repeating unit, its unit cell, becomes larger and more complex, the atomic-level picture provided by X-ray crystallography becomes less well-resolved (more "fuzzy") for a given number of observed reflections. Two limiting cases of X-ray crystallography—"small-molecule" and "macromolecular" crystallography—are often discerned. *Small-molecule crystallography* typically involves crystals with fewer than 100 atoms in their asymmetric unit; such crystal structures are usually so well resolved that the atoms can be discerned as isolated "blobs" of electron density. By contrast, *macromolecular crystallography* often involves tens of thousands of atoms in the unit cell. Such crystal structures are generally less well-resolved (more "smeared out"); the atoms and chemical bonds appear as tubes of electron density, rather than as isolated atoms. In general, small molecules are also easier to crystallize than macromolecules; however, X-ray crystallography has proven possible even for viruses with hundreds of thousands of atoms. Though normally x-ray crystallography can only be performed if the sample is in crystal form, new research has been done into sampling non-crystalline forms of samples.

Crystallization

Although crystallography can be used to characterize the disorder in an impure or irregular crystal, crystallography generally requires a pure crystal of high regularity to solve the structure of a complicated arrangement of atoms.

Pure, regular crystals can sometimes be obtained from natural or synthetic materials, such as samples of metals, minerals or other macroscopic materials. The regularity of such crystals can sometimes be improved with macromolecular crystal annealing and other methods. However, in many cases, obtaining a diffraction-quality crystal is the chief barrier to solving its atomic-resolution structure.

Fig. A protein crystal seen under a microscope. Crystals used in X-ray crystallography may be smaller than a millimeter across.

Small-molecule and macromolecular crystallography differ in the range of possible techniques used to produce diffraction-quality crystals. Small molecules generally have few degrees of conformational freedom, and may be crystallized by a wide range of methods, such as chemical vapor deposition and recrystallization. By contrast, macromolecules generally have many degrees of freedom and their crystallization must be carried out to maintain a stable structure. For example, proteins and larger RNA molecules cannot be crystallized if their tertiary structure has been unfolded; therefore, the range of crystallization conditions is restricted to solution conditions in which such molecules remain folded.

Protein crystals are almost always grown in solution. The most common approach is to lower the solubility of its component molecules very gradually; if this is done too quickly, the molecules will precipitate from solution, forming a useless dust or amorphous gel on the bottom of the container. Crystal growth in solution is characterized by two steps: *nucleation* of a microscopic crystallite (possibly having only 100 molecules), followed by *growth* of that crystallite, ideally to a diffraction-quality crystal.

The solution conditions that favor the first step (nucleation) are not always the same conditions that favor the second step (subsequent growth). The crystallographer's goal is to identify solution conditions that favor the development of a single, large crystal, since larger crystals offer improved resolution of the molecule. Consequently, the solution conditions should *disfavor* the first step (nucleation) but *favor* the second (growth), so that only one large crystal forms per droplet. If nucleation is favored too much, a shower of small crystallites will form in the droplet, rather than one large crystal; if favored too little, no crystal will form whatsoever.

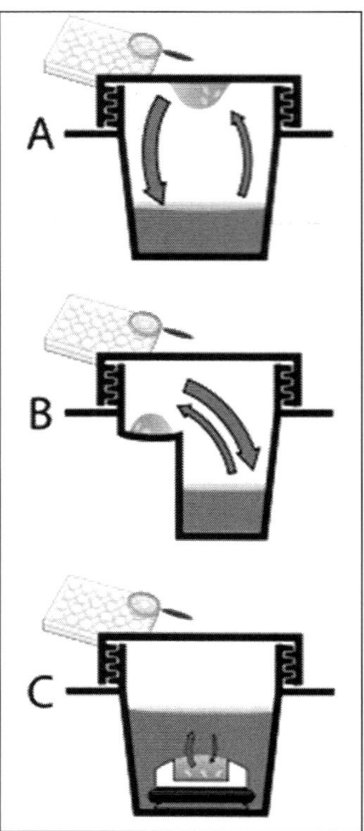

Fig. Three methods of preparing crystals,
A: Hanging drop. B: Sitting drop. C: Microdialysis

It is extremely difficult to predict good conditions for nucleation or growth of well-ordered crystals. In practice, favorable conditions are identified by *screening*; a very large batch of the molecules is prepared, and a wide variety of crystallization solutions are tested. Hundreds, even thousands, of solution conditions are generally tried before finding the successful one. The various conditions can use one or more physical mechanisms to lower the solubility of the molecule; for example, some may change the pH, some contain salts of the Hofmeister series or chemicals that lower the dielectric constant of the solution, and still others contain large polymers such as polyethylene glycol that drive the molecule out of solution by entropic effects.

It is also common to try several temperatures for encouraging crystallization, or to gradually lower the temperature so that the solution becomes supersaturated. These methods require large amounts of the target molecule, as they use high concentration of the molecule(s) to be crystallized. Due to the difficulty in obtaining such large quantities (milligrams) of crystallization-grade protein, robots have been developed that are capable of accurately dispensing crystallization trial drops that are in the order of 100 nanoliters in volume. This means that 10-fold less protein is used per experiment when compared to crystallization trials set up by hand (in the order of 1 microliter).

Several factors are known to inhibit or mar crystallization. The growing crystals are generally held at a constant temperature and protected from shocks or vibrations that might disturb their crystallization. Impurities in the molecules or in the crystallization solutions are often inimical to crystallization. Conformational flexibility in the molecule also tends to make crystallization less likely, due to entropy. Ironically, molecules that tend to self-assemble into regular helices are often unwilling to assemble into crystals. Crystals can be marred by twinning, which can occur when a unit cell can pack equally favorably in multiple orientations; although recent advances in computational methods may allow solving the structure of some twinned crystals. Having failed to crystallize a target molecule, a crystallographer may try again with a slightly modified version of the molecule; even small changes in molecular properties can lead to large differences in crystallization behavior.

Data collection

Mounting the crystal

Fig. Play media

Animation showing the five motions possible with a four-circle kappa goniometer. The rotations about each of the four angles φ, κ, ω and 2θ leave the crystal within the X-ray beam, but change the crystal orientation. The detector (red box) can be slid closer or further away from the crystal, allowing higher resolution data to be taken (if closer) or better discernment of the Bragg peaks (if further away).

The crystal is mounted for measurements so that it may be held in the X-ray beam and rotated. There are several methods of mounting. In the past, crystals were loaded into glass capillaries with the crystallization solution (the mother liquor). Nowadays, crystals of small molecules are typically attached with oil or glue to a glass fiber or a loop, which is made of nylon or plastic and attached to a solid rod. Protein crystals are scooped up by a loop, then flash-frozen with liquid nitrogen. This freezing reduces the radiation damage of the X-rays, as well as the noise in the Bragg peaks due to thermal motion (the Debye-Waller effect). However, untreated protein crystals often crack if flash-frozen; therefore, they are generally pre-soaked in a cryoprotectant solution before freezing. Unfortunately, this pre-soak may itself cause the crystal to crack, ruining it for crystallography. Generally, successful cryo-conditions are identified by trial and error.

The capillary or loop is mounted on a goniometer, which allows it to be positioned accurately within the X-ray beam and rotated. Since both the crystal and the beam are often very small, the crystal must be centered within the beam to within ~25 micrometers accuracy, which is aided by a camera focused on the crystal. The most common type of goniometer is the "kappa goniometer", which offers three angles of rotation: the ω angle, which rotates about an axis perpendicular to the beam; the κ angle, about an axis at ~50° to the ω axis; and, finally, the φ angle about the loop/capillary axis. When the κ angle is zero, the ω and φ axes are aligned. The ê rotation allows for convenient mounting of the crystal, since the arm in which the crystal is mounted may be swung out towards the crystallographer. The oscillations carried out during data collection (mentioned below) involve the ω axis only. An older type of goniometer is the four-circle goniometer, and its relatives such as the six-circle goniometer.

X-ray Sources

Rotating Anode

Small scale can be done on a local X-ray tube source, typically coupled with an image plate detector. These have the advantage of being (relatively) inexpensive and easy to maintain, and allow for quick screening and collection of samples. However, the wavelength light produced is limited by anode material, typically copper. Further, intensity is limited by the power applied and cooling capacity available to avoid melting the anode.

In such systems, electrons are boiled off of a cathode and accelerated through a strong electric potential of ~50 kV; having reached a high speed, the electrons collide with a metal plate, emitting *bremsstrahlung* and some strong spectral lines corresponding to the excitation of inner-shell electrons of the metal.

The most common metal used is copper, which can be kept cool easily, due to its high thermal conductivity, and which produces strong K_α and K_β lines. The K_β line is sometimes suppressed with a thin (~10μm) nickel foil. The simplest and cheapest variety of sealed X-ray tube has a stationary anode (the Crookes tube) and run with ~2 kW of electron beam power. The more expensive variety has a rotating-anode type source that run with ~14 kW of e-beam power.

X-rays are generally filtered (by use of X-Ray Filters) to a single wavelength (made monochromatic) and collimated to a single direction before they are allowed to strike the crystal. The filtering not only simplifies the data analysis, but also removes radiation that degrades the crystal without contributing useful information. Collimation is done either with a collimator (basically, a long tube) or with a clever arrangement of gently curved mirrors. Mirror systems are preferred for small crystals (under 0.3 mm) or with large unit cells (over 150 Å)

Synchrotron Radiation

Synchrotron radiation are some of the brightest lights on earth. It is the single most powerful tool available to X-ray crystallographers. It is made of X-ray beams generated in large machines called synchrotrons. These machines accelerate electrically charged particles, often electrons, to nearly the speed of light and confine them in a (roughly) circular loop using magnetic fields.

Synchrotrons are generally national facilities, each with several dedicated beamlines where data is collected without interruption. Synchrotrons were originally designed for use by high-energy physicists studying subatomic particles and cosmic phenomena. The largest component of each synchrotron is its electron storage ring. This ring is actually not a perfect circle, but a many-sided polygon. At each corner of the polygon, or sector, precisely aligned magnets bend the electron stream. As the electrons' path is bent, they emit bursts of energy in the form of X-rays.

Using synchrotron radiation frequently has specific requirements for X-ray crystallography. The intense ionizing radiation can cause radiation damage to samples, particularly macromolecular crystals. Cryo crystallography protects the sample from radiation damage, by freezing the crystal at liquid nitrogen temperatures (~100 K). However, synchrotron radiation frequently has the advantage of user selectable wavelengths, allowing for anomalous scattering experiments which maximizes anomalous signal. This is critical in experiments such as SAD and MAD.

Free Electron Laser

Recently, free electron lasers have been developed for use in X-ray crystallography. These are the brightest X-ray sources currently available; with the X-rays coming in femtosecond bursts. The intensity of the source is such that atomic resolution diffraction patterns can be resolved for crystals otherwise too small for collection. However, the intense light source also destroys the sample, requiring multiple crystals to be shot. As each crystal is randomly oriented in the beam, hundreds of thousands of individual diffraction images must be collected in order to get a complete data-set. This method, serial femtosecond crystallography, has been used in solving the structure of a number of protein crystal structures, sometimes noting differences with equivalent structures collected from synchrotron sources.

Recording the reflections

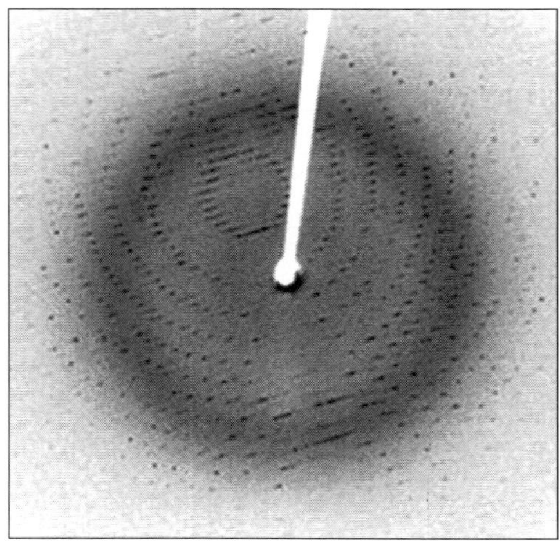

Fig. An X-ray diffraction pattern of a crystallized enzyme. The pattern of spots (*reflections*) and the relative strength of each spot (*intensities*) can be used to determine the structure of the enzyme.

When a crystal is mounted and exposed to an intense beam of X-rays, it scatters the X-rays into a pattern of spots or *reflections* that can be observed on a screen behind the crystal. A similar pattern may be seen by shining a laser pointer at a compact disc. The relative intensities of these spots provide the information to determine the arrangement of molecules within the crystal in atomic detail.

The intensities of these reflections may be recorded with photographic film, an area detector or with a charge-coupled device (CCD) image sensor. The peaks at small angles correspond to low-resolution data, whereas those at high angles represent high-resolution data; thus, an upper limit on the eventual resolution of the structure can be determined from the first few images. Some measures of diffraction quality can be determined at this point, such as the mosaicity of the crystal and its overall disorder, as observed in the peak widths. Some pathologies of the crystal that would render it unfit for solving the structure can also be diagnosed quickly at this point.

One image of spots is insufficient to reconstruct the whole crystal; it represents only a small slice of the full Fourier transform. To collect all the necessary information, the crystal must be rotated step-by-step through 180°, with an image recorded at every step; actually, slightly more than 180° is required to cover reciprocal space, due to the curvature of the Ewald sphere. However, if the crystal has a higher symmetry, a smaller angular range such as 90° or 45° may be recorded. The rotation axis should be changed at least once, to avoid developing a "blind spot" in reciprocal space close to the rotation axis. It is customary to rock the crystal slightly (by 0.5–2°) to catch a broader region of reciprocal space.

Multiple data sets may be necessary for certain phasing methods. For example, MAD phasing requires that the scattering be recorded at least three (and usually four, for redundancy) wavelengths of the incoming X-ray radiation. A single crystal may degrade too much during the collection of one data set, owing to radiation damage; in such cases, data sets on multiple crystals must be taken.

Data analysis

Crystal symmetry, unit cell, and image scaling

The recorded series of two-dimensional diffraction patterns, each corresponding to a different crystal orientation, is converted into a three-dimensional model of the electron density; the conversion uses the mathematical technique of Fourier transforms, which is explained below. Each spot corresponds to a different type of variation in the electron density; the crystallographer must determine *which* variation corresponds to *which* spot (*indexing*), the relative strengths of the spots in different images (*merging and scaling*) and how the variations should be combined to yield the total electron density (*phasing*).

Data processing begins with *indexing* the reflections. This means identifying the dimensions of the unit cell and which image peak corresponds to which position in reciprocal space. A byproduct of indexing is to determine the symmetry of the crystal, i.e., its *space group*. Some space groups can be eliminated from the beginning. For example, reflection symmetries cannot be observed in chiral molecules; thus, only 65 space groups of 230 possible are allowed for protein molecules which are almost always chiral. Indexing is generally accomplished using an *autoindexing* routine. Having assigned symmetry, the data is then *integrated*. This converts the hundreds of images containing the thousands of reflections into a single file, consisting of (at the very least) records of the Miller index of each reflection, and an intensity for each reflection (at this state the file often also includes error estimates and measures of partiality (what part of a given reflection was recorded on that image)).

A full data set may consist of hundreds of separate images taken at different orientations of the crystal. The first step is to merge and scale these various images, that is, to identify which peaks appear in two or more images (*merging*) and to scale the relative images so that they have a consistent intensity scale. Optimizing the intensity scale is critical because the relative intensity of the peaks is the key information from which the structure is determined. The repetitive technique of crystallographic data collection and the often high symmetry of crystalline materials cause the diffractometer to record many symmetry-equivalent reflections multiple times. This allows calculating the symmetry-related R-factor, a reliability index based upon how similar are the measured intensities of symmetry-equivalent reflections, thus assessing the quality of the data.

Initial phasing

The data collected from a diffraction experiment is a reciprocal space representation of the crystal lattice. The position of each diffraction 'spot' is governed by the size and shape of the unit cell, and the inherent symmetry within the crystal. The intensity of each diffraction 'spot' is recorded, and this intensity is proportional to the square of the *structure factor* amplitude. The structure factor is a complex number containing information relating to both the amplitude and phase of a wave. In order to obtain an interpretable *electron density map*, both amplitude and phase must be known (an electron density map allows a crystallographer to build a starting model of the molecule). The phase cannot be directly recorded during a diffraction experiment: this is known as the phase problem. Initial phase estimates can be obtained in a variety of ways:

- *Ab initio* phasing or direct methods – This is usually the method of choice for small molecules (<1000 non-hydrogen atoms), and has been used successfully to solve the phase problems for small proteins. If the resolution of the data is better than 1.4 Å (140 pm), direct methods can be used to obtain phase information, by exploiting known phase relationships between certain groups of reflections.
- Molecular replacement – if a related structure is known, it can be used as a search model in molecular replacement to determine the orientation and position of the molecules within the unit cell. The phases obtained this way can be used to generate *electron density maps*.
- Anomalous X-ray scattering (*MAD or SAD phasing*) – the X-ray wavelength may be scanned past an absorption edge of an atom, which changes the scattering in a known way. By recording full sets of reflections at three different wavelengths (far below, far above and in the middle of the absorption edge) one can solve for the substructure of the anomalously diffracting atoms and hence the structure of the whole molecule. The most popular method of incorporating anomalous scattering atoms into proteins is to express the protein in a methionine auxotroph (a host incapable of synthesizing methionine) in a media rich in seleno-methionine, which contains selenium atoms. A MAD experiment can then be conducted around the absorption

edge, which should then yield the position of any methionine residues within the protein, providing initial phases.

- Heavy atom methods (multiple isomorphous replacement) – If electron-dense metal atoms can be introduced into the crystal, direct methods or Patterson-space methods can be used to determine their location and to obtain initial phases. Such heavy atoms can be introduced either by soaking the crystal in a heavy atom-containing solution, or by co-crystallization (growing the crystals in the presence of a heavy atom). As in MAD phasing, the changes in the scattering amplitudes can be interpreted to yield the phases. Although this is the original method by which protein crystal structures were solved, it has largely been superseded by MAD phasing with selenomethionine.

Model building and phase refinement

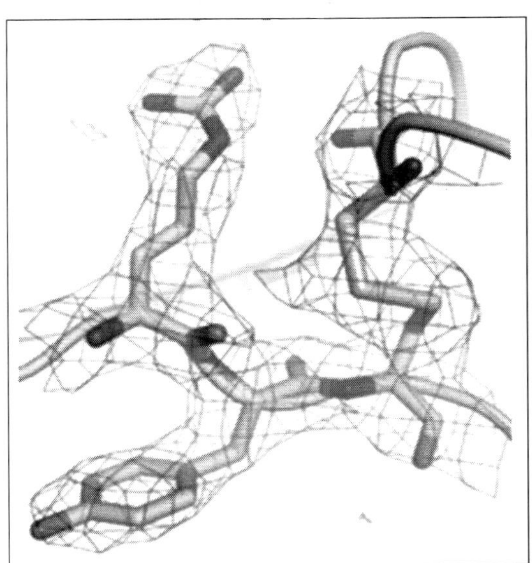

Fig. A protein crystal structure at 2.7 Å resolution. The mesh encloses the region in which the electron density exceeds a given threshold. The straight segments represent chemical bonds between the non-hydrogen atoms of an arginine (upper left), a tyrosine (lower left), a disulfide bond (upper right, in yellow), and some peptide groups (running left-right in the middle). The two curved green tubes represent spline fits to the polypeptide backbone.

Further information: Molecular modeling

Having obtained initial phases, an initial model can be built. This model can be used to refine the phases, leading to an improved model, and so on. Given a model of some atomic positions, these positions and their respective Debye-Waller factors (or B-factors, accounting for the thermal motion of the atom) can be refined to fit the observed diffraction data, ideally yielding a better set of phases. A new model can then be fit to the new electron density map and a further round of refinement is carried out. This continues until the correlation between the diffraction data and the model is maximized. The agreement is measured by an R-factor defined as

$$R = \frac{\sum_{\text{all reflections}} |F_o - F_c|}{\sum_{\text{all reflections}} |F_o|}$$

where F is the structure factor. A similar quality criterion is R_{free}, which is calculated from a subset (~10%) of reflections that were not included in the structure refinement. Both R factors depend on the resolution of the data. As a rule of thumb, R_{free} should be approximately the resolution in angstroms divided by 10; thus, a data-set with 2 Å resolution should yield a final R_{free} ~ 0.2. Chemical bonding features such as stereochemistry, hydrogen bonding and distribution of bond lengths and angles are complementary measures of the model quality. Phase bias is a serious problem in such iterative model building. *Omit maps* are a common technique used to check for this.

It may not be possible to observe every atom of the crystallized molecule – it must be remembered that the resulting electron density is an average of all the molecules within the crystal. In some cases, there is too much residual disorder in those atoms, and the resulting electron density for atoms existing in many conformations is smeared to such an extent that it is no longer detectable in the electron density map. Weakly scattering atoms such as hydrogen are routinely invisible. It is also possible for a single atom to appear multiple times in an electron density map, e.g., if a protein sidechain has multiple (<4) allowed conformations. In still other cases, the crystallographer may detect that the covalent structure deduced for the molecule was incorrect, or changed. For example, proteins may be cleaved or undergo post-translational modifications that were not detected prior to the crystallization.

Deposition of the structure

Once the model of a molecule's structure has been finalized, it is often deposited in a crystallographic database such as the Cambridge Structural Database (for small molecules), the Inorganic Crystal Structure Database (ICSD) (for inorganic compounds) or the Protein Data Bank (for protein structures). Many structures obtained in private commercial ventures to crystallize medicinally relevant proteins are not deposited in public crystallographic databases.

DIFFRACTION THEORY

The main goal of X-ray crystallography is to determine the density of electrons *f*(r) throughout the crystal, where r represents the three-dimensional position vector within the crystal. To do this, X-ray scattering is used to collect data about its Fourier transform F(q), which is inverted mathematically to obtain the density defined in real space, using the formula,

$$f(\mathbf{r}) = \frac{1}{(2\pi)^3} \int F(\mathbf{q}) e^{i\mathbf{q}\cdot\mathbf{r}} d\mathbf{q}$$

where the integral is taken over all values of q. The three-dimensional real vector q represents a point in reciprocal space, that is, to a particular oscillation in the electron density as one moves in the direction in which q points. The length of q corresponds to 2π divided by the wavelength of the oscillation. The corresponding formula for a Fourier transform will be used below

$$F(\mathbf{q}) = \int f(\mathbf{r}) e^{-i\mathbf{q}\cdot\mathbf{r}} d\mathbf{r}$$

where the integral is summed over all possible values of the position vector r within the crystal.

The Fourier transform F(q) is generally a complex number, and therefore has a magnitude $|F(q)|$ and a phase φ(q) related by the equation

$$F(\mathbf{q}) = |F(\mathbf{q})| e^{i\phi(\mathbf{q})}$$

The intensities of the reflections observed in X-ray diffraction give us the magnitudes $|F(q)|$ but not the phases φ(q). To obtain the phases, full sets of reflections are collected with known alterations to the scattering, either by modulating the wavelength past a certain absorption edge or by adding strongly scattering (i.e., electron-dense) metal atoms such as mercury. Combining the magnitudes and phases yields the full Fourier transform F(q), which may be inverted to obtain the electron density f(r).

Crystals are often idealized as being *perfectly* periodic. In that ideal case, the atoms are positioned on a perfect lattice, the electron density is perfectly periodic, and the Fourier transform F(q) is zero except when q belongs to the reciprocal lattice (the so-called *Bragg peaks*). In reality, however, crystals are not perfectly periodic; atoms vibrate about their mean position, and there may be disorder of various types, such as mosaicity, dislocations, various point defects, and heterogeneity in the conformation of crystallized molecules. Therefore, the Bragg peaks have a finite width and there may be significant *diffuse scattering*, a continuum of scattered X-rays that fall between the Bragg peaks.

Intuitive understanding by Bragg's law

An intuitive understanding of X-ray diffraction can be obtained from the Bragg model of diffraction. In this model, a given reflection is associated with a set of evenly spaced sheets running through the crystal, usually passing through the centers of the atoms of the crystal lattice. The orientation of a particular set of sheets is identified by its three Miller indices (h, k, l), and let their spacing be noted by d. William Lawrence Bragg proposed a model in which the incoming X-rays are scattered specularly (mirror-like) from each plane; from that assumption, X-rays scattered from adjacent planes will combine constructively (constructive interference) when the angle λ between the plane and the X-ray results in a path-length difference that is an integer multiple n of the X-ray wavelength λ.

$$2d\sin\theta = n\lambda$$

A reflection is said to be *indexed* when its Miller indices (or, more correctly, its reciprocal lattice vector components) have been identified from the known wavelength and the scattering angle 2θ. Such indexing gives the unit-cell parameters, the lengths and angles of the unit-cell, as well as its space group. Since Bragg's law does not interpret the relative intensities of the reflections, however, it is generally inadequate to solve for the arrangement of atoms within the unit-cell; for that, a Fourier transform method must be carried out.

Scattering as a Fourier transform

The incoming X-ray beam has a polarization and should be represented as a vector wave; however, for simplicity, let it be represented here as a scalar wave. We also ignore the complication of the time dependence of the wave and just concentrate on the wave's spatial dependence. Plane waves can be represented by a wave vector k_{in}, and so the strength of the incoming wave at time $t=0$ is given by

$$Ae^{ikinr}$$

At position r within the sample, let there be a density of scatterers $f(r)$; these scatterers should produce a scattered spherical wave of amplitude proportional to the local amplitude of the incoming wave times the number of scatterers in a small volume dV about r,

$$\text{amplitude of scattered wave} = Ae^{ik\bullet r}Sf(r)dV$$

where S is the proportionality constant.

Let's consider the fraction of scattered waves that leave with an outgoing wave-vector of k_{out} and strike the screen at r_{screen}. Since no energy is lost (elastic, not inelastic scattering), the wavelengths are the same as are the magnitudes of the wave-vectors $|k_{in}| = |k_{out}|$. From the time that the photon is scattered at r until it is absorbed at r_{screen}, the photon undergoes a change in phase

$$e^{ikout\bullet}(r_{screen}-r)$$

The net radiation arriving at r_{screen} is the sum of all the scattered waves throughout the crystal,

$$AS \int d\mathbf{r} f(\mathbf{r}) e^{i\mathbf{k}_{in}\cdot\mathbf{r}} e^{i\mathbf{k}_{out}\cdot(\mathbf{r}_{screen}-\mathbf{r})} = ASe^{i\mathbf{k}_{out}\cdot\mathbf{r}_{screen}} \int d\mathbf{r} f(\mathbf{r}) e^{i(\mathbf{k}_{in}-\mathbf{k}_{out})\cdot\mathbf{r}}$$

which may be written as a Fourier transform

$$ASe^{i\mathbf{k}_{out}\cdot\mathbf{r}_{screen}} \int d\mathbf{r} f(\mathbf{r}) e^{-i\mathbf{q}\cdot\mathbf{r}} = ASe^{i\mathbf{k}_{out}\cdot\mathbf{r}_{screen}} F(\mathbf{q})$$

where $q = k_{out} - k_{in}$. The measured intensity of the reflection will be square of this amplitude

$$A^2S^2|F(\mathbf{q})|^2$$

Friedel and Bijvoet mates

For every reflection corresponding to a point q in the reciprocal space, there is another reflection of the same intensity at the opposite point -q. This opposite reflection is known as the *Friedel mate* of the original reflection. This symmetry results from the mathematical fact that the density of electrons *f*(r) at a position r is always a real number. As noted above, *f*(r) is the inverse transform of its Fourier transform *F*(q); however, such an inverse transform is a complex number in general. To ensure that *f*(r) is real, the Fourier transform *F*(q) must be such that the Friedel mates *F*(–q) and *F*(q) are complex conjugates of one another. Thus, *F*(–q) has the same magnitude as *F*(q) but they have the opposite phase, i.e., φ(q) = –φ(q)

$$F(-\mathbf{q}) = |F(-\mathbf{q})|\, e^{i\phi(-\mathbf{q})} = F^*(\mathbf{q}) = |F(\mathbf{q})| e^{-i\phi(\mathbf{q})}$$

The equality of their magnitudes ensures that the Friedel mates have the same intensity $|F|$. This symmetry allows one to measure the full Fourier transform from only half the reciprocal space, e.g., by rotating the crystal slightly more than 180° instead of a full 360° revolution. In crystals with significant symmetry, even more reflections may have the same intensity (Bijvoet mates); in such cases, even less of the reciprocal space may need to be measured. In favorable cases of high symmetry, sometimes only 90° or even only 45° of data are required to completely explore the reciprocal space.

The Friedel-mate constraint can be derived from the definition of the inverse Fourier transform

$$f(\mathbf{r}) = \int \frac{d\mathbf{q}}{(2\pi)^3} F(\mathbf{q}) e^{i\mathbf{q}\cdot\mathbf{r}} = \int \frac{d\mathbf{q}}{(2\pi)^3} |F(\mathbf{q})|\, e^{i\phi(\mathbf{q})} e^{i\mathbf{q}\cdot\mathbf{r}}$$

Since Euler's formula states that e = cos(x) + i sin(x), the inverse Fourier transform can be separated into a sum of a purely real part and a purely imaginary part

$$f(\mathbf{r}) = \int \frac{d\mathbf{q}}{(2\pi)^3} |F(\mathbf{q})| e^{i(\phi+\mathbf{q}\cdot\mathbf{r})} = \int \frac{d\mathbf{q}}{(2\pi)^3} |F(\mathbf{q})| \cos(\phi + \mathbf{q}\cdot\mathbf{r}) + i \int \frac{d\mathbf{q}}{(2\pi)^3} |F(\mathbf{q})| \sin(\phi + \mathbf{q}\cdot\mathbf{r}) = I_{\cos} + iI_{\sin}$$

The function f(r) is real if and only if the second integral $I_{\sin}$ is zero for all values of r. In turn, this is true if and only if the above constraint is satisfied

$$I_{\sin} = \int \frac{d\mathbf{q}}{(2\pi)^3} |F(\mathbf{q})| \sin(\phi + \mathbf{q}\cdot\mathbf{r}) = \int \frac{d\mathbf{q}}{(2\pi)^3} |F(-\mathbf{q})| \sin(-\phi - \mathbf{q}\cdot\mathbf{r}) = -I_{\sin}$$

since $I_{\sin}$ = "$I_{\sin}$ implies that $I_{\sin}$=0.

Ewald's sphere

Each X-ray diffraction image represents only a slice, a spherical slice of reciprocal space, as may be seen by the Ewald sphere construction. Both k_{out} and k_{in} have the same length, due to the elastic scattering, since the wavelength has not changed. Therefore, they may be represented as two radial vectors in a sphere in reciprocal space, which shows the values of q that are sampled in a given diffraction image. Since there is a slight spread in the incoming wavelengths of the incoming X-ray beam, the values of $|F(q)|$ can be measured only for q vectors located between the two spheres corresponding to those radii. Therefore, to obtain a full set of Fourier transform data, it is necessary to rotate the crystal through slightly more than 180°, or sometimes less if sufficient symmetry is present. A full 360° rotation is not needed because of a symmetry intrinsic to the Fourier transforms of real functions (such as the electron density), but "slightly more" than 180° is needed to cover all of reciprocal space within a given resolution because of the curvature of the Ewald sphere. In practice, the crystal is rocked by a small amount (0.25-1°) to incorporate reflections near the boundaries of the spherical Ewald shells.

Patterson function

A well-known result of Fourier transforms is the autocorrelation theorem, which states that the autocorrelation c(r) of a function f(r)

$$c(\mathbf{r}) = \int d\mathbf{x} f(\mathbf{x}) f(\mathbf{x}+\mathbf{r}) = \int \frac{d\mathbf{q}}{(2\pi)^3} C(\mathbf{q}) e^{i\mathbf{q}\cdot\mathbf{r}}$$

has a Fourier transform C(q) that is the squared magnitude of F(q)

$$C(\mathbf{q}) = |F(\mathbf{q})|^2$$

Therefore, the autocorrelation function c(r) of the electron density (also known as the *Patterson function*) can be computed directly from the reflection intensities, without computing the phases. In principle, this could be used to determine the crystal structure directly; however, it is difficult to realize in practice. The autocorrelation function corresponds to the distribution of vectors between atoms in the crystal; thus, a crystal of *N* atoms in its unit cell may have *N(N-1)* peaks in its Patterson function. Given the inevitable errors in measuring the intensities, and the mathematical difficulties of reconstructing atomic positions from the interatomic vectors, this technique is rarely used to solve structures, except for the simplest crystals.

Advantages of a crystal

In principle, an atomic structure could be determined from applying X-ray scattering to non-crystalline samples, even to a single molecule. However, crystals offer a much stronger signal due to their periodicity. A crystalline sample is by definition periodic; a crystal is composed of many unit cells repeated indefinitely in three independent directions. Such periodic systems have a Fourier transform that is concentrated at periodically repeating points in reciprocal space known as *Bragg peaks*; the Bragg peaks correspond to the reflection spots observed in the diffraction image. Since the amplitude at these reflections grows linearly with the number *N* of scatterers, the observed *intensity* of these spots should grow quadratically, like *N*. In other words, using a crystal concentrates the weak scattering of the individual unit cells into a much more powerful, coherent reflection that can be observed above the noise. This is an example of constructive interference.

In a liquid, powder or amorphous sample, molecules within that sample are in random orientations. Such samples have a continuous Fourier spectrum that uniformly spreads its amplitude thereby reducing the measured signal intensity, as is observed in SAXS.

More importantly, the orientational information is lost. Although theoretically possible, it is experimentally difficult to obtain atomic-resolution structures of complicated, asymmetric molecules from such rotationally averaged data. An intermediate case is fiber diffraction in which the subunits are arranged periodically in at least one dimension.

NUCLEAR MAGNETIC RESONANCE

Nuclear magnetic resonance (NMR) is a physical phenomenon in which nuclei in a magnetic field absorb and re-emit electromagnetic radiation. This energy is at a specific resonance frequency which depends on the strength of the magnetic field and the magnetic properties of the isotope of the atoms; in practical applications, the frequency is similar to VHF and UHF television broadcasts (60–1000 MHz). NMR allows the observation of specific quantum mechanical magnetic properties of the atomic nucleus. Many scientific techniques exploit NMR phenomena to study molecular physics, crystals, and non-crystalline materials through NMR spectroscopy. NMR is also routinely used in advanced medical imaging techniques, such as in magnetic resonance imaging (MRI).

All isotopes that contain an odd number of protons and/or of neutrons (see Isotope) have an intrinsic magnetic moment and angular momentum, in other words a nonzero spin, while all nuclides with even numbers of both have a total spin of zero. The most commonly studied nuclei are 1H and 13C, although nuclei from isotopes of many other elements (e.g. 2H, 6Li, 10B, 11B, 14N, 15N, 17O, 19F, 23Na, 29Si, 31P, 35Cl, 113Cd, 129Xe, 195Pt) have been studied by high-field NMR spectroscopy as well.

A key feature of NMR is that the resonance frequency of a particular substance is directly proportional to the strength of the applied magnetic field. It is this feature that is exploited in imaging techniques; if a sample is placed in a non-uniform magnetic field then the resonance frequencies of the sample's nuclei depend on where in the field they are located. Since the resolution of the imaging technique depends on the magnitude of magnetic field gradient, many efforts are made to develop increased field strength, often using superconductors. The effectiveness of NMR can also be improved using hyperpolarization, and/or using two-dimensional, three-dimensional and higher-dimensional multi-frequency techniques.

The principle of NMR usually involves two sequential steps:

- The alignment (polarization) of the magnetic nuclear spins in an applied, constant magnetic field B_0.
- The perturbation of this alignment of the nuclear spins by employing an electro-magnetic, usually radio frequency (RF) pulse. The required perturbing frequency is dependent upon the static magnetic field (H_0) and the nuclei of observation.

The two fields are usually chosen to be perpendicular to each other as this maximizes the NMR signal strength. The resulting response by the total magnetization (M) of the nuclear spins is the phenomenon that is exploited in NMR spectroscopy and magnetic resonance imaging. Both use intense applied magnetic fields (H_0) in order to achieve dispersion and very high stability to deliver spectral resolution, the details of which are described by chemical shifts, the Zeeman effect, and Knight shifts (in metals).

NMR phenomena are also utilized in low-field NMR, NMR spectroscopy and MRI in the Earth's magnetic field (referred to as Earth's field NMR), and in several types of magnetometers.

Nuclear magnetic resonance was first described and measured in molecular beams by Isidor Rabi in 1938, by extending the Stern–Gerlach experiment, and in 1944, Rabi was awarded the Nobel Prize in Physics for this work. In 1946, Felix Bloch and Edward Mills Purcell expanded the technique for use on liquids and solids, for which they shared the Nobel Prize in Physics in 1952.

Yevgeny Zavoisky likely observed nuclear magnetic resonance in 1941, well before Felix Bloch and Edward Mills Purcell, but dismissed the results as not reproducible.

Purcell had worked on the development of radar during World War II at the Massachusetts Institute of Technology's Radiation Laboratory. His work during that project on the production and detection of radio frequency power and on the absorption of such RF power by matter laid the foundation for Rabi's discovery of NMR.

Rabi, Bloch, and Purcell observed that magnetic nuclei, like 1H and 31P, could absorb RF energy when placed in a magnetic field and when the RF was of a frequency specific to the identity of the nuclei. When this absorption occurs, the nucleus is described as being *in resonance*. Different atomic nuclei within a molecule resonate at different (radio) frequencies for the same magnetic field strength. The observation of such magnetic resonance frequencies of the nuclei present in a molecule allows any trained user to discover essential chemical and structural information about the molecule.

The development of NMR as a technique in analytical chemistry and biochemistry parallels the development of electromagnetic technology and advanced electronics and their introduction into civilian use.

THEORY OF NUCLEAR MAGNETIC RESONANCE

Nuclear spin and magnets

All nucleons, that is neutrons and protons, composing any atomic nucleus, have the intrinsic quantum property of spin. The overall spin of the nucleus is determined by the spin quantum number *S*. If the number of both the protons and neutrons in a given nuclide are even then $S = 0$, i.e. there is no overall spin. Then, just as electrons pair up in atomic orbitals, so do even numbers of protons or even numbers of neutrons (which are also spin-D_2 particles and hence fermions) pair up giving zero overall spin.

However, a proton and neutron will have lower energy when their spins are parallel, not anti-parallel. Parallel spin alignment does not infringe upon the Pauli Exclusion Principle. The lowering of energy for parallel spins has to do with the quark structure of these two nucleons. Therefore, the spin ground state for the deuteron (the deuterium nucleus, or the H isotope of hydrogen)—that has only a proton and a neutron—corresponds to a spin value of 1, *not of zero*.

The single, isolated deuteron therefore exhibits an NMR absorption spectrum characteristic of a quadrupolar nucleus of spin 1, which in the "rigid" state at very low temperatures is a characteristic ('Pake') *doublet*, (not a singlet as for a single, isolated H, or any other isolated fermion or dipolar nucleus of spin 1/2). On the other hand, because of the Pauli Exclusion Principle, the tritium isotope of hydrogen must have a pair of anti-parallel spin neutrons (of total spin zero for the neutron-spin pair), plus a proton of spin 1/2. Therefore, the character of the tritium nucleus is again magnetic dipolar, *not quadrupolar*—

like its non-radioactive deuteron cousin—and the tritium nucleus total spin value is again 1/2, just like for the simpler, abundant hydrogen isotope, H nucleus (the *proton*). The NMR absorption (radio) frequency for tritium is however slightly higher than that of H because the tritium nucleus has a slightly higher gyromagnetic ratio than H. In many other cases of *non-radioactive* nuclei, the overall spin is also non-zero. For example, the 27Al nucleus has an overall spin value $S = D_2$.

A non-zero spin is thus always associated with a non-zero magnetic moment (μ) via the relation $\mu = \gamma S$, where γ is the gyromagnetic ratio. It is this magnetic moment that allows the observation of NMR absorption spectra caused by transitions between nuclear spin levels. Most nuclides (with some rare exceptions) that have both even numbers of protons and even numbers of neutrons, also have zero nuclear magnetic moments, and they also have zero magnetic dipole and quadrupole moments. Hence, such nuclides do not exhibit any NMR absorption spectra. Thus, 18O is an example of a nuclide that has no NMR absorption, whereas 13C, 31P, 35Cl and 37Cl are nuclides that do exhibit NMR absorption spectra. The last two nuclei are quadrupolar nuclei whereas the preceding two nuclei (13C and 31P) are dipolar ones.

Electron spin resonance (ESR) is a related technique in which transitions between electronic spin levels are detected rather than nuclear ones. The basic principles are similar but the instrumentation, data analysis, and detailed theory are significantly different. Moreover, there is a much smaller number of molecules and materials with unpaired electron spins that exhibit ESR (or electron paramagnetic resonance (EPR)) absorption than those that have NMR absorption spectra. ESR has much higher sensitivity than NMR does.

Values of spin angular momentum

The angular momentum associated with nuclear spin is quantized. This means both that the magnitude of angular momentum is quantized (i.e. S can only take on a restricted range of values), and also that the orientation of the associated angular momentum is quantized. The associated quantum number is known as the magnetic quantum number, m, and can take values from $+S$ to “S, in integer steps. Hence for any given nucleus, there are a total of $2S + 1$ angular momentum states.

The z-component of the angular momentum vector (S) is therefore $S_z = m\hbar$, where $\hbar$ is the reduced Planck constant. The z-component of the magnetic moment is simply:

$$\mu_z = \gamma S_z = \gamma m\hbar$$

Spin behavior in a magnetic field

Consider nuclei which have a spin of one-half, like 1H, 13C or 19F. The nucleus has two possible spin states: $m = -/_2$ or $m = /_2$ (also referred to as spin-up and spin-down, or sometimes α and β spin states, respectively). These states are degenerate, that is they have the same energy. Hence the number of atoms in these two states will be approximately equal at thermal equilibrium.

If a nucleus is placed in a magnetic field, however, the interaction between the nuclear magnetic moment and the external magnetic field mean the two states no longer have the same energy. The energy of a magnetic moment μ when in a magnetic field B_0 is given by:

$$E = -\boldsymbol{\mu} \cdot \mathbf{B_0} = -\mu_x B_{0x} - \mu_y B_{0y} - \mu_z B_{0z} .$$

Usually the *z* axis is chosen to be along B_0, and the above expression reduces to:

$$E = -\mu_z B_0 \ ,$$

or alternatively:

$$E = -\gamma m \hbar B_0 \ .$$

As a result the different nuclear spin states have different energies in a non-zero magnetic field. In less formal language, we can talk about the two spin states of a spin/$_2$ as being *aligned* either with or against the magnetic field. If ã is positive (true for most isotopes) then m =/$_2$ is the lower energy state.

The energy difference between the two states is:

$$\Delta E = \gamma \hbar B_0 \ ,$$

and this difference results in a small population bias toward the lower energy state.

Magnetic resonance by nuclei

Resonant absorption by nuclear spins will occur only when electromagnetic radiation of the correct frequency (e.g., equaling the Larmor precession rate) is being applied to match the energy difference between the nuclear spin levels in a constant magnetic field of the appropriate strength. The energy of an absorbed photon is then $E = h\nu_0$, where í$_0$ is the resonance radiofrequency that has to match (that is, it has to be equal to the Larmor precession frequency ν_L of the nuclear magnetization in the constant magnetic field B_0). Hence, a magnetic resonance absorption will only occur when $\Delta E = h\nu_0$, which is when $\nu_0 = \gamma B_0/(2\pi)$. Such magnetic resonance frequencies typically correspond to the radio frequency (or RF) range of the electromagnetic spectrum for magnetic fields up to roughly 20 T. It is this magnetic resonant absorption which is detected in NMR.

Nuclear shielding

It might appear from the above that all nuclei of the same nuclide (and hence the same γ) would resonate at the same frequency. This is not the case.

The most important perturbation of the NMR frequency for applications of NMR is the "shielding" effect of the surrounding shells of electrons. Electrons, similar to the nucleus, are also charged and rotate with a spin to produce a magnetic field opposite to the magnetic field produced by the nucleus. In general, this electronic shielding reduces the magnetic field *at the nucleus* (which is what determines the NMR frequency).

As a result the energy gap is reduced, and the frequency required to achieve resonance is also reduced. This shift in the NMR frequency due to the electronic molecular orbital coupling to the external magnetic field is called chemical shift, and it explains why NMR is able to probe the chemical structure of molecules, which depends on the electron density distribution in the corresponding molecular orbitals. If a nucleus in a specific chemical group is shielded to a higher degree by a higher electron density of its surrounding molecular orbital, then its NMR frequency will be shifted "upfield" (that is, a lower chemical shift), whereas if it is less shielded by such surrounding electron density, then its NMR frequency will be shifted "downfield" (that is, a higher chemical shift).

Unless the local symmetry of such molecular orbitals is very high (leading to "isotropic" shift), the shielding effect will depend on the orientation of the molecule with respect to the external field (B_0). In solid-state NMR spectroscopy, magic angle spinning is required to average out this orientation dependence in order to obtain values close to the average chemical shifts. This is unnecessary in conventional NMR investigations of molecules, since rapid "molecular tumbling" averages out the chemical shift anisotropy (CSA). In this case, the term "average" chemical shift (ACS) is used.

Relaxation

The process called population relaxation refers to nuclei that return to the thermodynamic state in the magnet. This process is also called T_1, "spin-lattice" or "longitudinal magnetic" relaxation, where T_1 refers to the mean time for an individual nucleus to return to its thermal equilibrium state of the spins. Once the nuclear spin population is relaxed, it can be probed again, since it is in the initial, equilibrium (mixed) state.

The precessing nuclei can also fall out of alignment with each other (returning the net magnetization vector to a non-precessing field) and stop producing a signal. This is called T_2 or *transverse relaxation*. Because of the difference in the actual relaxation mechanisms involved (for example, inter-molecular vs. intra-molecular magnetic dipole-dipole interactions), T_1 is usually (except in rare cases) longer than T_2 (that is, slower spin-lattice relaxation, for example because of smaller dipole-dipole interaction effects). In practice, the value of T_2^* which is the actually observed decay time of the observed NMR signal, or free induction decay, (to 1/e of the initial amplitude immediately after the resonant RF pulse)— also depends on the static magnetic field inhomogeneity, which is quite significant. (There is also a smaller but significant contribution to the observed FID shortening from the RF inhomogeneity of the resonant pulse). In the corresponding FT-NMR spectrum—meaning the Fourier transform of the free induction decay—the T_2^* time is inversely related to the width of the NMR signal in frequency units. Thus, a nucleus with a long T_2 relaxation time gives rise to a very sharp NMR peak in the FT-NMR spectrum for a very homogeneous ("well-shimmed") static magnetic field, whereas nuclei with shorter T_2 values give rise to broad FT-NMR peaks even when the magnet is shimmed well. Both T_1 and T_2 depend on the rate of molecular motions as well as the gyromagnetic ratios of both the resonating and their strongly interacting, next-neighbor nuclei that are not at resonance.

A Hahn echo decay experiment can be used to measure the dephasing time, as shown in the animation below. The size of the echo is recorded for different spacings of the two pulses. This reveals the decoherence which is not refocused by the π pulse. In simple cases, an exponential decay is measured which is described by the T_2 time.

NMR SPECTROSCOPY

NMR spectroscopy is one of the principal techniques used to obtain physical, chemical, electronic and structural information about molecules due to either the chemical shift, Zeeman effect, or the Knight shift effect, or a combination of both, on the resonant frequencies of the nuclei present in the sample. It is a powerful technique that can provide detailed information on the topology, dynamics and three-dimensional structure of molecules in solution and the solid state. Thus, structural and dynamic information is obtainable (with or without "magic angle" spinning (MAS)) from NMR studies of quadrupolar nuclei (that is, those nuclei with spin S $>/_2$) even in the presence of magnetic "dipole-dipole" interaction broadening (or simply, dipolar broadening) which is always much smaller than the quadrupolar interaction strength because it is a magnetic vs. an electric interaction effect.

Additional structural and chemical information may be obtained by performing double-quantum NMR experiments for quadrupolar nuclei such as 2H. Also, nuclear magnetic resonance is one of the techniques that has been used to design quantum automata, and also build elementary quantum computers.

Continuous-wave (CW) spectroscopy

In its first few decades, nuclear magnetic resonance spectrometers used a technique known as continuous-wave spectroscopy (CW spectroscopy). Although NMR spectra could be, and have been, obtained using a fixed magnetic field and sweeping the frequency of the electromagnetic radiation, this more typically involved using a fixed frequency source and varying the current (and hence magnetic field) in an electromagnet to observe the resonant absorption signals. This is the origin of the counterintuitive, but still common, "high field" and "low field" terminology for low frequency and high frequency regions respectively of the NMR spectrum.

CW spectroscopy is inefficient in comparison with Fourier analysis techniques (see below) since it probes the NMR response at individual frequencies in succession. Since the NMR signal is intrinsically weak, the observed spectrum suffers from a poor signal-to-noise ratio. This can be mitigated by signal averaging i.e. adding the spectra from repeated measurements. While the NMR signal is constant between scans and so adds linearly, the random noise adds more slowly – proportional to the square-root of the number of spectra (see random walk). Hence the overall signal-to-noise ratio increases as the square-root of the number of spectra measured.

Fourier-transform spectroscopy

Most applications of NMR involve full NMR spectra, that is, the intensity of the NMR signal as a function of frequency. Early attempts to acquire the NMR spectrum more efficiently than simple CW methods involved illuminating the target simultaneously with more than one frequency. A revolution in NMR occurred when short pulses of radio-frequency radiation began to be used—centered at the middle of the NMR spectrum. In simple terms, a short pulse of a given "carrier" frequency "contains" a range of frequencies centered about the carrier frequency, with the range of excitation (bandwidth) being inversely proportional to the pulse duration, i.e. the Fourier transform of a short pulse contains contributions from all the frequencies in the neighborhood of the principal frequency. The restricted range of the NMR frequencies made it relatively easy to use short (millisecond to microsecond) radio frequency pulses to excite the entire NMR spectrum.

Applying such a pulse to a set of nuclear spins simultaneously excites all the single-quantum NMR transitions. In terms of the net magnetization vector, this corresponds to tilting the magnetization vector away from its equilibrium position (aligned along the external magnetic field). The out-of-equilibrium magnetization vector precesses about the external magnetic field vector at the NMR frequency of the spins. This oscillating magnetization vector induces a current in a nearby pickup coil, creating an electrical signal oscillating at the NMR frequency.

This signal is known as the free induction decay (FID), and it contains the vector sum of the NMR responses from all the excited spins. In order to obtain the frequency-domain NMR spectrum (NMR absorption intensity vs. NMR frequency) this time-domain signal (intensity vs. time) must be Fourier transformed. Fortunately the development of Fourier Transform NMR coincided with the development of digital computers and the digital Fast Fourier

Transform. Fourier methods can be applied to many types of spectroscopy. (See the full article on Fourier transform spectroscopy.)

Richard R. Ernst was one of the pioneers of pulse NMR, and he won a Nobel Prize in chemistry in 1991 for his work on Fourier Transform NMR and his development of multi-dimensional NMR (see below).

Multi-dimensional NMR Spectroscopy

The use of pulses of different shapes, frequencies and durations in specifically designed patterns or *pulse sequences* allows the spectroscopist to extract many different types of information about the molecule. Multi-dimensional nuclear magnetic resonance spectroscopy is a kind of FT NMR in which there are at least two pulses and, as the experiment is repeated, the pulse sequence is systematically varied. In *multidimensional nuclear magnetic resonance* there will be a sequence of pulses and, at least, one variable time period. In three dimensions, two time sequences will be varied. In four dimensions, three will be varied.

There are many such experiments. In one, these time intervals allow (amongst other things) magnetization transfer between nuclei and, therefore, the detection of the kinds of nuclear-nuclear interactions that allowed for the magnetization transfer. Interactions that can be detected are usually classified into two kinds. There are *through-bond* interactions and *through-space* interactions, the latter usually being a consequence of the nuclear Overhauser effect. Experiments of the nuclear Overhauser variety may be employed to establish distances between atoms, as for example by 2D-FT NMR of molecules in solution.

Although the fundamental concept of 2D-FT NMR was proposed by Jean Jeener from the Free University of Brussels at an International Conference, this idea was largely developed by Richard Ernst who won the 1991 Nobel prize in Chemistry for his work in FT NMR, including multi-dimensional FT NMR, and especially 2D-FT NMR of small molecules. Multi-dimensional FT NMR experiments were then further developed into powerful methodologies for studying biomolecules in solution, in particular for the determination of the structure of biopolymers such as proteins or even small nucleic acids.

In 2002 Kurt Wüthrich shared the Nobel Prize in Chemistry (with John Bennett Fenn and Koichi Tanaka) for his work with protein FT NMR in solution.

Solid-state NMR spectroscopy

This technique complements X-ray crystallography in that it is frequently applicable to molecules in a liquid or liquid crystal phase, whereas crystallography, as the name implies, is performed on molecules in a solid phase. Though nuclear magnetic resonance is used to study solids, extensive atomic-level molecular structural detail is especially challenging to obtain in the solid state. There is little signal averaging by thermal motion in the solid state, where most molecules can only undergo restricted vibrations and rotations at room temperature, each in a slightly different electronic environment, therefore exhibiting a different NMR absorption peak. Such a variation in the electronic environment of the resonating nuclei results in a blurring of the observed spectra—which is often only a broad Gaussian band for non-quadrupolar spins in a solid- thus making the interpretation of such "dipolar" and "chemical shift anisotropy" (CSA) broadened spectra either very difficult or impossible.

Professor Raymond Andrew at the University of Nottingham in the UK pioneered the development of high-resolution solid-state nuclear magnetic resonance. He was the first to report the introduction of the MAS (magic angle sample spinning; MASS) technique that allowed him to achieve spectral resolution in solids sufficient to distinguish between chemical groups with either different chemical shifts or distinct Knight shifts. In MASS, the sample is spun at several kilohertz around an axis that makes the so-called magic angle θ_m (which is ~54.74°, where $\cos\theta_m = 1/3$) with respect to the direction of the static magnetic field B_0; as a result of such magic angle sample spinning, the chemical shift anisotropy bands are averaged to their corresponding average (isotropic) chemical shift values. The above expression involving $\cos\theta_m$ has its origin in a calculation that predicts the magnetic dipolar interaction effects to cancel out for the specific value of θ_m called the magic angle. One notes that correct alignment of the sample rotation axis as close as possible to θ_m is essential for cancelling out the dipolar interactions whose strength for angles sufficiently far from θ_m is usually greater than ~10 kHz for C-H bonds in solids, for example, and it is thus greater than their CSA values.

There are different angles for the sample spinning relative to the applied field for the averaging of quadrupole interactions and paramagnetic interactions, correspondingly ~30.6° and ~70.1°

A concept developed by Sven Hartmann and Erwin Hahn was utilized in transferring magnetization from protons to less sensitive nuclei (popularly known as cross-polarization) by M.G. Gibby, Alex Pines and John S. Waugh. Then, Jake Schaefer and Ed Stejskal demonstrated also the powerful use of cross-polarization under MASS conditions (CP-MAS) which is now routinely employed to measure high resolution spectra of low-abundance and low-sensitivity nuclei, namely carbon-13, in solids.

Sensitivity

Because the intensity of nuclear magnetic resonance signals and, hence, the sensitivity of the technique depends on the strength of the magnetic field the technique has also advanced over the decades with the development of more powerful magnets. Advances made in audio-visual technology have also improved the signal-generation and processing capabilities of newer instruments. As noted above, the sensitivity of nuclear magnetic resonance signals is also dependent on the presence of a magnetically susceptible nuclide and, therefore, either on the natural abundance of such nuclides or on the ability of the experimentalist to artificially enrich the molecules, under study, with such nuclides.

The most abundant naturally occurring isotopes of hydrogen and phosphorus (for example) are both magnetically susceptible and readily useful for nuclear magnetic resonance spectroscopy. In contrast, carbon and nitrogen have useful isotopes but which occur only in very low natural abundance.

Other limitations on sensitivity arise from the quantum-mechanical nature of the phenomenon. For quantum states separated by energy equivalent to radio frequencies, thermal energy from the environment causes the populations of the states to be close to equal. Since incoming radiation is equally likely to cause stimulated emission (a transition from the upper to the lower state) as absorption, the NMR effect depends on an excess of nuclei in the lower states. Several factors can reduce sensitivity, including

- Increasing temperature, which evens out the population of states. Conversely, low temperature NMR can sometimes yield better results than room-temperature NMR, providing the sample remains liquid.
- Saturation of the sample with energy applied at the resonant radiofrequency. This manifests in both CW and pulsed NMR; in the first case (CW) this happens by using too much continuous power that keeps the upper spin levels completely populated; in the second case (pulsed), each pulse (that is at least a 90° pulse) leaves the sample saturated, and four to five times the (longitudinal) relaxation time (5 T_1) must pass before the next pulse or pulse sequence can be applied. For single pulse experiments, shorter RF pulses that tip the magnetization by less than 90° can be used, which loses some intensity of the signal, but allows for shorter *recycle delays*. The optimum there is called an *Ernst angle*, after the Nobel laureate. Especially in solid state NMR, or in samples with very few nuclei with spins > 0, (diamond with the natural 1% of Carbon-13 is especially troublesome here) the longitudinal relaxation times can be on the range of hours, while for proton-NMR they are more on the range of one second.
- Non-magnetic effects, such as electric-quadrupole coupling of spin-1 and spin-$/_2$ nuclei with their local environment, which broaden and weaken absorption peaks. 14N, an abundant spin-1 nucleus, is difficult to study for this reason. High resolution NMR instead probes molecules using the rarer 15N isotope, which has spin-$/_2$.

Isotopes

Many isotopes of chemical elements can be used for NMR analysis. Commonly used nuclei:

- Is the nucleus most sensitive to NMR signal (apart from 3H which is not commonly used due to its instability and radioactivity). Proton NMR produces narrow chemical shift with sharp signals. Fast acquisition of quantitative results (peak integrals in stoichiometric ratio) is possible due to short relaxation time. The 1H signal has been the sole diagnostic nucleus used for clinical magnetic resonance imaging.
- 2H, a spin 1 nucleus commonly utilized as signal-free medium in the form of deuterated solvents during proton NMR, to avoid signal interference from hydrogen-containing solvents in measurement of 1H solutes. Also used in determining the behavior of lipids in lipid membranes and other solids or liquid crystals as it is a relatively non-perturbing label which can selectively replace 1H. Alternatively, 2H can be detected in media specially labeled with 2H. Deuterium resonance is commonly used in high-resolution NMR spectroscopy to monitor drifts in the magnetic field strength (lock) and to improve the homogeneity of the external magnetic field.
- 3He, is very sensitive to NMR. There is a very low percentage in natural helium, and subsequently has to be purified from 4He. It is used mainly in studies of endohedral fullerenes, where its chemical inertness is beneficial to ascertaining the structure of the entrapping fullerene.
- 11B, more sensitive than 10B, yields sharper signals. Quartz tubes must be used as borosilicate glass interferes with measurement.
- 13C spin-1/2, is widely used, despite its relative paucity in naturally occurring carbon (approximately 1%). It is stable to nuclear decay. Since there is a low percentage in natural carbon, spectrum acquisition on samples which have not been experimentally enriched in 13C takes a long time. Frequently used for labeling of compounds in synthetic and metabolic studies. Has low sensitivity and wide chemical shift, yields sharp signals. Low percentage makes it useful by preventing spin-spin couplings and makes the spectrum appear less crowded. Slow relaxation means that spectra are not integrable unless long acquisition times are used.
- 14N, spin-1, medium sensitivity nucleus with wide chemical shift. Its large quadrupole moment interferes in acquisition of high resolution spectra, limiting usefulness to smaller molecules and functional groups with a high degree of symmetry such as the headgroups of lipids.
- 15N, spin-1/2, relatively commonly used. Can be used for labeling

compounds. Nucleus very insensitive but yields sharp signals. Low percentage in natural nitrogen together with low sensitivity requires high concentrations or expensive isotope enrichment.

- 17O, spin-5/2, low sensitivity and very low natural abundance (0.037%), wide chemical shifts range (up to 2000 ppm). Quadrupole moment causing a line broadening. Used in metabolic and biochemical studies in studies of chemical equilibria.
- 19F, spin-1/2, relatively commonly measured. Sensitive, yields sharp signals, has wide chemical shift.
- 31P, spin-1/2, 100% of natural phosphorus. Medium sensitivity, wide chemical shifts range, yields sharp lines. Spectra tend to have a moderate amount of noise. Used in biochemical studies and in coordination chemistry where phosphorus containing ligands are involved.
- 35Cl and 37Cl, broad signal. 35Cl significantly more sensitive, preferred over 37Cl despite its slightly broader signal. Organic chlorides yield very broad signals, its use is limited to inorganic and ionic chlorides and very small organic molecules.
- 43Ca, used in biochemistry to study calcium binding to DNA, proteins, etc. Moderately sensitive, very low natural abundance.
- 195Pt, used in studies of catalysts and complexes.

APPLICATIONS

Medicine

The application of nuclear magnetic resonance best known to the general public is magnetic resonance imaging for medical diagnosis and magnetic resonance microscopy in research settings, however, it is also widely used in chemical studies, notably in NMR spectroscopy such as proton NMR, carbon-13 NMR, deuterium NMR and phosphorus-31 NMR. Biochemical information can also be obtained from living tissue (e.g. human brain tumors) with the technique known as in vivo magnetic resonance spectroscopy or chemical shift NMR Microscopy.

These studies are possible because nuclei are surrounded by orbiting electrons, which are charged particles that generate small, local magnetic fields that add to or subtract from the external magnetic field, and so will partially shield the nuclei. The amount of shielding depends on the exact local environment. For example, a hydrogen bonded to an oxygen will be shielded differently from a hydrogen bonded to a carbon atom. In addition, two hydrogen nuclei can interact via a process known as spin-spin coupling, if they are on the same molecule, which will split the lines of the spectra in a recognizable way.

As one of the two major spectroscopic techniques used in metabolomics, NMR is used to generate metabolic fingerprints from biological fluids to obtain information about disease states or toxic insults.

Chemistry

By studying the peaks of nuclear magnetic resonance spectra, chemists can determine the structure of many compounds. It can be a very selective technique, distinguishing among many atoms within a molecule or collection of molecules of the same type but which differ only in terms of their local chemical environment. NMR spectroscopy is used to unambiguously identify known and novel compounds, and as such, is usually required by scientific journals for identity confirmation of synthesized new compounds. See the articles on carbon-13 NMR and proton NMR for detailed discussions.

By studying T_2 information, a chemist can determine the identity of a compound by comparing the observed nuclear precession frequencies to known frequencies. Further structural data can be elucidated by observing *spin-spin coupling*, a process by which the precession frequency of a nucleus can be influenced by the magnetization transfer from nearby chemically bound nuclei. Spin-spin coupling is observed in NMR of hydrogen-1 (1H NMR), since its natural abundance is nearly 100%; isotope enrichment is required for most other elements.

Because the nuclear magnetic resonance *timescale* is rather slow, compared to other spectroscopic methods, changing the temperature of a T_2*experiment can also give information about fast reactions, such as the Cope rearrangement or about structural dynamics, such as ring-flipping in cyclohexane. At low enough temperatures, a distinction can be made between the axial and equatorial hydrogens in cyclohexane.

An example of nuclear magnetic resonance being used in the determination of a structure is that of buckminsterfullerene (often called "buckyballs", composition C_{60}). This now famous form of carbon has 60 carbon atoms forming a sphere. The carbon atoms are all in identical environments and so should see the same internal H field. Unfortunately, buckminsterfullerene contains no hydrogen and so 13C nuclear magnetic resonance has to be used. 13C spectra require longer acquisition times since carbon-13 is not the common isotope of carbon (unlike hydrogen, where 1H is the common isotope). However, in 1990 the spectrum was obtained by R. Taylor and co-workers at the University of Sussex and was found to contain a single peak, confirming the unusual structure of buckminsterfullerene.

Purity determination (w/w NMR)

NMR is primarily used for structural determination, however it can also be used for purity determination, providing that the structure and molecular weight of the compound is known. This technique requires the use of an internal standard of a known purity. Typically this standard will have a high molecular weight to facilitate accurate weighing, but relatively few protons so as to give a clear peak for later integration e.g. 1,2,3,4-tetrachloro-5-nitrobenzene. Accurately weighed portions of both the standard and sample are combined and analysed by NMR. Suitable peaks are selected for both compounds and the purity of the sample determined via the following equation.

$$Purity = \frac{Wt(Std) \times n[H](Std) \times MW(Spl)}{Wt(Spl) \times MW(Std) \times n[H](Spl)} \times P$$

Where:
Wt(Std): Weight of internal standard
Wt(Spl): Weight of sample
n(Std): The integrated area of the peak selected for comparison in the standard, corrected for the number of protons in that functional group
n(Spl): The integrated area of the peak selected for comparison in the sample, corrected for the number of protons in that functional group
MW(Std): Molecular weight of standard
MW(Spl): Molecular weight of sample
P: Purity of internal standard

Non-destructive testing

Nuclear magnetic resonance is extremely useful for analyzing samples non-destructively. Radio waves and static magnetic fields easily penetrate many types of matter and anything that is not inherently ferromagnetic. For example, various expensive biological samples, such as nucleic acids, including RNA and DNA, or proteins, can be studied using nuclear magnetic resonance for weeks or months before using destructive biochemical experiments. This also makes nuclear magnetic resonance a good choice for analyzing dangerous samples.

Acquisition of dynamic information

In addition to providing static information on molecules by determining their 3D structures in solution, one of the remarkable advantages of NMR over X-ray crystallography is that it can be used to obtain important dynamic information.

Data acquisition in the petroleum industry

Another use for nuclear magnetic resonance is data acquisition in the petroleum industry for petroleum and natural gas exploration and recovery. A borehole is drilled into rock and sedimentary strata into which nuclear magnetic resonance logging equipment is lowered. Nuclear magnetic resonance analysis of these boreholes is used to measure rock porosity, estimate permeability from pore size distribution and identify pore fluids (water, oil and gas). These instruments are typically low field NMR spectrometers.

Flow probes for NMR spectroscopy

Recently, real-time applications of NMR in liquid media have been developed using specifically designed flow probes (flow cell assemblies) which can replace standard tube probes. This has enabled techniques that can incorporate the use of high performance liquid chromatography (HPLC) or other continuous flow sample introduction devices.

Process control

NMR has now entered the arena of real-time process control and process optimization in oil refineries and petrochemical plants. Two different types of NMR analysis are utilized to provide real time analysis of feeds and products in order to control and optimize unit operations. Time-domain NMR (TD-NMR) spectrometers operating at low field (2–20 MHz for 1H) yield free induction decay data that can be used to determine absolute hydrogen content values, rheological information, and component composition. These spectrometers are used in mining, polymer production, cosmetics and food manufacturing as well as coal analysis. High resolution FT-NMR spectrometers operating in the 60 MHz range with shielded permanent magnet systems yield high resolution 1H NMR spectra of refinery and petrochemical streams. The variation observed in these spectra with changing physical and chemical properties is modeled using chemometrics to yield predictions on unknown samples. The prediction results are provided to control systems via analogue or digital outputs from the spectrometer.

Earth's field NMR

In the Earth's magnetic field, NMR frequencies are in the audio frequency range, or the very low frequency and ultra low frequency bands of the radio frequency spectrum. Earth's field NMR (EFNMR) is typically stimulated by applying a relatively strong dc magnetic field pulse to the sample and, after the end of the pulse, analyzing the resulting low frequency alternating magnetic field that occurs in the Earth's magnetic field due to free induction decay (FID). These effects are exploited in some types of magnetometers, EFNMR spectrometers, and MRI imagers. Their inexpensive portable nature makes these instruments valuable for field use and for teaching the principles of NMR and MRI.

An important feature of EFNMR spectrometry compared with high-field NMR is that some aspects of molecular structure can be observed more clearly at low fields and low frequencies, whereas other aspects observable at high fields are not observable at low fields.

This is because:

- Electron-mediated heteronuclear J-couplings (spin-spin couplings) are field independent, producing clusters of two or more frequencies separated by several Hz, which are more easily observed in a fundamental resonance of about 2 kHz. "Indeed it appears that enhanced resolution is possible due to the long spin relaxation times and high field homogeneity which prevail in EFNMR."
- Chemical shifts of several ppm are clearly separated in high field NMR spectra, but have separations of only a few millihertz at proton EFNMR frequencies, so are usually lost in noise etc.

Zero Field NMR

In Zero Field NMR all magnetic fields are shielded such that magnetic fields below nT (nano-Tesla) are achieved and the nuclear precession frequencies of all nuclei are close to zero and indistinguishable. Under those circumstances the observed spectra are no-longer dictated by chemical shifts but primarily by J-coupling interactions which are independent of the external magnetic field. Since inductive detection schemes are not sensitive at very low frequencies, on the order of the J-couplings (typically between 0 and 1000 Hz), alternative detection schemes are used. Specifically, sensitive magnetometers turn out to be good detectors for Zero Field NMR. A zero magnetic field environment does not provide any polarization hence it is the combination of zero-field NMR with hyperpolarization schemes that makes zero field NMR attractive.

ELECTRON MICROSCOPE

An electron microscope is a microscope that uses a beam of accelerated electrons as a source of illumination. Because the wavelength of an electron can be up to 100,000 times shorter than that of visible light photons, the electron microscope has a higher resolving power than a light microscope and can reveal the structure of smaller objects. A transmission electron microscope can achieve better than 50 pm resolution and magnifications of up to about 10,000,000x whereas most light microscopes are limited by diffraction to about 200 nm resolution and useful magnifications below 2000x.

The transmission electron microscope uses electrostatic and electromagnetic lenses to control the electron beam and focus it to form an image. These electron optical lenses are analogous to the glass lenses of an optical light microscope.

Electron microscopes are used to investigate the ultrastructure of a wide range of biological and inorganic specimens including microorganisms, cells, large molecules, biopsy samples, metals, and crystals. Industrially, the electron microscope is often used for quality control and failure analysis. Modern electron microscopes produce electron micrographs using specialized digital cameras and frame grabbers to capture the image.

The first electromagnetic lens was developed in 1926 by Hans Busch.

According to Dennis Gabor, the physicist Leó Szilárd tried in 1928 to convince Busch to build an electron microscope, for which he had filed a patent.

German physicist Ernst Ruska and the electrical engineer Max Knoll constructed the prototype electron microscope in 1931, capable of four-hundred-power magnification; the apparatus was the first demonstration of the principles of electron microscopy. Two years later, in 1933, Ruska built an electron microscope that exceeded the resolution attainable with an optical (light) microscope. Moreover, Reinhold Rudenberg, the scientific director of Siemens-Schuckertwerke, obtained the patent for the electron microscope in May 1931.

In 1932, Ernst Lubcke of Siemens & Halske built and obtained images from a prototype electron microscope, applying concepts described in the Rudenberg patent applications. Five years later (1937), the firm financed the work of Ernst Ruska and Bodo von Borries, and employed Helmut Ruska (Ernst's brother) to develop applications for the microscope, especially with biological specimens. Also in 1937, Manfred von Ardenne pioneered the scanning electron microscope. The first *practical* electron microscope was constructed in 1938, at the University of Toronto, by Eli Franklin Burton and students Cecil Hall, James Hillier, and Albert Prebus; and Siemens produced the first *commercial* transmission electron microscope (TEM) in 1939. Although contemporary electron microscopes are capable of two million-power magnification, as scientific instruments, they remain based upon Ruska's prototype.

Bibliography

A V N Swamy: *Fundamentals of Biochemical Engineering*, BS Publication, Delhi, 2007.

Afroz Alam, Sharad Vats and Abhishek Tripathi: *Plant Physiology*, Narendra Publishing House, Delhi, 2015.

Ajit Kumar Ghosh: *Flowering : Physiological, Biochemical and Molecular Aspects*, Cyber Tech Publication, Delhi, 2010.

Charles Hardin, Jennifer Edwards, Andrew Riell, William Miller and Dominique Robertson: *Cloning Gene Expression and Protein Purification*, Oxford University Press, Delhi, 2008.

Edwin Oxlade: *Plant Physiology*, Viva Books, Delhi, 2010.

G. Tripathi: *Cellular and Biochemical Science*, I.K. International Publication, Delhi, 2010.

G. Whitmore: *Metabolism, Biosynthesis and Biochemical Energetics*, Sarup Publication, Delhi, 2002.

Iqbal Hussain: *Plant Physiology*, Oxford Book Company, Delhi, 2008.

J N Govil and Lalit Tiwari: *Recent Progress in Medicinal Plants Vol. 35: Phytoconstituents and Biochemical Processes*, Studium Press India, 2013.

J. Christopher: *Biochemical Elements*, Anmol Publication, Delhi, 2006.

James Bailey: *Biochemical Engg Fundamentals*, Tata Mcgrawhill, 1987.

James M Bower and Hamid Bolouri: *Computational Modeling of Genetic and Biochemical Networks*, MIT Press, 2004.

Mahendra R. Awode: *Chemical Thermodynamics (A Treatise on the Application of Thermodynamics of Chemical and Biochemical Systems)*, Dattsons, 2002.

Neeru Mathur: *Plant Physiology*, RBSA Publishers, Jaipur, 2010.

P. Sudhakar, P. Latha and P.V. Reddy: *Phenotyping Crop Plants for Physiological and Biochemical Traits*, BS Publications, Delhi, 2014.

P.N. Panday, S.K. Sharan and P.K. Mishra: *Silk Culture : A Biochemical Approach*, APH Publication, Delhi, 2005.

R. Ranjan, S.P. Bohra and M. Jeet Asija: *Plant Senescence: Physiological Biochemical and Molecular Aspects*, Agrobios Publication, Delhi, 2001.

R.M. Michael: *The Biochemical Basis of Sports Performance*, Cyber Tech Publications, Delhi, 2012.

Renuka Desai: *Plant Physiology*, Adhyayan Publication, Delhi, 2008.

S.S. Purohit and R. Ranjan: *Flowering : Physiological Biochemical and Molecular Aspects*, Agrobios Publication, Delhi, 2002.

S.S. Purohit and R. Ranjan: *Flowering: Physiological, Biochemical and Molecular Aspects*, Agrobios Publication, Delhi, 2002.

S.S. Purohit and R. Ranjan: *Photosynthesis: Physiological, Biochemical and Molecular Aspects*, Agrobios Publication, Delhi, 2002.

Susheela M. Das and Abha Bhardwaj: *Plant Physiology*, Wisdom Press, Delhi, 2012.

Index

A

Agglomeration 124
Analogous 125
Angiosperms 181, 183
Antimicrobial 79
Appropriate 71
Assumption 92
Asymmetrical 32, 33, 34, 35

B

Bacteriophage 58
Biological Molecules 237, 245, 253, 254

C

Carbohydrates 31, 32, 33
Cell Types 35
Cellulose 21, 23
Chemical Composition 31
Chemical Information 89
Chiral Chromatography 228
Chromatography Terms 219
Chromosome 89
Circular DNA 89
Conservative 64, 66
Considerable 99, 105, 107, 113, 118
Crystallography 62
Cytoplasm 32, 33, 35

D

Data Analysis 216, 217, 255, 257, 268
Deoxyribonucleic Acid 57, 63
DNA Synthesis 63
Dynamic Information 271, 279

E

Electrolytes 100, 101, 124, 125, 126, 127
Electrophoresis 66
Equilibrium 101, 110, 122
Eukaryotic 57
Exchange Chromatography 225
Excision Pair 65

F

Filterability 117
Flocculation 124, 125, 126, 127, 128, 129
Floriculturists 93
Fluorescent 66
Fork 59

G

Gel electrophoresis 210, 211, 217
Gelation 129, 130
Gelcondition 130
Genetic 56, 58, 60, 61, 62, 63
Glycoproteins 31, 32, 33, 34, 36

H

Helix 67, 69, 70, 75, 80, 81, 83
Horticulturists 93
Hysteresis 131, 132

I

Imbibition 112, 113
Initial Phasing 258
Insulin 78, 79
Interpretations 95

L

Lagging 64
Liquid Chromatography 219, 221, 223, 224, 226, 227, 228, 280
Lubricating 98

M

Macromolecules 37, 39
Magnetic Resonance 246, 265, 266, 267, 269, 271, 273, 274, 275, 276, 277, 278, 279
Membrane Fluidity 37, 39
Membrane Proteins 32, 37
Microfilaments 34
Mitochondria 33
Molecules 32, 35, 36, 37, 38, 39, 40, 41
Mutations 58, 64, 65

N

Nuclear Shielding 269

P

Patterson Function 264, 265
Peptidoglycan 79, 80
Permeability 112
Permeability Barrier 21
Phagocytes 29
Phosphorylation 22, 78, 81, 82
Photoreceptors 40, 41
Planar Chromatography 222
Polymerase 61, 63, 64, 65
Primary Antibody 213, 214, 215, 218, 232, 233, 234, 235
Process Control 280
Protein Structure 40
Protoplasm 92, 112, 113, 115, 116, 119, 131, 132
Pseudomonads 28
Punctuated 30

R

Reaction 65
Ribosome Factories 75

S

Scattering Techniques 247
Secondary Antibody 214, 215, 216, 229, 233, 234, 235, 236
Semiconservative 64, 66
Solubility 97, 98
Special Techniques 226
Streptococcus Pyogenes 29
Syneresis 132

T

Terminology 98
Thecations 100
Thixotropy 132
Thymine 57, 59
Thyroperoxidase 78
Tissue Preparation 210, 211
Transcription 60, 62, 63

U

Unraveling 60, 62

V

Viscosity 113, 120, 121, 130, 132

W

Western Blot 210, 213, 215, 216, 217, 218